COMPLEX ANALYSIS AND APPLICATIONS

COMPLEX ANALYSIS AND APPLICATIONS

COMPLEX ANALYSIS AND APPLICATIONS

Proceedings of the 13th International Conference on Finite or Infinite Dimensional Complex Analysis and Applications

Shantou University, China 8 – 12 August 2005

editors

Yuefei Wang
Academy of Mathematics and System Sciences, Beijing, China

Shengjian Wu
Peking University, Beijing, China

Hasi Wulan
Shantou University, Guangdong, China

Lo Yang
Academy of Mathematics and System Sciences, Beijing, China

World Scientific

NEW JERSEY • LONDON • SINGAPORE • BEIJING • SHANGHAI • HONG KONG • TAIPEI • CHENNAI

Published by

World Scientific Publishing Co. Pte. Ltd.
5 Toh Tuck Link, Singapore 596224
USA office: 27 Warren Street, Suite 401-402, Hackensack, NJ 07601
UK office: 57 Shelton Street, Covent Garden, London WC2H 9HE

British Library Cataloguing-in-Publication Data
A catalogue record for this book is available from the British Library.

COMPLEX ANALYSIS AND APPLICATIONS
Proceedings of the 13th International Conference on Finite or Infinite Dimensional Complex Analysis and Applications

ISBN 981-256-868-9

Printed in Singapore by World Scientific Printers (S) Pte Ltd

PREFACE

This proceedings consists of articles by participants in the 13th International Conference on Finite or Infinite Dimensional Complex Analysis and Applications, which was held at Shantou University, China, August 8–12, 2005. More than 90 participants from 17 countries or regions attended this conference. The conference was supported by the LI KA SHING Foundation, National Natural Science Foundation of China, Chinese Mathematical Society, Academy of Mathematics and Systems Science of Chinese Academy of Sciences and Shantou University.

The proceedings contains all plenary lectures and some selected talks at the conference. These contributions mainly concentrated in the following topics:

- Applied Complex Analysis
- Clifford Analysis
- Complex Dynamical Systems
- Complex Function Spaces and Operator Theory
- Complex Numerical Analysis
- Qusiconformal Mapping, Teichmüller Theory and Klainian Groups
- Several Complex Variables
- Value Distribution Theory and Complex Differential Equations
- Related Topics

The book offers not only reviews of these fields for researches but also perspectives on ways they may proceed, together with a comprehensive bibliography of books and papers. The book also serves to inform readers of the main research directions pursued outside their own institutions.

We would like to thank all of the contributors and referees. We thank all members of The International Advisory Board and Organizing committee of the conference for their efforts. We wish to thank the listed institutions for their support. On this basis, many participants did get financial help including receiving a copy of the proceedings.

Thanks are due to Ms. Yali Liu who was heavily involved with the conference organization and to all the other helpers from the Department of Mathematics, Shantou University.

The Editors

CONTENTS

COMPLEX BOUNDARY VALUE PROBLEMS IN A QUARTER PLANE

H. BEGEHR

I. Math. Inst.
FU Berlin Arnimallee 3
14195 Berlin, Germany
Email: begehr@math.fu-berlin.de

G. HARUTYUNYAN

Fac. Computer Sci. Appl. Math.
Yerevan State Univ. Alex Manoogian 1
375049 Yerevan, Armenia
Email: gohar@ysu.am

Dirichlet and Neumann boundary value problems are considered for the inhomogeneous Cauchy-Riemann equation in a quarter plane. Solvability conditions and solutions are given in explicit form.

Keywords: Inhomogeneous Cauchy-Riemann equation, quarter plane, Dirichlet boundary value problem, Neumann boundary value problem.

2000 Mathematics Subject Classification: 30E25, 35C15, 35F15

1. Introduction

The Cauchy-Pompeiu representation formula for functions in the upper right quarter plane $\mathbb{Q}_1$ of the complex plane $\mathbb{C}$ follows from the Gauss theorem for regular domains and a limiting process, see [1].

Cauchy-Pompeiu representation Any $w \in C^1(\mathbb{Q}_1;\mathbb{C}) \cap C(\bar{\mathbb{Q}}_1;\mathbb{C})$, where

$$\mathbb{Q}_1 = \{z \in \mathbb{C} : 0 < \text{ Re } z, 0 < \text{Im } z\},$$

for which for some $0 < \delta$ the function $(1+r)^\delta M(r,w)$ with

$$M(r,w) = \max\{|w(z)| : |z| = r, 0 \leq \text{ Re } z, 0 \leq \text{ Im } z\}$$

is bounded in $\mathbb{R}^+$ and $w_{\bar{z}} \in L_1(\mathbb{Q}_1; \mathbb{C})$ is representable as

$$w(z) = \frac{1}{2\pi i}\int_0^{+\infty} w(t)\frac{dt}{t-z} - \frac{1}{2\pi i}\int_0^{+\infty} w(it)\frac{dt}{t+iz} - \frac{1}{\pi}\int_{\mathbb{Q}_1} w_{\bar{\zeta}}(\zeta)\frac{d\xi d\eta}{\zeta - z}. \quad (1)$$

Introducing the harmonic Green function for $\mathbb{Q}_1$

$$G_1(z,\zeta) = \log\left|\frac{\bar{\zeta}^2 - z^2}{\zeta^2 - z^2}\right|^2 = \log\left|\frac{(\bar{\zeta}-z)(\bar{\zeta}+z)}{(\zeta-z)(\zeta+z)}\right|^2$$

this representation can be altered into

$$w(z) = \frac{1}{\pi}\int_{\mathbb{Q}_1} w(\zeta)\left[\frac{1}{(\bar{\zeta}+z)^2} - \frac{1}{(\bar{\zeta}-z)^2}\right] d\xi d\eta - \frac{1}{\pi}\int_{\mathbb{Q}_1} w_{\bar{\zeta}}(\zeta)\partial_z G_1(z,\zeta) d\xi d\eta, \quad (2)$$

see [1]. For respective representations in the upper half plane see [8], in the unit disc see e.g. [2,3,4,5] and in arbitrary regular domains see [9]. In [1] formula (1) is adjusted to Schwarz boundary data and the solution to the Schwarz boundary value problem is given for the inhomogeneous Cauchy-Riemann and Bitsadze equations.

Here the Dirichlet and Neumann boundary value problems are treated for the Cauchy-Riemann equation. These problems are known to be overdetermined so that solvability conditins have to be determined, see e.g [4,5,6,7].

2. The Dirichlet problem

The complex Gauss theorem in the form

$$\frac{1}{2\pi i}\int_{\partial D} w(z)dz = \frac{1}{\pi}\int_D w_{\bar{z}}(z)dxdy$$

for functions $w \in C^1(D;\mathbb{C}) \cap C(\bar{D};\mathbb{C})$ for bounded domains D in the complex plane $\mathbb{C}$ with piecewise smooth boundary ∂D, see e.g. [2], applied for

$$\mathbb{Q}_{1R} = \{z = x+iy : |z| < R, 0 < x, 0 < y\}$$

besides leading to (1) also supplies for $w \in C^1(\mathbb{Q}_1;\mathbb{C}) \cap C(\bar{\mathbb{Q}}_1;\mathbb{C})$ with in $\mathbb{R}^+$ bounded $(1+r)^\delta M(r,w)$ and $w_{\bar{z}} \in L_{p,2}(\mathbb{Q};\mathbb{C}), 2 < p$, and $z \in \mathbb{Q}_1$

$$\frac{1}{2\pi i}\int_0^{+\infty} w(t)\frac{dt}{t-\bar{z}} - \frac{1}{2\pi i}\int_0^{+\infty} w(it)\frac{dt}{t+i\bar{z}} - \frac{1}{\pi}\int_{\mathbb{Q}_1} w_{\bar{\zeta}}(\zeta)\frac{d\xi d\eta}{\zeta-\bar{z}} = 0, \quad (3)$$

$$\frac{1}{2\pi i}\int_0^{+\infty} w(t)\frac{dt}{t+\bar{z}} - \frac{1}{2\pi i}\int_0^{+\infty} w(it)\frac{dt}{t-i\bar{z}} - \frac{1}{\pi}\int_{\mathbb{Q}_1} w_{\bar{\zeta}}(\zeta)\frac{d\xi d\eta}{\zeta+\bar{z}} = 0, \quad (4)$$

$$\frac{1}{2\pi i}\int_0^{+\infty} w(t)\frac{dt}{t+z} - \frac{1}{2\pi i}\int_0^{+\infty} w(it)\frac{dt}{t-iz} - \frac{1}{\pi}\int_{\mathbb{Q}_1} w_{\bar{\zeta}}(\zeta)\frac{d\xi d\eta}{\zeta+z} = 0. \quad (5)$$

Properly combined with (1) they lead to solutions to Schwarz problems, see [1]. But they together with (1) also lead to solvability conditions and solutions to the Dirichlet problem for the Cauchy-Riemann equation

Theorem 1 *The Dirichlet problem*

$$w_{\bar{z}} = f \text{ in } \mathbb{Q}_1, w = \gamma_1 \text{ for } 0 \le x, y = 0, w = \gamma_2 \text{ for } 0 \le y, x = 0$$

for $f \in L_{p,2}(\mathbb{Q}_1;\mathbb{C}), 2 < p, \gamma_1, \gamma_2 \in C(\mathbb{R};\mathbb{C})$ *such that* $(1+t)^\delta\gamma_1(t), (1+t)^\delta\gamma_2(t)$ *are bounded for some* $0 < \delta$ *and satisfying the compatibility condition* $\gamma_1(0) = \gamma_2(0)$ *is uniquely weakly solvable in the class* $C^1(\mathbb{Q}_1;\mathbb{C}) \cap C(\bar{\mathbb{Q}}_1;\mathbb{C})$ *if and only if*

$$\frac{1}{2\pi i}\int_0^{+\infty} \gamma_1(t)\frac{dt}{t-\bar{z}} - \frac{1}{2\pi i}\int_0^{+\infty} \gamma_2(t)\frac{dt}{t+i\bar{z}} - \frac{1}{\pi}\int_{\mathbb{Q}_1} f(\zeta)\frac{d\xi d\eta}{\zeta-\bar{z}} = 0, \quad (6)$$

$$\frac{1}{2\pi i}\int_0^{+\infty} \gamma_1(t)\frac{dt}{t+\bar{z}} - \frac{1}{2\pi i}\int_0^{+\infty} \gamma_2(t)\frac{dt}{t-i\bar{z}} - \frac{1}{\pi}\int_{\mathbb{Q}_1} f(\zeta)\frac{d\xi d\eta}{\zeta+\bar{z}} = 0, \quad (7)$$

$$\frac{1}{2\pi i}\int_0^{+\infty} \gamma_1(t)\frac{dt}{t+z} - \frac{1}{2\pi i}\int_0^{+\infty} \gamma_2(t)\frac{dt}{t-iz} - \frac{1}{\pi}\int_{\mathbb{Q}_1} f(\zeta)\frac{d\xi d\eta}{\zeta+z} = 0. \quad (8)$$

The solution is

$$w(z) = \frac{1}{2\pi i}\int_0^{+\infty} \gamma_1(t)\frac{dt}{t-z} - \frac{1}{2\pi i}\int_0^{+\infty} \gamma_2(t)\frac{dt}{t+iz} - \frac{1}{\pi}\int_{\mathbb{Q}_1} f(\zeta)\frac{d\xi d\eta}{\zeta-z}. \quad (9)$$

Remark *The boundedness condition on the boundary data γ_1, γ_2 at infinity can be weakened to $\gamma_1, \gamma_2 \in L_2(\mathbb{R}^+; \mathbb{C})$.*

Proof If this Dirichlet problem has a solution in the class $C^1(\mathbb{Q}_1; \mathbb{C}) \cap C(\bar{\mathbb{Q}}_1; \mathbb{C})$ satisfying $w_{\bar{z}} \in L_{p,2}(\mathbb{Q}_1; \mathbb{C})$ for some $2 < p$ then according to (1) it is of the form (9). On the other hand (9) can be verified to be a solution. From the assumptions on γ_1, γ_2 and f it follows by direct computation and from [10], p. 43, that $(1+r)^\rho M(r, w)$ is bounded on $\mathbb{R}^+$ with $\rho = \min\{\delta, \frac{p-2}{p}\}$. Moreover, w obviously satisfies the differential equation. For checking the boundary conditions the solvability conditions (6) to (8) are needed. Adding (7) to (9) and subtracting (6) and (8) leads to

$$w(z) = \frac{2}{\pi} \int_0^{+\infty} \gamma_1(t) \frac{y}{|t-z|^2} \frac{t^2 + |z|^2}{|t+z|^2} dt + \frac{2i}{\pi} \int_0^{+\infty} \gamma_2(t) \frac{y}{|t-iz|^2} \frac{t^2 - |z|^2}{|t-iz|^2} dt$$

$$-\frac{2}{\pi} \int_{\mathbb{Q}_1} f(\zeta) \left(\frac{z}{\zeta^2 - z^2} - \frac{\bar{z}}{\zeta^2 - \bar{z}^2} \right) d\xi d\eta. \tag{10}$$

Adding (8) to (9) and subtracting (6) and (7) shows

$$w(z) = \frac{2}{\pi i} \int_0^{+\infty} \gamma_1(t) \frac{x}{|t-z|^2} \frac{t^2 - |z|^2}{|t+z|^2} dt + \frac{2}{\pi i} \int_0^{+\infty} \gamma_2(t) \frac{x}{|t+iz|^2} \frac{t^2 + |z|^2}{|t-iz|^2} dt$$

$$-\frac{2}{\pi} \int_{\mathbb{Q}_1} f(\zeta) \left(\frac{z}{\zeta^2 - z^2} + \frac{\bar{z}}{\zeta^2 - \bar{z}^2} \right) d\xi d\eta. \tag{11}$$

Observing that

$$\frac{1}{2\pi i} \frac{z - \bar{z}}{|t-z|^2} = \frac{1}{\pi} \frac{y}{|t-z|^2}, \ z = x + iy, \ x, t \in \mathbb{R}, \ 0 < y$$

is the Poisson kernel for the upper half plane, see e.g. [8], from (10) for $t_0 \in \mathbb{R}^+$

$$\lim_{z \to t_0} w(z) = \lim_{z \to t} \frac{2}{\pi} \int_0^{+\infty} \gamma_1(t) \frac{y}{|t-z|^2} \frac{t^2 + |z|^2}{|t+z|^2} dt = \gamma_1(t_0)$$

and from (11) likewise for $t_0 \in \mathbb{R}^+$

$$\lim_{z \to it_0} w(z) = \lim_{z \to it_0} \frac{2}{\pi i} \int_0^{+\infty} \gamma_2(t) \frac{x}{|t+iz|^2} \frac{t^2 + |z|^2}{|t-iz|^2} dt = \gamma_2(t_0).$$

The solution to the Dirichlet problem can be used to find the solution and the solvability conditions to the Neumann problem for the inhomogeneous Cauchy-Riemann equation.

3. The Neumann problem

The normal derivative on the real axis is $\partial_y = i(\partial_z - \partial_{\bar{z}})$ while $\partial_x = \partial_z + \partial_{\bar{z}}$ is the normal derivative on the imaginary axis. With respect to the upper right quadrant $\mathbb{Q}_1$ these are derivatives with respect to the inner normals. For simplicity at first the homogeneous Cauchy-Riemann equation is considered.

Theorem 2 *The Neumann problem*

$$\partial_y w = \gamma_1 \text{ for } 0 \leq x, y = 0, \ \partial_x w = \gamma_2 \text{ for } 0 \leq y, x = 0, \ w(0) = c$$

for analytic functions in $\mathbb{Q}_1$ *being continuous in* $\bar{\mathbb{Q}}_1$, *is uniquely solvable for* $\gamma_1, \gamma_2 \in C(\mathbb{R}^+; \mathbb{C}), c \in \mathbb{C}$, *such that for some* $0 < \delta$ *the functions* $(1 + |t|)^\delta \gamma_1(t), (1 + |t|)^\delta \gamma_2(t)$ *are bounded, if and only if*

$$\frac{1}{2\pi} \int_0^{+\infty} \gamma_1(t) \frac{dt}{t - \bar{z}} + \frac{1}{2\pi i} \int_0^{+\infty} \gamma_2(t) \frac{dt}{t + i\bar{z}} = 0, \tag{12}$$

$$\frac{1}{2\pi} \int_0^{+\infty} \gamma_1(t) \frac{dt}{t + \bar{z}} + \frac{1}{2\pi i} \int_0^{+\infty} \gamma_2(t) \frac{dt}{t - i\bar{z}} = 0, \tag{13}$$

$$\frac{1}{2\pi} \int_0^{+\infty} \gamma_1(t) \frac{dt}{t + z} + \frac{1}{2\pi i} \int_0^{+\infty} \gamma_2(t) \frac{dt}{t - iz} = 0. \tag{14}$$

The solution is

$$w(z) = c + \frac{1}{2\pi} \int_0^{+\infty} \gamma_1(t) \log \left| \frac{t^2 - z^2}{t^2} \right|^2 dt + \frac{1}{2\pi} \int_0^{+\infty} \gamma_2(t) \log \left| \frac{t^2 + z^2}{t^2} \right|^2 dt. \tag{15}$$

Proof For an analytic function the boundary conditions are

$$\partial_y w = i w' = \gamma_1, \ \partial_x w = w' = \gamma_2.$$

These are Dirichlet conditions for the analytic function w'. Hence, according to Theorem 1, the solution is

$$w'(z) = -\frac{1}{2\pi}\int_0^{+\infty} \gamma_1(t)\frac{dt}{t-z} - \frac{1}{2\pi i}\int_0^{+\infty} \gamma_2(t)\frac{dt}{t+iz} \tag{16}$$

if and only if (12)-(14) are satisfied. Integrating (16) shows (15) or also

$$w(z) = w\left(\frac{1+i}{\sqrt{2}}\right) + \frac{1}{2\pi}\int_0^{+\infty} \gamma_1 \log\frac{|t^2-z^2|^2}{1+t^4} + \frac{1}{2\pi}\int_0^{+\infty} \gamma_2(t)\log\frac{|t^2+z^2|^2}{1+t^4}dt.$$

To verify that (15) in fact is the solution under the solvability conditions (12) to (14) at first w has to be recognized as an analytic function. Differentiation with respect to $\bar{z}$ shows

$$\begin{aligned} w_{\bar{z}}(z) &= -\frac{1}{2\pi}\int_0^{+\infty} \gamma_1(t)\frac{2\bar{z}}{t^2-\bar{z}^2}dt + \frac{1}{2\pi}\int_0^{+\infty} \gamma_2(t)\frac{2\bar{z}}{t^2+\bar{z}^2}dt, \\ &= -\frac{1}{2\pi}\int_0^{+\infty} \gamma_1(t)\left(\frac{1}{t-\bar{z}} - \frac{1}{t-\bar{z}}\right)dt \\ &\quad + \frac{1}{2\pi i}\int_0^{+\infty} \gamma_2(t)\left(\frac{1}{t-i\bar{z}} - \frac{1}{t+i\bar{z}}\right)dt = 0 \end{aligned}$$

according to (12) and (13). Similarly,

$$\begin{aligned} w'(z) &= -\frac{1}{2\pi}\int_0^{+\infty} \gamma_1(t)\left(\frac{1}{t-z} - \frac{1}{t+z}\right)dt \\ &\quad + \frac{1}{2\pi i}\int_0^{+\infty} \gamma_2(t)\left(\frac{1}{t-iz} - \frac{1}{t+iz}\right)dt \\ &= -\frac{1}{2\pi}\int_0^{+\infty} \gamma_1(t)\frac{dt}{t-z} - \frac{1}{2\pi i}\int_0^{+\infty} \gamma_2(t)\frac{dt}{t+iz} \end{aligned}$$

because of (14).

Theorem 3 *The Neumann problem*

$$w_{\bar{z}} = f \text{ in } \mathbb{Q}_1, \ w(0) = c,$$
$$\partial_y w = \gamma_1 \text{ for } 0 \le x, y = 0, \ \partial_x w = \gamma_2 \text{ for } 0 \le y, x = 0,$$

where $f \in L_{p,2}(\mathbb{Q}_1;\mathbb{C}) \cap C^\alpha(\mathbb{Q}_1;\mathbb{C})$ *for* $2 < p, 0 < \alpha < 1, \gamma_1, \gamma_2 \in C(\mathbb{R}^+;\mathbb{C})$ *such that* $(1+t)^\delta \gamma_1(t), (1+t)^\delta \gamma_2(t), (1+t)^\delta f(t), (1+t)^\delta f(it)$ *are bounded for some* $0 < \delta, c \in \mathbb{C}$ *are uniquely weakly solvable if and only if*

$$\frac{1}{2\pi}\int_0^{+\infty} (\gamma_1(t) + if(t))\frac{dt}{t-\bar{z}}$$
$$+\frac{1}{2\pi i}\int_0^{+\infty} (\gamma_2(t) - f(it))\frac{dt}{t+i\bar{z}} + \frac{1}{\pi}\int_{\mathbb{Q}_1} f(\zeta)\frac{d\xi d\eta}{\zeta - \bar{z}} = 0, \tag{17}$$

$$\frac{1}{2\pi}\int_0^{+\infty} (\gamma_1(t) + if(t))\frac{dt}{t+\bar{z}}$$
$$+\frac{1}{2\pi i}\int_0^{+\infty} (\gamma_2(t) - f(it))\frac{dt}{t-i\bar{z}} + \frac{1}{\pi}\int_{\mathbb{Q}_1} f(\zeta)\frac{d\xi d\eta}{\zeta + \bar{z}} = 0, \tag{18}$$

$$\frac{1}{2\pi}\int_0^{+\infty} (\gamma_1(t) + if(t))\frac{dt}{t+z}$$
$$+\frac{1}{2\pi i}\int_0^{+\infty} (\gamma_2(t) - f(it))\frac{dt}{t-iz} + \frac{1}{\pi}\int_{\mathbb{Q}_1} f(\zeta)\frac{d\xi d\eta}{\zeta + z} = 0. \tag{19}$$

The solution is

$$w(z) = c + \frac{1}{2\pi}\int_0^{+\infty} (\gamma_1(t) + if(t)) \log\left|\frac{t^2 - z^2}{t^2}\right|^2 dt$$
$$+\frac{1}{2\pi}\int_0^{+\infty} (\gamma_2(t) - f(it)) \log\left|\frac{t^2 + z^2}{t^2}\right|^2 dt$$
$$-\frac{z}{\pi}\int_{\mathbb{Q}_1} \frac{f(\zeta)}{\zeta}\,\frac{d\xi d\eta}{\zeta - z}. \tag{20}$$

Proof Consider the function

$$\varphi = w - Tf, \; Tf(z) = -\frac{1}{\pi} \int\limits_{\mathbb{Q}_1} f(\zeta) \frac{d\xi d\eta}{\zeta - z}.$$

It is analytic in $\mathbb{Q}_1$ satisfying

$$\varphi' = -i\partial_y(w - Tf) = -i\gamma_1 - \Pi f + f \text{ for } 0 \leq x, y = 0,$$
$$\varphi' = \partial_x(w - Tf) = \gamma_2 - \Pi f - f \text{ for } 0 \leq y, x = 0$$

with

$$\Pi f(z) = 1/\pi \int_{\mathbb{Q}_1} f(\zeta) \frac{d\xi d\eta}{(\zeta - z)^2}.$$

According to [10], p. 63 Πf is Hölder continuous in $\bar{\mathbb{Q}}_1$. Although the asymptotic behaviour of Πf at infinity is not known the following conclusions follow because of the Fubini theorem.

Thus on the basis of Theorem 1 this Dirichlet problem is solvable if and only if

$$\frac{1}{2\pi i} \int\limits_0^{+\infty} [i\gamma_1(t) + \Pi f(t) - f(t)] \frac{dt}{t - \bar{z}}$$

$$+ \frac{1}{2\pi i} \int\limits_0^{+\infty} [\gamma_2(t) - \Pi f(it) - f(it)] \frac{dt}{t + i\bar{z}} = 0, \tag{21}$$

$$\frac{1}{2\pi i} \int\limits_0^{+\infty} [\gamma_1(t) + \Pi f(t) - f(t)] \frac{dt}{t + \bar{z}}$$

$$+ \frac{1}{2\pi i} \int\limits_0^{+\infty} [\gamma_2(t) - \Pi f(it) - f(it)] \frac{dt}{t - i\bar{z}} = 0, \tag{22}$$

$$\frac{1}{2\pi i} \int\limits_0^{+\infty} [\gamma_1(t) + \Pi f(t) - f(t)] \frac{dt}{t + z}$$

$$+ \frac{1}{2\pi i} \int\limits_0^{+\infty} [\gamma_2(t) - \Pi f(it) - f(it)] \frac{dt}{t - iz} = 0. \tag{23}$$

The solution is

$$\varphi'(z) = -\frac{1}{2\pi i}\int_0^{+\infty}[i\gamma_1(t) + \Pi f(t) - f(t)]\frac{dt}{t-z}$$

$$-\frac{1}{2\pi i}\int_0^{+\infty}[\gamma_2(t) - \Pi f(it) - f(it)]\frac{dt}{t+iz}. \tag{24}$$

For $z \in \mathbb{Q}_1$

$$\frac{1}{2\pi i}\int_0^{+\infty}\Pi f(t)\frac{dt}{t-\bar{z}} - \frac{1}{2\pi i}\int_0^{+\infty}\Pi f(it)\frac{dt}{t+i\bar{z}}$$

$$= \frac{1}{\pi}\int_{\mathbb{Q}_1} f(\zeta)\left[\frac{1}{2\pi i}\int_0^{+\infty}\frac{dt}{(t-\zeta)(t-\bar{z})} - \frac{1}{2\pi i}\int_0^{+\infty}\frac{idt}{(it-\zeta)(it-\bar{z})}\right]d\xi d\eta$$

$$= \frac{1}{\pi}\int_{\mathbb{Q}_1} f(\zeta)\frac{1}{2\pi i}\int_{\partial\mathbb{Q}_1}\frac{d\tilde{\zeta}}{(\tilde{\zeta}-\zeta)(\tilde{\zeta}-\bar{z})}d\xi d\eta = \frac{1}{\pi}\int_{\mathbb{Q}_1} f(\zeta)\frac{d\xi d\eta}{\zeta-\bar{z}} \tag{25}$$

and similarly

$$\frac{1}{2\pi i}\int_0^{+\infty}\Pi f(t)\frac{dt}{t+\bar{z}} - \frac{1}{2\pi i}\int_0^{+\infty}\Pi f(it)\frac{dt}{t-iz}$$

$$= \frac{1}{\pi}\int_{\mathbb{Q}_1} f(\zeta)\frac{d\xi d\eta}{\zeta+\bar{z}}, \tag{26}$$

$$\frac{1}{2\pi i}\int_0^{+\infty}\Pi f(t)\frac{dt}{t+z} - \frac{1}{2\pi i}\int_0^{+\infty}\Pi f(it)\frac{dt}{t-iz}$$

$$= \frac{1}{\pi}\int_{\mathbb{Q}_1} f(\zeta)\frac{d\xi d\eta}{t+z} \tag{27}$$

and

$$\frac{1}{2\pi i}\int_0^{+\infty}\Pi f(t)\frac{dt}{t-z} - \frac{1}{2\pi i}\int_0^{+\infty}\Pi f(it)\frac{dt}{t+iz} = 0. \tag{28}$$

This shows (17), (18), (19) and

$$\varphi'(z) = -\frac{1}{2\pi i}\int_0^{+\infty}[i\gamma_1(t) - f(t)]\frac{dt}{t-z} - \frac{1}{2\pi i}\int_0^{+\infty}[\gamma_2(t) - f(it)]\frac{dt}{t+iz}.$$

Integrating this last equation shows

$$\varphi(z) = \varphi(0) + \frac{1}{2\pi i}\int_0^{+\infty}[i\gamma_1(t) - f(t)]\log\left|\frac{t^2 - z^2}{t^2}\right|^2 dt$$

$$+ \frac{1}{2\pi}\int_0^{+\infty}[\gamma_2(t) - f(it)]\log\left|\frac{t^2 + z^2}{t^2}\right|^2 dt$$

so that

$$w(z) = \varphi(t) + Tf(z) = w(0) + Tf(z) - Tf(0)$$

which is (20).

Dirichtlet and Neumann problems are usually solved for the Poisson equation. They can also be considered for the Bitsadze equation, see [4], part II, in the case of the unit disc. For the quater plane $\mathbb{Q}_1$ these problems can be treated in a similar manner as in this paper.

References

1. S.A. Abdyumanapov, H. Begehr, A.B. Tungatarov, *Some Schwarz problems in a quarter plane,* Preprint, FU Berlin, 2005.
2. H. Begehr, *Complex analytic methods for partial differential equations. An introductory text,* Singapore: World-Sci., 1994.
3. H. Begehr, Orthogonal decompositions of the function space $L_2(\bar{D};\mathbb{C})$,*J. Reine Angew. Math.* **549** (2002):191-219.
4. H. Begehr, Boundary value problems in complex analysis, *I; II, Bol. Asoc. Mat. Venezolana* **12** (2005):65–85; 217–250.
5. H. Begehr, Basic boundary value problems in complex analysis, *Proc. Workshop Recent Trends in Appl. Complex Analysis, METU, Ankara,* 2004, to appear.
6. H. Begehr, G. Harutyunyan, Robin boundary value problem for the Cauchy-Riemann operator,*Complex Var., Theory Appl.* **50**(2005):1125–1136.
7. H. Begehr, J.C. Vanegas, Iterated Neumann problem for the higher order Poisson equation, *Math. Nachr.* **279** (2006): 38–57.
8. E. Gärtner, *Basic complex boundary value problems in the upper half plane,* Preprint, FU Berlin, 2005.
9. A. Krausz, *Intergraldarstellungen mit Greenschen Funktionen höherer Ordnung in Gebieten und Polygebieten,* Dissertation, FU Berlin, 2005; http://www.diss.fu-berlin.de/2005/128/ .
10. I.N. Vekua, *Generalized analytic functions,* Pergamon, Oxford, 1967.

SOME EQUATIONS OVER 4-DIMENSIONAL CLIFFORD ALGEBRA

WENSHENG CAO
School of Mathematical Sciences
Peking University
Beijing 100871, P. R. China
and
Institute of Mathematics
Hunan University of Science and Technology
Xiangtan, Hunan 411201, P. R. China
E-mail: wscao@math.pku.edu.cn

Some real matrix representations of 4-dimensional Clifford numbers are presented and some properties of 4-dimensional Clifford numbers are obtained. As an application, an explicit expression of the general solution of equation $axb = d$ and $ax = xb$ are obtained in terms of a and b in 4-dimensional Clifford algebra $C\ell_{0,3}$. As a by-product, we obtain the necessary and sufficient conditions for two Clifford numbers to be similar.

Keywords: Quaternion, Clifford number, matrix representation, equation.

2000 Mathematics Subject Classification: 30F40; 15A33; 22E55

1. Introduction

From the general definition of Clifford algebra [1], the basis of 4-dimensional Clifford algebra $C\ell_{0,3}$ is $e_i, i = 0, \cdots, 7$ and the multiplication rules for the basis of $C\ell_{0,3}$ are as follows.

For $a = a_0 + a_1e_1 + a_2e_2 + a_3e_3 + a_4e_4 + a_5e_5 + a_6e_6 + a_7e_7 \in C\ell_{0,3}$,we define $|a| = \sqrt{\sum_{i=0}^{7} a_i^2}$ will be the Euclidean norm of a.We point out that when $n > 3$, $|a^2| = |a|^2$ will fail to hold and there exist zero-divisors. For

Supported by NSFs (No: 10271043), (No: 10571048) of China and China Postdoctoral Science Foundation.

1	e_1	e_2	e_3	e_4	e_5	e_6	e_7
e_1	-1	e_3	-e_2	e_5	-e_4	e_7	-e_6
e_2	-e_3	-1	e_1	e_6	-e_7	-e_4	e_5
e_3	e_2	-e_1	-1	e_7	e_6	-e_5	-e_4
e_4	-e_5	-e_6	e_7	-1	e_1	e_2	-e_3
e_5	e_4	-e_7	-e_6	-e_1	-1	e_3	e_2
e_6	e_7	e_4	e_5	-e_2	-e_3	-1	-e_1
e_7	-e_6	e_5	-e_4	-e_3	e_2	-e_1	1

instance:

$$|(1+e_7)^2| = 2\sqrt{2}, \text{while } |1+e_7|^2 = 2, \tag{1}$$

$$(1+e_7)(1-e_7) = 0. \tag{2}$$

In this paper we will concentrate on the properties of Clifford numbers and some equations in $C\ell_{0,3}$. Our first task is to investigate the properties of Clifford numbers by using the real quaternions and their matrix representations in Section 2. Based on the matrix representation of Clifford numbers, we will give the concepts of Moore-Penrose inverse of Clifford numbers and obtain the explicit solutions to equation $axb = d$ in Section 3. The nontrivial task of solving equation $ax = xb$ is treated in Section 4.

2. The real matrix representation of elements in $C\ell_{0,3}$

We will represent 4-dimensional Clifford numbers with two quaternions. At first, we recall some properties of division ring of real quaternions $\mathbb{H}$ and their matrix representations.

It is well known that the real quaternion algebra $\mathbb{H}$ is algebraically isomorphic to the matrix algebra corresponding to the bijective map

$$L : q = q_0 + q_1e_1 + q_2e_2 + q_3e_3 \in \mathbb{H} \mapsto L(q) := \begin{pmatrix} q_0 & -q_1 & -q_2 & -q_3 \\ q_1 & q_0 & -q_3 & q_2 \\ q_2 & q_3 & q_0 & -q_1 \\ q_3 & -q_2 & q_1 & q_0 \end{pmatrix}.$$

Denote $\overrightarrow{x} = (x_0, x_1, x_2, x_3)^T$ for $x = x_0 + x_1e_1 + x_2e_2 + x_3e_3 \in \mathbb{H}$, where A^T stands for the transposed matrix of A. Then the following equations hold.

$$\overrightarrow{qx} = L(q)\,\overrightarrow{x}\,,\ \overrightarrow{xq} = R(q)\,\overrightarrow{x}\,, \tag{3}$$

where $R(q) = E_4L(q)^TE_4$ and $E_4 = diag(1,-1,-1,-1)$. For more details of the properties of quaternions and related topics, see [2,3,4,5] and the references therein.

We can write

$$a = a_0 + a_1e_1 + a_2e_2 + a_3e_3 + a_4e_4 + a_5e_5 + a_6e_6 + a_7e_7 = a_h + a_He_4, \quad (4)$$

where $a_h = a_0 + a_1e_1 + a_2e_2 + a_3e_3, a_H = a_4 + a_5e_1 + a_6e_2 + a_7e_3 \in \mathbb{H}$. We define

$$\bar{a} = a_0 - a_1e_1 - a_2e_2 - a_3e_3 - a_4e_4 - a_5e_5 - a_6e_6 + a_7e_7 \quad (5)$$

and

$$\mathcal{C}\mathtt{re}(a) = \frac{1}{2}(a+\bar{a}) = a_0 + a_7e_7,\ \mathcal{C}\mathtt{im}(a) = a - \mathcal{C}\mathtt{re}(a).$$

Note that $\mathbb{R} + \mathbb{R}e_7$ is the center field of $C\ell_{0,3}$. For $a \in C\ell_{0,3}$, we define $\mathcal{T} : C\ell_{0,3} \mapsto \mathbb{R}$ by

$$\mathcal{T}(a) = a_0a_7 + a_2a_5 - a_1a_6 - a_3a_4. \quad (6)$$

Then

$$a\bar{a} = \bar{a}a = |a|^2 + 2\mathcal{T}(a)e_7 \quad (7)$$

and

$$|ab|^2 = |a|^2|b|^2 + 4\mathcal{T}(a)\mathcal{T}(b),\ \mathcal{T}(ab) = |a|^2\mathcal{T}(b) + |b|^2\mathcal{T}(a),\ \mathcal{T}(a) = \mathcal{T}(\bar{a}).$$

Based on the real matrix representations of quaternions, we can introduce the real matrix representations of numbers in $C\ell_{0,3}$ as follows.

Definition 2.1. Let $a = a_h + a_He_4 \in C\ell_{0,3}$. Then the real 8-by-8 real matrix

$$\omega(a) := \begin{pmatrix} L(a_h) & -L(a_H)P_4 \\ L(a_H)P_4 & L(a_h) \end{pmatrix} \quad (8)$$

is called the left matrix representation of a over $\mathbb{R}$, where $P_4 = diag(1,-1,-1,1)$. The real 8-by-8 real matrix

$$\nu(a) := K_8\omega(a)^TK_8 \quad (9)$$

is called the right matrix representation of a, where $K_8 = diag(-1,1,1,1,1,1,1,-1)$.

We list some properties of the representation matrix as following propositions.

Proposition 2.1. *Denote* $\overrightarrow{x} = (x_0, x_1, x_2, x_3, x_4, x_5, x_6, x_7)^T$ *for* $x = x_0 + x_1e_1 + x_2e_2 + x_3e_3 + x_4e_4 + x_5e_5 + x_6e_6 + x_7e_7$. *Then*

$$\overrightarrow{ax} = \omega(a)\overrightarrow{x} \quad \text{and} \quad \overrightarrow{xa} = \nu(a)\overrightarrow{x}. \tag{10}$$

Proposition 2.2. *Let* $a, b \in C\ell_{0,3}$ *and* $\lambda \in \mathbb{R}$. *Then*

(1) $a = b \Longleftrightarrow \omega(a) = \omega(b) \Longleftrightarrow \nu(a) = \nu(b)$;
(2) $\omega(a+b) = \omega(a)+\omega(b), \omega(\lambda a) = \lambda\omega(a), \omega(ab) = \omega(a)\omega(b), \omega(1) = I_8$;
(3) $\nu(a+b) = \nu(a)+\nu(b), \nu(\lambda a) = \lambda\nu(a), \nu(ab) = \nu(b)\nu(a), \nu(1) = I_8$;
(4) $\omega(\bar{a}) = \omega(a)^T, \nu(\bar{a}) = \nu(a)^T, \omega(a)\nu(b) = \nu(b)\omega(a)$;

where I_8 *denotes the* 8×8 *unit matrix.*

Proposition 2.3. *The eigenvalues of right matrix representation of* a, $\omega(a)$, *are given by*

$$\lambda_{1,2} = a_0 + a_7 \pm i\sqrt{|a|^2 + 2T(a) - (a_0 + a_7)^2}$$

and

$$\lambda_{3,4} = a_0 - a_7 \pm i\sqrt{|a|^2 - 2T(a) - (a_0 - a_7)^2},$$

where $i = \sqrt{-1}$ *and each eigenvalue occurs with algebraic multiplicity 2.*

In order to find the solution of $ax = xb$ in Section 4, we need the following propositions which proof can be verify directly.

Proposition 2.4. *The eigenvalues of* $\omega(a) - \nu(b)$ *can be expressed by*

$$\begin{aligned}\mu_{1,2} = {} & (a_0 + a_7) - (b_0 + b_7) + i(\sqrt{(a_1 - a_6)^2 + (a_3 - a_4)^2 + (a_2 + a_5)^2} \\ & \pm \sqrt{(b_1 - b_6)^2 + (b_3 - b_4)^2 + (b_2 + b_5)^2}\,),\end{aligned}$$

$$\begin{aligned}\mu_{3,4} = {} & (a_0 + a_7) - (b_0 + b_7) - i(\sqrt{(a_1 - a_6)^2 + (a_3 - a_4)^2 + (a_2 + a_5)^2} \\ & \pm \sqrt{(b_1 - b_6)^2 + (b_3 - b_4)^2 + (b_2 + b_5)^2}\,),\end{aligned}$$

$$\begin{aligned}\mu_{5,6} = {} & (a_0 - a_7) - (b_0 - b_7) + i(\sqrt{(a_1 + a_6)^2 + (a_3 + a_4)^2 + (a_2 - a_5)^2} \\ & \pm \sqrt{(b_1 + b_6)^2 + (b_3 + b_4)^2 + (b_2 - b_5)^2}\,),\end{aligned}$$

$$\begin{aligned}\mu_{7,8} = {} & (a_0 - a_7) - (b_0 - b_7) - i(\sqrt{(a_1 + a_6)^2 + (a_3 + a_4)^2 + (a_2 - a_5)^2} \\ & \pm \sqrt{(b_1 + b_6)^2 + (b_3 + b_4)^2 + (b_2 - b_5)^2}\,).\end{aligned}$$

Proposition 2.5. *The eigenvalues of $\omega(a) + \nu(b)$ can be expressed by*

$$\mu_{1,2} = (a_0 + a_7) + (b_0 + b_7) + i(\sqrt{(a_1 - a_6)^2 + (a_3 - a_4)^2 + (a_2 + a_5)^2} \pm \sqrt{(b_1 - b_6)^2 + (b_3 - b_4)^2 + (b_2 + b_5)^2}\,),$$

$$\mu_{3,4} = (a_0 + a_7) + (b_0 + b_7) - i(\sqrt{(a_1 - a_6)^2 + (a_3 - a_4)^2 + (a_2 + a_5)^2} \pm \sqrt{(b_1 - b_6)^2 + (b_3 - b_4)^2 + (b_2 + b_5)^2}\,),$$

$$\mu_{5,6} = (a_0 - a_7) + (b_0 - b_7) + i(\sqrt{(a_1 + a_6)^2 + (a_3 + a_4)^2 + (a_2 - a_5)^2} \pm \sqrt{(b_1 + b_6)^2 + (b_3 + b_4)^2 + (b_2 - b_5)^2}\,),$$

$$\mu_{7,8} = (a_0 - a_7) + (b_0 - b_7) - i(\sqrt{(a_1 + a_6)^2 + (a_3 + a_4)^2 + (a_2 - a_5)^2} \pm \sqrt{(b_1 + b_6)^2 + (b_3 + b_4)^2 + (b_2 - b_5)^2}\,).$$

3. The Moore-Penrose inverse of elements in $C\ell_{0,3}$

Definition 3.1. Let $a \in C\ell_{0,3}$. If there exists an $x \in C\ell_{0,3}$ such that $ax = 1$, then x is called the inverse of a, which is denoted by a^{-1}.

For $a \in C\ell_{0,3}$, we define $p : C\ell_{0,3} \mapsto \mathbb{R}$ by

$$p(a) = |a|^4 - 4\mathcal{T}(a)^2. \tag{11}$$

Proposition 3.1. *If $p(a) = |a|^4 - 4\mathcal{T}(a)^2 \neq 0$, then a is invertible and*

$$a^{-1} = \frac{(|a|^2 - 2\mathcal{T}(a)e_7)}{p(a)}\bar{a}.$$

Lemma 3.1. *The following equations*

$$ax = 0, \ \bar{a}x = 0, \ \bar{a}ax = 0 \tag{12}$$

have the same solutions. Moreover, if a is not invertible then the general solution is

$$x = (|a|^2 - 2\mathcal{T}(a)e_7)p, \ \forall p \in C\ell_{0,3}. \tag{13}$$

Definition 3.2. The Moore-Penrose inverse of $a \in C\ell_{0,3}$, denoted by a^+, is a Clifford number x such that

$$axa = a, \ xax = x, \ \overline{ax} = ax. \tag{14}$$

Note that if x is the Moore-Penrose inverse of a then $\overline{xa} = xa$ and thus $\omega(x)$ is the unique Moore-Penrose inverse of $\omega(a)$.

If $a = 0$ then $a^+ = 0$ and if a is invertible then $a^+ = a^{-1} = \frac{(\bar{a}a)^*}{p(a)}\bar{a}$. For non-zero element a, by $\bar{a}a = a\bar{a} = |a|^2 + 2\mathcal{T}(a)e_7$, we have

$$a\frac{\bar{a}}{2|a|^2}a = a, \quad \frac{\bar{a}}{2|a|^2}a\frac{\bar{a}}{2|a|^2} = \frac{\bar{a}}{2|a|^2}, \quad \overline{a\frac{\bar{a}}{2|a|^2}} = a\frac{\bar{a}}{2|a|^2}, \tag{15}$$

which implies that the Moore-Penrose inverse of a (which is not invertible) is

$$a^+ = \frac{\bar{a}}{2|a|^2}. \tag{16}$$

It is easy to verify that $\omega(a^+) = \omega(a)^+$ and the following proposition.

Proposition 3.2. *If $a \in C\ell_{0,3}$ is not invertible, then a satisfies one of the following two cases:*
(1) $a\bar{a} = |a|^2(1 + e_7)$ and $a = ae_7$; (2) $a\bar{a} = |a|^2(1 - e_7)$ and $a = -ae_7$.

Examples: $a = e_2 + e_5$, $b = e_2 + e_3 + e_5$, $ab = -2 + e_1 - e_6 - 2e_7$, $ba = -2 - e_1 + e_6 - 2e_7$, $a^+ = -\frac{1}{4}(e_2 + e_5)$, $b^{-1} = -\frac{1}{5}(e_2 + 3e_3 + 2e_4 + e_5)$, $ae_7 = a$.

By the theory of linear equation, we have the following two theorems.

Theorem 3.1. *Let $a \in C\ell_{0,3}$ be a nonzero element and not invertible. Then the linear equation $ax = d$ has a solution if and only if $\frac{a\bar{a}}{2|a|^2}d = d$. In that case, the general solution can be expressed by*

$$x = \frac{\bar{a}d}{2|a|^2} + (2|a|^2 - a\bar{a})p, \ \forall\ p \in C\ell_{0,3}.$$

Theorem 3.2. *Let $a, b \in C\ell_{0,3}$ be nonzero elements and not invertible. Then the linear equation $axb = d$ has a solution if and only if $\frac{ab\overline{ab}}{4|a|^2|b|^2}d = d$. In that case, the general solution can be expressed by*

$$x = \frac{\bar{a}d\bar{b}}{4|a|^2|b|^2} + (4|a|^2|b|^2 - ab\overline{ab})p, \ \forall\ p \in C\ell_{0,3}.$$

4. Linear equation $ax = xb$ over $C\ell_{0,3}$

Due to the non-commutativity and existence of non-division elements of $C\ell_{0,3}$, it is non-trivial to find the solutions of equation in $C\ell_{0,3}$.

We say $a, b \in C\ell_{0,3}$ satisfy *condition 1* if

$$a_0 + a_7 = b_0 + b_7 \text{ and } |a|^2 + 2\mathcal{T}(a) = |b|^2 + 2\mathcal{T}(b) \tag{17}$$

and *condition 2* if

$$a_0 - a_7 = b_0 - b_7 \text{ and } |a|^2 - 2\mathcal{T}(a) = |b|^2 - 2\mathcal{T}(b). \tag{18}$$

By the equivalently of $(\omega(a) - \nu(b))\overrightarrow{x} = 0$ and the linear equation

$$ax = xb, \tag{19}$$

Proposition 2.4 implies that

Proposition 4.1. *Equation (19) has a nonzero solution if and only if a, b satisfy condition 1 or condition 2.*

In order to find the general solution of Eq.(19), we list some properties of its solutions.

Proposition 4.2. *The solution x of Eq.(19) with $|a| \neq |b|$ satisfies*
(1) $x = xe_7$ provided condition 1 holds; (2) $x = -xe_7$ provided condition 2 holds.

Proposition 4.3. *Equation (19) with $|a| = |b|$ has a nonzero solution if and only if one of the following three cases holds: (1) $\mathcal{T}(a) = \mathcal{T}(b), a_0 + a_7 = b_0 + b_7, a_0 \neq b_0$; (2) $\mathcal{T}(a) = \mathcal{T}(b), a_0 - a_7 = b_0 - b_7, a_0 \neq b_0$; (3) $\mathcal{T}(a) = \mathcal{T}(b), a_0 = b_0, a_7 = b_7$. Moreover, the solution x of Eq.(19) satisfies (4) $x = xe_7$ provided (1) holds ; (5) $x = -xe_7$ provided (2) holds.*

Proposition 4.4. *Equation (19) has a solution which is invertible if and only if*

$$\mathcal{C}\mathtt{re}(a) = \mathcal{C}\mathtt{re}(b), |a| = |b|, \text{ and } \mathcal{T}(a) = \mathcal{T}(b).$$

Remark 4.1. We say two Clifford numbers a and b are *similar* if there exists invertible Clifford number $q \in C\ell_{0,3}$ such that $a = qbq^{-1}$. The above proposition provides the necessary and sufficient conditions for two Clifford numbers to be similar, which generalizes the corresponding concept of quaternions.

We are now ready to consider the general solution of Eq.(19). This will be divided into three cases.

Case 1: $\mathcal{C}\mathtt{im}(a) = 0$ or $\mathcal{C}\mathtt{im}(b) = 0$

The case 1, say $\mathcal{C}\mathtt{im}(b) = 0$, implies that Eq.(19) is equivalent to $(a - b)x = 0$. By Lemma 3.1, we can summarize these results as the following theorem.

Theorem 4.1. *If case 1 and condition 1 or condition 2 hold, then the general solution of Eq.(19) is*

$$x = (|a - b|^2 - 2\mathcal{T}(a - b)e_7)p, \ \forall \ p \in C\ell_{0,3}.$$

Case 2: $|a| \neq |b|$, $\mathcal{C}\mathtt{im}(a) \neq 0$ *and* $\mathcal{C}\mathtt{im}(b) \neq 0$

Theorem 4.2. *If case 2 and condition 1 hold, then*

(1) if $|\mathcal{C}\mathtt{im}(a)|^2 = -2\mathcal{T}(\mathcal{C}\mathtt{im}(a))$ *and* $|\mathcal{C}\mathtt{im}(b)|^2 = -2\mathcal{T}(\mathcal{C}\mathtt{im}(b))$ *then the general solution can be expressed by*

$$x = (1 + e_7)p, \quad \forall p \in C\ell_{0,3}; \tag{20}$$

(2) otherwise the general solution can be expressed by

$$x = (1 + e_7)(\mathcal{C}\mathtt{im}(a)p + p\mathcal{C}\mathtt{im}(b)), \quad \forall \ p \in C\ell_{0,3}. \tag{21}$$

If case 2 and condition 2 hold, then

(3) if $|\mathcal{C}\mathtt{im}(a)|^2 = 2\mathcal{T}(\mathcal{C}\mathtt{im}(a))$ *and* $|\mathcal{C}\mathtt{im}(b)|^2 = 2\mathcal{T}(\mathcal{C}\mathtt{im}(b))$ *then the general solution can be expressed by*

$$x = (1 - e_7)p, \quad \forall \ p \in C\ell_{0,3}; \tag{22}$$

(4) otherwise the general solution can be expressed by

$$x = (1 - e_7)(\mathcal{C}\mathtt{im}(a)p + p\mathcal{C}\mathtt{im}(b)), \quad \forall \ p \in C\ell_{0,3}. \tag{23}$$

Proof. At first, we assume that condition 1 holds. By Proposition 4.2, if x_0 is a nonzero solution of $ax = xb$ then $(1 - e_7)x_0 = 0$. It follows from $ax_0 - x_0 b = (\mathcal{C}\mathtt{re}(a) - \mathcal{C}\mathtt{re}(b))x_0 + \mathcal{C}\mathtt{im}(a)x_0 - x_0\mathcal{C}\mathtt{im}(b)$ that

$$\mathcal{C}\mathtt{im}(a)x_0 = x_0\mathcal{C}\mathtt{im}(b).$$

If $|\mathcal{C}\mathtt{im}(a)|^2 = -2\mathcal{T}(\mathcal{C}\mathtt{im}(a))$ and $|\mathcal{C}\mathtt{im}(b)|^2 = -2\mathcal{T}(\mathcal{C}\mathtt{im}(b))$ hold, then each solution of Eq.(19) satisfies $\mathcal{C}\mathtt{im}(a)x = x\mathcal{C}\mathtt{im}(b) = 0$. By Lemma 3.1, the general solution can be expressed as

$$x = (1 + e_7)p, \quad \forall \ p \in C\ell_{0,3}.$$

This completes the proof of Theorem 4.2 (1).

For the remaining cases, since

$$\mathcal{C}\mathtt{im}(a)^2 - \mathcal{C}\mathtt{im}(b)^2 = (\mathcal{C}\mathtt{re}(a)^2 - \mathcal{C}\mathtt{re}(b)^2) + (|b|^2 - |a|^2 + 2(\mathcal{T}(b) - \mathcal{T}(a))e_7),$$

we have $(1+e_7)(\mathcal{C}\mathtt{im}(a)^2 - \mathcal{C}\mathtt{im}(b)^2) = 0$. Thus, $x = (1+e_7)(\mathcal{C}\mathtt{im}(a)p + p\mathcal{C}\mathtt{im}(b))$, which satisfies $ax - xb = 0$, is a solution of Eq.(19) for each p in $\mathcal{C}\ell_{0,3}$.

Suppose that x_0 is any solution of $ax = xb$ of the cases in Theorem 4.2 (2). Let $p = \mathcal{C}\mathtt{im}(a)^+ x_0/4$ provided $\mathcal{C}\mathtt{im}(a)$ is invertible or $|\mathcal{C}\mathtt{im}(a)|^2 = 2\mathcal{T}(\mathcal{C}\mathtt{im}(a))$, and let $p = x_0\mathcal{C}\mathtt{im}(b)^+/4$ provided $\mathcal{C}\mathtt{im}(b)$ is invertible or $|\mathcal{C}\mathtt{im}(b)|^2 = 2\mathcal{T}(\mathcal{C}\mathtt{im}(b))$. Then $x_0 = (1+e_7)(\mathcal{C}\mathtt{im}(a)p + p\mathcal{C}\mathtt{im}(b))$ for such p.

The above statements show that, for cases of Theorem 4.2 (2), any solution of $ax = xb$ can be expressed by (21). This completes the proof of Theorem 4.2 (2). By similar arguments, we can obtain the proof of Theorem 4.2 (3),(4). □

Case 3: $|a| = |b|$, $\mathcal{C}\mathtt{im}(a) \neq 0$ *and* $\mathcal{C}\mathtt{im}(b) \neq 0$

Theorem 4.3. *If case 3 and the hypotheses in Proposition 4.3 (1) hold, then*

(1) if $|\mathcal{C}\mathtt{im}(a)|^2 = -2\mathcal{T}(\mathcal{C}\mathtt{im}(a))$ *and* $|\mathcal{C}\mathtt{im}(b)|^2 = -2\mathcal{T}(\mathcal{C}\mathtt{im}(b))$ *then the general solution can be expressed by*

$$x = (1+e_7)p, \quad \forall p \in \mathcal{C}\ell_{0,3}; \tag{24}$$

(2) if $|\mathcal{C}\mathtt{im}(a)|^2 \neq -2\mathcal{T}(\mathcal{C}\mathtt{im}(a))$ *or* $|\mathcal{C}\mathtt{im}(b)|^2 \neq -2\mathcal{T}(\mathcal{C}\mathtt{im}(b))$ *then the general solution can be expressed by*

$$x = (1+e_7)(\mathcal{C}\mathtt{im}(a)p + p\mathcal{C}\mathtt{im}(b)), \quad \forall\, p \in \mathcal{C}\ell_{0,3}. \tag{25}$$

If case 3 and the hypotheses in Proposition 4.3 (2) hold, then

(3) if $|\mathcal{C}\mathtt{im}(a)|^2 = 2\mathcal{T}(\mathcal{C}\mathtt{im}(a))$ *and* $|\mathcal{C}\mathtt{im}(b)|^2 = 2\mathcal{T}(\mathcal{C}\mathtt{im}(b))$ *then the general solution can be expressed by*

$$x = (1-e_7)p, \quad \forall\, p \in \mathcal{C}\ell_{0,3}; \tag{26}$$

(4) if $|\mathcal{C}\mathtt{im}(a)|^2 \neq 2\mathcal{T}(\mathcal{C}\mathtt{im}(a))$ *or* $|\mathcal{C}\mathtt{im}(b)|^2 \neq 2\mathcal{T}(\mathcal{C}\mathtt{im}(b))$ *then the general solution can be expressed by*

$$x = (1-e_7)(\mathcal{C}\mathtt{im}(a)p + p\mathcal{C}\mathtt{im}(b)), \quad \forall\, p \in \mathcal{C}\ell_{0,3}. \tag{27}$$

If case 3 and the hypotheses in Proposition 4.3 (3) hold, then

(5) if $\mathcal{C}\mathtt{im}(a)$ *is invertible then the general solution can be expressed by*

$$x = \mathcal{C}\mathtt{im}(a)p + p\mathcal{C}\mathtt{im}(b), \quad \forall\, p \in \mathcal{C}\ell_{0,3}; \tag{28}$$

(6) if $\mathcal{C}\mathtt{im}(a)$ *is not invertible and*

(a) *if* $|\mathcal{C}\mathtt{im}(a)|^2 = -2\mathcal{T}(\mathcal{C}\mathtt{im}(a))$, *then the general belongs to the direct sum*

$$Q_1 \oplus Q_2 = \{x = x_1 + x_2 : x_1 \in Q_1, x_2 \in Q_2\},$$

where $Q_1 = \{x : x = \mathcal{C}\mathtt{im}(a)p + p\mathcal{C}\mathtt{im}(b), \forall p \in C4\}$ *and* $Q_2 = \{x : x = (1 + e_7)p, \forall\ p \in C\ell_{0,3}\}$.

(b) *if* $|\mathcal{C}\mathtt{im}(a)|^2 = 2\mathcal{T}(\mathcal{C}\mathtt{im}(a))$, *then the general belongs to the direct sum*

$$Q_1 \oplus Q_2 = \{x = x_1 + x_2 : x_1 \in Q_1, x_2 \in Q_2\},$$

where $Q_1 = \{x : x = \mathcal{C}\mathtt{im}(a)p + p\mathcal{C}\mathtt{im}(b), \forall p \in C4\}$ *and* $Q_2 = \{x : x = (1 - e_7)p, \forall\ p \in C\ell_{0,3}\}$.

Proof. The proofs of the Theorem 4.3 (1)-(4) follow from the similar reasoning as that in the proof of Theorem 4.2.

If the hypotheses in Proposition 4.3 (3) hold then $\mathcal{C}\mathtt{im}(a)^2 - \mathcal{C}\mathtt{im}(b)^2 = \mathcal{C}\mathtt{re}(a)^2 - \mathcal{C}\mathtt{re}(b)^2 = 0$. Thus $x = \mathcal{C}\mathtt{im}(a)p + p\mathcal{C}\mathtt{im}(b)$, which satisfies $ax - xb = (\mathcal{C}\mathtt{im}(a)^2 - \mathcal{C}\mathtt{im}(b)^2)p = 0$, is a solution of Eq.(19) for each p in $C\ell_{0,3}$. Under the hypotheses in Proposition 4.3 (3), x is a solution of Eq.(19) if and only if $\mathcal{C}\mathtt{im}(a)x = x\mathcal{C}\mathtt{im}(b)$.

If $\mathcal{C}\mathtt{im}(a)$ is invertible, then any solution x_0 can be expressed by $\mathcal{C}\mathtt{im}(a)p + p\mathcal{C}\mathtt{im}(b)$ for $p = \mathcal{C}\mathtt{im}(a)^{-1}x_0/2$. This completes the proof of Theorem 4.3 (5).

If $\mathcal{C}\mathtt{im}(a)$ is not invertible, say $|\mathcal{C}\mathtt{im}(a)|^2 = -2\mathcal{T}(\mathcal{C}\mathtt{im}(a))$, then $|\mathcal{C}\mathtt{im}(b)|^2 = -2\mathcal{T}(\mathcal{C}\mathtt{im}(b))$. Thus $x = (1 + e_7)p$, which satisfies $\mathcal{C}\mathtt{im}(a)x = x\mathcal{C}\mathtt{im}(b)$, is a solutions of Eq.(19) for each p in $C\ell_{0,3}$.

Let $Q_1 = \{x : x = (1 + e_7)p,\ \forall\ p \in C\ell_{0,3}\}$ and $Q_2 = \{x : x = \mathcal{C}\mathtt{im}(a)p + p\mathcal{C}\mathtt{im}(b),\ \forall\ p \in C\ell_{0,3}\}$. We can prove that $Q_1 \cap Q_2 = \{0\}$.

By Propositions 2.4 and 2.5, we know that $rank[\omega(a) - \nu(b)] = 2$ and $rank[\omega(\mathcal{C}\mathtt{im}(a)) + \nu(\mathcal{C}\mathtt{im}(b))] = 2$. Thus all the solutions of Eq.(19) belong to direct sum $Q_1 \oplus Q_2$ of Q_1 and Q_2.

Similarly, we can obtain the proof of the case $|\mathcal{C}\mathtt{im}(a)|^2 = 2\mathcal{T}(\mathcal{C}\mathtt{im}(a))$. This completes the proofs of Theorem 4.3 (6). □

Remark 4.2. By proposition 3.2, if $a \in C\ell_{0,3}$ is not invertible, then $a = \frac{1}{2}(1 + e_7)a$ or $a = \frac{1}{2}(1 - e_7)a$. In both cases, we have $(1 - e_7)a = 0$ or $(1 + e_7)a = 0$. Thus, in the case (6) of Theorem 4.3, if $y = y_1 + y_2 \in Q_1 \oplus Q_2$ with $0 \neq y_1 \in Q_1$ and $0 \neq y_2 \in Q_2$ then y is invertible.

References

1. L.V. Ahlfors, *Möbius transformations and Clifford numbers, in Differential Geometry and Complex Analysis,* ed I. Chavel & H.M. Farkas, 65-73, Springer 1985.
2. W. Cao, J.R. Parker and X. Wang, On the classification of quaternion Mobuis transformation, *Math. Proc. Cambridge. Philos. Soc.,* **137**(2004): 349-361.
3. J. Groß, G. Trenkler and S.-O. Troschke, Quaternions: further contributions to a matrix oriented approach, *Linear Algebra Appl.,* **326**(2001): 205-213.
4. K. Gürlebeck and W. Sprößig, *Quaternionic analysis and elliptic boundary value problems,* Birkhäuser, Basel, 1990.
5. Y. Tian, Matrix representations of octonions and their applications, *Adv. Appl. Clifford Algebras,* **10**(2000):61-90.

A CHANGE OF SCALE FORMULA FOR WIENER INTEGRALS OF UNBOUNDED FUNCTIONS OVER WIENER PATHS IN ABSTRACT WIENER SPACE*

K. S. CHANG

Department of Mathematics
Yonsei University
Seoul 120-749, Korea
E-mail: kunchang@yonsei.ac.kr

D. H. CHO†

Department of Mathematics
Kyonggi University
Kyonggido Suwon 443-760, Korea
E-mail: j94385@kyonggi.ac.kr

B. S. KIM

School of Liberal Arts
Seoul National University of Technology
Seoul 139-743, Korea
E-mail: mathkbs@snut.ac.kr

T. S. SONG

Department of Mathematics
Yonsei University
Seoul 120-749, Korea
E-mail: teukseob@dreamwiz.com

I. YOO

Department of Mathematics
Yonsei University
Kangwondo 220-710, Korea
E-mail: iyoo@dragon.yonsei.ac.kr

In 1987, Cameron and Storvick introduced change of scale formulas for Wiener integrals of bounded functions in a Banach algebra $\mathcal{S}$ on classical Wiener space. In 1994, Yoo and Skoug extended these results to abstract Wiener space, and recently Chang, Kim, Song and Yoo obtained a change of scale formula of various functions which need not be bounded or continuous. Let $C_0(\mathbb{B})$ denote the space of all abstract Wiener space-valued continuous functions on $[0,T]$ which vanish at 0. In this paper, we establish a change of scale formula for Wiener integrals of functions on $C_0(\mathbb{B})$ which need not be bounded or continuous. Using our formula, we obtain the above change of scale formulas for Wiener integrals of bounded or unbounded functions on classical and abstract Wiener spaces, as corollaries.

Keywords: Analytic Feynman integral, change of scale formula, Fresnel class, generalized Fresnel class, Wiener paths in abstract Wiener space.

2000 Mathematics Subject Classification: 28C20

1. Introduction and preliminaries

It has long been known that Wiener measure and Wiener measurability behave badly under change of scale transformation and under translation[1,2]. Cameron and Storvick expressed the analytic Feynman integral for a rather large class of functions as a limit of Wiener integrals[4]. As a result, they discovered a change of scale formula for Wiener integrals on the classical Wiener space[5]. In Refs. [12, 13, 15], Yoo, Yoon and Skoug extended these results to Yeh-Wiener space and to an abstract Wiener space. In particular, Yoo and Skoug established a change of scale formula for Wiener integrals of functions in Fresnel class $\mathcal{F}(\mathbb{B})$ on abstract Wiener space[12] and they developed this formula for a more generalized Fresnel class $\mathcal{F}_{A_1,A_2}$ than the class $\mathcal{F}(\mathbb{B})$ [13]. But the functions in both classes are bounded. Furthermore, in Ref. [14], Chang, Kim, Song and Yoo established a change of scale formula for Wiener integrals of functions of the form, for $x_1 \in \mathbb{B}$,

$$F_1(x_1) = G(x_1)\Psi((e_1,x_1)^\sim, \cdots, (e_n,x_1)^\sim) \tag{1}$$

for $G \in \mathcal{F}(\mathbb{B})$ and $\Psi = \psi + \phi$ where $\psi \in L_p(\mathbb{R}^n), 1 \le p < \infty$ and ϕ is the Fourier transform of a measure of bounded variation over $\mathbb{R}^n$. In the paper, they obtained the formula by direct calculations using the Wiener

*This research was supported by the Basic Science Research Institute Program, Korea Research Foundation under Grant KRF 2003-005-C00011.

†The second author was supported in part by Kyonggi University Research Grant.

integration formula on abstract Wiener space. Note that F_1 need not be continuous or bounded.

On the other hand, Chang, Cho and Yoo introduced the space $\mathcal{A}_{r,u}^{(p)}$ ($1 \le p \le \infty$) of cylinder type functions on $C_0(\mathbb{B})$, the space of abstract Wiener space-valued continuous functions $x(t)$ which are defined on $[0,T]$ with $x(0) = 0$, and they evaluated the conditional analytic Feynman integrals of the cylinder type functions[6]. In Ref. [8], Cho introduced the Banach algebra $\mathcal{F}(C_0(\mathbb{B});u)$ which is equivalent to both $\mathcal{F}(\mathbb{B})$ and $\mathcal{F}_{A_1,A_2}$, and evaluated the conditional analytic Feynman integral of functions in the class $\mathcal{F}(C_0(\mathbb{B});u)$.

In this paper, we establish a change of scale formula for Wiener integrals of possibly unbounded functions on $C_0(\mathbb{B})$ which have the form

$$G_r(x) = F(x)\Psi((e_1, x(u))^\sim, \cdots, (e_r, x(u))^\sim)$$

for $F \in \mathcal{F}(C_0(\mathbb{B});u)$ and $\Psi = f + \phi$ where $f \in L_p(\mathbb{R}^r), 1 \le p \le \infty$ and ϕ is the Fourier transform of a measure of bounded variation over $\mathbb{R}^r$. As a result, the change of scale formula for Wiener integrals of F_1 on $\mathbb{B}$ [14] is obtained by the formula over $C_0(\mathbb{B})$. This provides another method which drives the change of scale formula for Wiener integrals of the function F_1 given by (1).

Let $(\mathcal{H}, \mathbb{B}, m)$ be an abstract Wiener space[10]. Let $\{e_j : j = 1, 2, \cdots\}$ be a complete orthonormal set in the real separable Hilbert space $\mathcal{H}$ such that e_j's are in $\mathbb{B}^*$, the dual space of real separable Banach space $\mathbb{B}$. For each $h \in \mathcal{H}$ and $y \in \mathbb{B}$, define the stochastic inner product $(h, y)^\sim$ by

$$(h,y)^\sim = \begin{cases} \lim_{n\to\infty} \sum_{j=1}^{n} \langle h, e_j\rangle (y, e_j), & \text{if the limit exists;} \\ 0, & \text{otherwise,} \end{cases}$$

where $(\cdot, \cdot)$ denotes the dual pairing between $\mathbb{B}$ and $\mathbb{B}^*$ [9]. Note that for each $h (\neq 0)$ in $\mathcal{H}$, $(h, \cdot)^\sim$ is a Gaussian random variable on $\mathbb{B}$ with mean 0 and variance $|h|^2$; also $(h, y)^\sim$ is essentially independent of choice of the complete orthonormal set used in its definition and further, $(h, \lambda y)^\sim = (\lambda h, y)^\sim = \lambda (h, y)^\sim$ for all $\lambda \in \mathbb{R}$. It is well-known that if $\{h_1, h_2, \cdots, h_n\}$ is an orthogonal set in $\mathcal{H}$, then the random variables $(h_j, \cdot)^\sim$'s are independent. Moreover, if both h and y are in $\mathcal{H}$, then $(h, y)^\sim = \langle h, y\rangle$ where $\langle \cdot, \cdot \rangle$ denotes the inner product on $\mathcal{H}$.

Now, we introduce a useful integral formula which appears in the proofs of several results.

Lemma 1.1. *Let $\mathbb{C}_+ \equiv \{\lambda \in \mathbb{C} : \operatorname{Re}\lambda > 0\}$ and let $a \in \mathbb{C}_+$. Then for any* *real b, we have*

$$\int_{\mathbb{R}} \exp\{-au^2 + ibu\}du = \left(\frac{\pi}{a}\right)^{\frac{1}{2}} \exp\left\{-\frac{b^2}{4a}\right\}.$$

2. Wiener paths in abstract Wiener space

Let $C_0(\mathbb{B})$ denote the space of all continuous functions on $[0, T]$ into $\mathbb{B}$ which vanish at 0. Then $C_0(\mathbb{B})$ is a real separable Banach space with the norm $\|x\|_{C_0(\mathbb{B})} \equiv \sup_{0\le t\le T} \|x(t)\|_{\mathbb{B}}$. The minimal σ-field making the mapping $x \to x(t)$ measurable is $\mathcal{B}(C_0(\mathbb{B}))$, the Borel σ-field on $C_0(\mathbb{B})$. Further, Brownian motion in $\mathbb{B}$ induces a probability measure $m_{\mathbb{B}}$ on $(C_0(\mathbb{B}), \mathcal{B}(C_0(\mathbb{B})))$ which is mean-zero Gaussian[11].

A function defined on $C_0(\mathbb{B})$ is said to be Wiener measurable if it is measurable and the function is said to be Wiener integrable if it is integrable with respect to $m_{\mathbb{B}}$.

Now, we introduce Wiener integration theorem without proof. For the proof, see Ref. [11].

Theorem 2.1. *Let $\vec{t} = (t_1, t_2, \cdots, t_k)$ be given with $0 = t_0 < t_1 < t_2 < \cdots < t_k \le T$ and let $f : \mathbb{B}^k \to \mathbb{C}$ be a Wiener measurable function. Then, we have*

$$\int_{C_0(\mathbb{B})} f(x(t_1), x(t_2), \cdots, x(t_k))\, dm_{\mathbb{B}}(x)$$

$$\stackrel{*}{=} \int_{\mathbb{B}^k} f\left(\sqrt{t_1 - t_0}x_1, \sqrt{t_1 - t_0}x_1 + \sqrt{t_2 - t_1}x_2, \cdots, \sum_{j=1}^{k} \sqrt{t_j - t_{j-1}}x_j\right)$$

$$d\left(\prod_{j=1}^{k} m\right)(x_1, x_2, \cdots, x_k),$$

where by $\stackrel{*}{=}$ we mean that if either side exists, then both sides exist and they are equal.

A subset E of $C_0(\mathbb{B})$ is called a scale-invariant null set if $m_{\mathbb{B}}(\lambda E) = 0$ for any $\lambda > 0$. A property is said to hold scale-invariant almost everywhere (in abbreviation, s-a.e.) if it holds except for a scale-invariant null set. Let F be defined on $C_0(\mathbb{B})$ and let $F^\lambda(x) = F(\lambda^{-\frac{1}{2}}x)$ for $\lambda > 0$. Suppose that $E[F^\lambda]$ exists for any $\lambda > 0$, where the expectation is taken over $C_0(\mathbb{B})$ and it has the analytic extension $J^*_\lambda(F)$ on $\mathbb{C}_+$. Then we call $J^*_\lambda(F)$ the analytic Wiener integral of F over $C_0(\mathbb{B})$ with parameter λ and it is denoted by

$$E^{anw_\lambda}[F] = J^*_\lambda(F).$$

Moreover, if for a non-zero real q, $E^{anw_\lambda}[F]$ has a limit as λ approaches to $-iq$ through $\mathbb{C}_+$, then it is called the analytic Feynman integral of F over $C_0(\mathbb{B})$ with parameter q and denoted by

$$E^{anf_q}[F] = \lim_{\lambda \to -iq} E^{anw_\lambda}[F].$$

Similar definitions and notations can be understood on abstract Wiener space if F is defined on the space.

3. A change of scale formula over Wiener paths in abstract Wiener space

Throughout this paper, let $0 < u \le T$ be fixed, but arbitrarily. Let $\mathcal{H}$ be a real separable infinite dimensional Hilbert space, let r be a positive integer and let $\{e_1, \cdots, e_r\}$ be an orthonormal set in $\mathcal{H}$. Let

$$X_r(x_1) = ((e_1, x_1)^\sim, \cdots, (e_r, x_1)^\sim) \tag{2}$$

for x_1 in $\mathbb{B}$. For $1 \le p < \infty$, let $\mathcal{A}_{r,u}^{(p)}$ be the space of all cylinder type functions F_r defined on $C_0(\mathbb{B})$ of the form

$$F_r(x) = f(X_r(x(u))) \tag{3}$$

for s-a.e. x in $C_0(\mathbb{B})$ where $f : \mathbb{R}^r \to \mathbb{R}$ is in $L_p(\mathbb{R}^r)$. Let $\mathcal{A}_{r,u}^{(\infty)}$ be the space of all functions of the form (3) with $f \in L_\infty(\mathbb{R}^r)$, the space of essentially bounded functions on $\mathbb{R}^r$. Note that, without loss of generality, we can take f to be Borel measurable.

Let $\mathcal{M}(\mathcal{H})$ be the class of all $\mathbb{C}$-valued Borel measures on $\mathcal{H}$ with bounded variation and let $\mathcal{F}(C_0(\mathbb{B}); u)$ be the space of all s-equivalence classes of functions F which, for $\sigma \in \mathcal{M}(\mathcal{H})$, have the form

$$F(x) = \int_{\mathcal{H}} \exp\{i(h, x(u))^\sim\} d\sigma(h) \tag{4}$$

for $x \in C_0(\mathbb{B})$. Note that $\mathcal{F}(C_0(\mathbb{B}); u)$ is a Banach algebra which is equivalent to Fresnel class $\mathcal{F}(\mathbb{B})$ with the norm $\|F\| = \|\sigma\|$, the total variation of σ in $\mathcal{M}(\mathcal{H})$ [8].

For convenience, we introduce a useful notations using Gram-Schmidt orthonormalization process. Let $h \in \mathcal{H}$. Then we obtain an orthonormal set $\{e_1, \cdots, e_r, e_{r+1}\}$ as follows;

$$c_j(h) = \begin{cases} \langle h, e_j \rangle & \text{for } j = 1, \cdots, r; \\ (|h|^2 - \sum_{l=1}^r [c_l(h)]^2)^{\frac{1}{2}} & \text{for } j = r+1 \end{cases} \tag{5}$$

and

$$e_{r+1} = \frac{1}{c_{r+1}(h)}\Big(h - \sum_{j=1}^{r} c_j(h)e_j\Big)$$

if $c_{r+1}(h) \neq 0$. Then we have

$$h = \sum_{j=1}^{r+1} c_j(h)e_j \quad \text{and} \quad |h|^2 = \sum_{j=1}^{r+1}[c_j(h)]^2. \tag{6}$$

Note that (6) holds trivially for the case $c_{r+1}(h) = 0$. If we have an orthonormal set in $\mathcal{H}$ whose number of elements is larger than r, then we assume that the set contains $\{e_1, \cdots, e_r\}$ as a subset.

The following lemma is useful to prove our several results.

Lemma 3.1. *Let $a > 0$ and $h \in \mathcal{H}$. Further, let F_r be given by (3). Then we have*

$$\begin{aligned} &E[\exp\{i(h, ax(u))^{\sim}\}F_r(ax)] \\ &\overset{*}{=} \Big(\frac{1}{2\pi ua^2}\Big)^{\frac{r}{2}} \int_{\mathbb{R}^r} f(\vec{u}_r)\exp\Big\{\frac{ua^2}{2}\Big[\sum_{j=1}^{r}\Big(\frac{i}{ua^2}u_j + c_j(h)\Big)^2 - |h|^2\Big]\Big\}d\vec{u}_r \end{aligned} \tag{7}$$

with $\vec{u}_r = (u_1, \cdots, u_r)$, where by $\overset{*}{=}$ we mean that if either side exists, then both sides exist and they are equal, and $c_j(h)$ is given by (5).

Proof. By (3), (6) and Theorem 2.1, we have

$$\begin{aligned} &E[\exp\{i(h, ax(u))^{\sim}\}F_r(ax)] \\ &= \int_{C_0(\mathbb{B})} \exp\{ia(h, x(u))^{\sim}\}f(a((e_1, x(u))^{\sim}, \cdots, (e_r, x(u))^{\sim}))dm_{\mathbb{B}}(x) \\ &\overset{*}{=} \int_{\mathbb{B}} f(a\sqrt{u}((e_1, x_1)^{\sim}, \cdots, (e_r, x_1)^{\sim}))\exp\Big\{ia\sqrt{u}\sum_{j=1}^{r+1} c_j(h)(e_j, x_1)^{\sim}\Big\}dm(x_1). \end{aligned}$$

Let $\vec{v}_r = (v_1, \cdots, v_r)$ and $\vec{v}_{r+1} = (v_1, \cdots, v_r, v_{r+1})$. Since $(e_j, \cdot)^{\sim}$'s are Gaussian with mean 0 and variance 1 and they are independent, we have

$$\begin{aligned} &E[\exp\{i(h, ax(u))^{\sim}\}F_r(ax)] \\ &\overset{*}{=} \Big(\frac{1}{2\pi}\Big)^{\frac{r+1}{2}} \int_{\mathbb{R}^{r+1}} f(a\sqrt{u}\vec{v}_r)\exp\Big\{ia\sqrt{u}\sum_{j=1}^{r+1} c_j(h)v_j - \frac{1}{2}\sum_{j=1}^{r+1} v_j^2\Big\}d\vec{v}_{r+1} \end{aligned}$$

$$= (2\pi)^{\frac{1}{2}}\left(\frac{1}{2\pi}\right)^{\frac{r+1}{2}} \exp\left\{-\frac{ua^2[c_{r+1}(h)]^2}{2}\right\}\int_{\mathbb{R}^r} f(a\sqrt{u}\vec{v}_r)$$
$$\exp\left\{ia\sqrt{u}\sum_{j=1}^{r} c_j(h)v_j - \frac{1}{2}\sum_{j=1}^{r} v_j^2\right\}d\vec{v}_r$$

by Lemma 1.1. Now, let $\vec{u}_r = a\sqrt{u}\,\vec{v}_r$. Then, by (5) and the change of variable theorem, we have

$$E[\exp\{i(h, ax(u))^{\sim}\}F_r(ax)]$$
$$\stackrel{*}{=} \left(\frac{1}{2\pi ua^2}\right)^{\frac{r}{2}}\int_{\mathbb{R}^r} f(\vec{u}_r)\exp\left\{-\frac{ua^2}{2}\left(|h|^2 - \sum_{j=1}^{r}[c_j(h)]^2\right) + i\sum_{j=1}^{r} c_j(h)u_j\right.$$
$$\left. - \frac{1}{2ua^2}\sum_{j=1}^{r} u_j^2\right\}d\vec{u}_r$$
$$= \left(\frac{1}{2\pi ua^2}\right)^{\frac{r}{2}}\int_{\mathbb{R}^r} f(\vec{u}_r)\exp\left\{\frac{ua^2}{2}\left[\sum_{j=1}^{r}\left(\frac{i}{ua^2}u_j + c_j(h)\right)^2 - |h|^2\right]\right\}d\vec{u}_r$$

which is the desired result. □

Theorem 3.1. *Let G_r be given by*

$$G_r(x) = F(x)F_r(x) \tag{8}$$

for s-a.e. x in $C_0(\mathbb{B})$, where $F_r \in \mathcal{A}_{r,u}^{(p)}$ $(1 \le p \le \infty)$ and $F \in \mathcal{F}(C_0(\mathbb{B}); u)$ are given by (3) and (4), respectively. Then, for $\lambda \in \mathbb{C}_+$, $E^{anw_\lambda}[G_r]$ exists and it is given by

$$E^{anw_\lambda}[G_r]$$
$$= \left(\frac{\lambda}{2\pi u}\right)^{\frac{r}{2}}\int_{\mathcal{H}}\int_{\mathbb{R}^r} f(\vec{u}_r)\exp\left\{\frac{u}{2\lambda}\left[\sum_{j=1}^{r}\left(\frac{\lambda i}{u}u_j + c_j(h)\right)^2 - |h|^2\right]\right\}d\vec{u}_r d\sigma(h) \tag{9}$$

where $\vec{u}_r = (u_1, \cdots, u_r)$ and $c_j(h)$ is given by (5).

Proof. For $\lambda > 0$, we have

$$E[G_r^\lambda] = \int_{C_0(\mathbb{B})}\int_{\mathcal{H}} \exp\{i(h, \lambda^{-\frac{1}{2}}x(u))^{\sim}\}F_r^\lambda(x)\, d\sigma(h)dm_{\mathbb{B}}(x).$$

Now, we have

$$\begin{aligned}&\int_{\mathcal{H}}\int_{C_0(\mathbb{B})}|\exp\{i(h,\lambda^{-\frac{1}{2}}x(u))^{\sim}\}F_r^{\lambda}(x)|\,dm_{\mathbb{B}}(x)d|\sigma|(h)\\&=\|\sigma\|E[\,|F_r^{\lambda}|\,]\\&=\|\sigma\|\Big(\frac{\lambda}{2\pi u}\Big)^{\frac{r}{2}}\int_{\mathbb{R}^r}|f(\vec{u}_r)|\exp\Big\{-\frac{\lambda}{2u}\sum_{j=1}^{r}u_j^2\Big\}d\vec{u}_r<\infty\end{aligned}$$

by Hölder's inequality, where the last equality follows from (7) in Lemma 3.1 with $a=\lambda^{-\frac{1}{2}}$ and $h=0$. Again, by (7) in Lemma 3.1 with $a=\lambda^{-\frac{1}{2}}$ and Fubini's theorem, we have

$$\begin{aligned}&E[G_r^{\lambda}]\\&=\Big(\frac{\lambda}{2\pi u}\Big)^{\frac{r}{2}}\int_{\mathcal{H}}\int_{\mathbb{R}^r}f(\vec{u}_r)\exp\Big\{\frac{u}{2\lambda}\Big[\sum_{j=1}^{r}\Big(\frac{\lambda i}{u}u_j+c_j(h)\Big)^2-|h|^2\Big]\Big\}d\vec{u}_r d\sigma(h).\end{aligned}$$

By Bessel's inequality, for $\lambda\in\mathbb{C}_+$, we have

$$\begin{aligned}&|f(\vec{u}_r)|\,\Big|\exp\Big\{\frac{u}{2\lambda}\Big[\sum_{j=1}^{r}\Big(\frac{\lambda i}{u}u_j+c_j(h)\Big)^2-|h|^2\Big]\Big\}\Big|\\&=|f(\vec{u}_r)|\exp\Big\{-\frac{u\mathrm{Re}\,\lambda}{2|\lambda|^2}\Big(|h|^2-\sum_{j=1}^{r}[c_j(h)]^2\Big)-\frac{\mathrm{Re}\,\lambda}{2u}\sum_{j=1}^{r}u_j^2\Big\}\\&\le|f(\vec{u}_r)|\exp\Big\{-\frac{\mathrm{Re}\,\lambda}{2u}\sum_{j=1}^{r}u_j^2\Big\}\end{aligned}$$

and hence using Hölder's inequality, we have the theorem by Morera's theorem. □

For the case $p=1$ in Theorem 3.1, we have the following corollary by the dominated convergence theorem.

Corollary 3.1. *Let the assumptions and notations be given as in Theorem 3.1 except for $F_r\in\mathcal{A}_{r,u}^{(1)}$. Then, for a non-zero real q, $E^{anf_q}[G_r]$ exists and it is given by*

$$\begin{aligned}&E^{anf_q}[G_r]\\&=\Big(\frac{q}{2\pi ui}\Big)^{\frac{r}{2}}\int_{\mathcal{H}}\int_{\mathbb{R}^r}f(\vec{u}_r)\exp\Big\{\frac{iu}{2q}\Big[\sum_{j=1}^{r}\Big(\frac{q}{u}u_j+c_j(h)\Big)^2-|h|^2\Big]\Big\}d\vec{u}_r d\sigma(h).\end{aligned}$$

Let $\hat{\mathrm{M}}(\mathbb{R}^r)$ be the space of all functions ϕ defined on $\mathbb{R}^r$ by

$$\phi(\vec{u}_r) = \int_{\mathbb{R}^r} \exp\left\{ i \sum_{j=1}^{r} u_j\, t_j \right\} d\rho(t_1, \cdots, t_r), \tag{10}$$

where $\vec{u}_r = (u_1, \cdots, u_r) \in \mathbb{R}^r$ and ρ is a complex Borel measure of bounded variation over $\mathbb{R}^r$.

Theorem 3.2. *Let ϕ be given by (10). Let K_r be given by the right-hand side of (3) with replacing f by ϕ and let $G_r = FK_r$ where F is given by (4). Then, for $\lambda \in \mathbb{C}_+$, $E^{anw_\lambda}[G_r]$ exists and it is given by*

$$E^{anw_\lambda}[G_r] = \int_{\mathcal{H}} \int_{\mathbb{R}^r} \exp\left\{ -\frac{u}{2\lambda} \left[|h|^2 + 2 \sum_{j=1}^{r} c_j(h) t_j + \sum_{j=1}^{r} t_j^2 \right] \right\} d\rho(\vec{t}_r) d\sigma(h)$$

where $\vec{t}_r = (t_1, \cdots, t_r)$ and $c_j(h)$ is given by (5).

Proof. Since ϕ is bounded, we have $K_r \in \mathcal{A}_{r,u}^{(\infty)}$. By Theorem 3.1, $E^{anw_\lambda}[G_r]$ exists for $\lambda \in \mathbb{C}_+$ and it is given by

$$\begin{aligned} &E^{anw_\lambda}[G_r] \\ &= \left(\frac{\lambda}{2\pi u} \right)^{\frac{r}{2}} \int_{\mathcal{H}} \int_{\mathbb{R}^r} \phi(\vec{u}_r) \exp\left\{ \frac{u}{2\lambda} \left[\sum_{j=1}^{r} \left(\frac{\lambda i}{u} u_j + c_j(h) \right)^2 - |h|^2 \right] \right\} d\vec{u}_r d\sigma(h) \end{aligned}$$

from (9), where $\vec{u}_r = (u_1, \cdots, u_r)$. Hence, by Fubini's theorem, we have

$$\begin{aligned} &E^{anw_\lambda}[G_r] \\ &= \left(\frac{\lambda}{2\pi u} \right)^{\frac{r}{2}} \int_{\mathcal{H}} \int_{\mathbb{R}^r} \int_{\mathbb{R}^r} \exp\left\{ i \sum_{j=1}^{r} t_j u_j + \frac{u}{2\lambda} \left[\sum_{j=1}^{r} \left(\frac{\lambda i}{u} u_j + c_j(h) \right)^2 - |h|^2 \right] \right\} \\ &\quad d\vec{u}_r d\rho(\vec{t}_r) d\sigma(h) \\ &= \left(\frac{\lambda}{2\pi u} \right)^{\frac{r}{2}} \int_{\mathcal{H}} \int_{\mathbb{R}^r} \int_{\mathbb{R}^r} \exp\left\{ -\frac{\lambda}{2u} \sum_{j=1}^{r} u_j^2 + i \sum_{j=1}^{r} [t_j + c_j(h)] u_j \right. \\ &\quad \left. + \frac{u}{2\lambda} \left[\sum_{j=1}^{r} [c_j(h)]^2 - |h|^2 \right] \right\} d\vec{u}_r d\rho(\vec{t}_r) d\sigma(h) \end{aligned}$$

where $\vec{t_r} = (t_1, \cdots, t_r)$. By Lemma 1.1, we have

$$\begin{aligned}
&E^{anw_\lambda}[G_r] \\
&= \int_{\mathcal{H}} \int_{\mathbb{R}^r} \exp\left\{ -\frac{u}{2\lambda} \left[\sum_{j=1}^{r} [[t_j + c_j(h)]^2 - [c_j(h)]^2] + |h|^2 \right] \right\} d\rho(\vec{t_r}) d\sigma(h) \\
&= \int_{\mathcal{H}} \int_{\mathbb{R}^r} \exp\left\{ -\frac{u}{2\lambda} \left[|h|^2 + 2\sum_{j=1}^{r} c_j(h) t_j + \sum_{j=1}^{r} t_j^2 \right] \right\} d\rho(\vec{t_r}) d\sigma(h)
\end{aligned}$$

which completes the proof. □

The following corollary is an immediate consequence of Theorem 3.2, by Bessel's inequality and the dominated convergence theorem.

Corollary 3.2. *Let the assumptions and notations be given as in Theorem 3.2. Then, for a non-zero real q, $E^{anf_q}[G_r]$ exists and it is given by*

$$E^{anf_q}[G_r] = \int_{\mathcal{H}} \int_{\mathbb{R}^r} \exp\left\{ \frac{u}{2qi} \left[|h|^2 + 2\sum_{j=1}^{r} c_j(h) t_j + \sum_{j=1}^{r} t_j^2 \right] \right\} d\rho(\vec{t_r}) d\sigma(h).$$

From the previous theorems and corollaries, we have the following corollary by the linearity of analytic Wiener and Feynman integrals over Wiener paths in abstract Wiener space.

Corollary 3.3. *Let F be given by (4) and F_r, K_r be given as in Theorems 3.1, 3.2, respectively. Then, for $\lambda \in \mathbb{C}_+$, $E^{anw_\lambda}[F(F_r + K_r)]$ exists and it is given by*

$$\begin{aligned}
&E^{anw_\lambda}[F(F_r + K_r)] \\
&= \int_{\mathcal{H}} \left[\left(\frac{\lambda}{2\pi u} \right)^{\frac{r}{2}} \int_{\mathbb{R}^r} f(\vec{u}_r) \exp\left\{ \frac{u}{2\lambda} \left[\sum_{j=1}^{r} \left(\frac{\lambda i}{u} u_j + c_j(h) \right)^2 - |h|^2 \right] \right\} d\vec{u}_r \right. \\
&\quad \left. + \int_{\mathbb{R}^r} \exp\left\{ -\frac{u}{2\lambda} \left[|h|^2 + 2\sum_{j=1}^{r} c_j(h) t_j + \sum_{j=1}^{r} t_j^2 \right] \right\} d\rho(\vec{t_r}) \right] d\sigma(h)
\end{aligned}$$

where $\vec{u}_r = (u_1, \cdots, u_r)$, $\vec{t_r} = (t_1, \cdots, t_r)$ and $c_j(h)$ is given by (5). In particular, if $F_r \in \mathcal{A}_{r,u}^{(1)}$, then, for a non-zero real q, $E^{anf_q}[F(F_r + K_r)]$

exists and it is given by

$$\begin{aligned}
&E^{anf_q}[F(F_r+K_r)]\\
&=\int_{\mathcal{H}}\Big[\Big(\frac{q}{2\pi ui}\Big)^{\frac{r}{2}}\int_{\mathbb{R}^r} f(\vec{u}_r)\exp\Big\{\frac{iu}{2q}\Big[\sum_{j=1}^{r}\Big(\frac{q}{u}u_j+c_j(h)\Big)^2-|h|^2\Big]\Big\}d\vec{u}_r\\
&+\int_{\mathbb{R}^r}\exp\Big\{\frac{u}{2qi}\Big[|h|^2+2\sum_{j=1}^{r}c_j(h)t_j+\sum_{j=1}^{r}t_j^2\Big]\Big\}d\rho(\vec{t}_r)\Big]d\sigma(h).
\end{aligned}$$

Lemma 3.2. *Let $\{e_1,\cdots,e_n\}$ be an orthonormal set in $\mathcal{H}$ with $n>r$. For $\lambda\in\mathbb{C}_+$ and $h\in\mathcal{H}$, let*

$$K=\int_{C_0(\mathbb{B})}\exp\Big\{\frac{1-\lambda}{2u}\sum_{j=1}^{n}[(e_j,x(u))^\sim]^2+i(h,x(u))^\sim\Big\}F_r(x)dm_{\mathbb{B}}(x)$$

where $F_r\in\mathcal{A}_{r,u}^{(p)}(1\le p\le\infty)$ is given by (3). Then we have

$$\begin{aligned}
K=\lambda^{-\frac{n}{2}}\Big(\frac{\lambda}{2\pi u}\Big)^{\frac{r}{2}}&\exp\Big\{\frac{(\lambda-1)u}{2\lambda}\sum_{j=1}^{n}[c_j(h)]^2-\frac{u}{2}|h|^2\Big\}\\
\times\int_{\mathbb{R}^r}&f(\vec{u}_r)\exp\Big\{\frac{u}{2\lambda}\sum_{j=1}^{r}\Big(\frac{\lambda i}{u}u_j+c_j(h)\Big)^2\Big\}d\vec{u}_r,
\end{aligned}$$

where $\vec{u}_r=(u_1,\cdots,u_r)$ and $c_j(h)$ is given by (5) with replacing r by n.

Proof. Using Gram-Schmidt orthonormalization process, (5), (6) and Theorem 2.1, we have

$$\begin{aligned}
&K\\
&=\int_{\mathbb{B}}\exp\Big\{\frac{1-\lambda}{2}\sum_{j=1}^{n}[(e_j,x_1)^\sim]^2+i\sqrt{u}\sum_{j=1}^{n+1}c_j(h)(e_j,x_1)^\sim\Big\}f(\sqrt{u}((e_1,x_1)^\sim,\\
&\cdots,(e_r,x_1)^\sim))dm(x_1)
\end{aligned}$$

$$\begin{aligned}
&=\Big(\frac{1}{2\pi}\Big)^{\frac{n+1}{2}}\int_{\mathbb{R}^{n+1}}f(\sqrt{u}\vec{v}_r)\exp\Big\{\frac{1-\lambda}{2}\sum_{j=1}^{n}v_j^2+i\sqrt{u}\sum_{j=1}^{n+1}c_j(h)v_j-\frac{1}{2}\sum_{j=1}^{n+1}v_j^2\Big\}\\
&d\vec{v}_{n+1}
\end{aligned}$$

where $\vec{v}_r=(v_1,\cdots,v_r)$ and $\vec{v}_{n+1}=(v_1,\cdots,v_r,v_{r+1},\cdots,v_n,v_{n+1})$, since $(e_j,\cdot)^\sim$'s are Gaussian with mean 0 and variance 1. Let $\vec{u}_{n+1}=$

$(u_1,\cdots,u_n,u_{n+1}) = \sqrt{u}\vec{v}_{n+1}$ and $\vec{u}_r = \sqrt{u}\vec{v}_r$. By the change of variable theorem and Lemma 1.1, we have

$$K = \left(\frac{1}{2\pi u}\right)^{\frac{n+1}{2}} \int_{\mathbb{R}^{n+1}} f(\vec{u}_r) \exp\left\{\frac{1-\lambda}{2u}\sum_{j=1}^{n} u_j^2 \right.$$

$$\left. +i\sum_{j=1}^{n+1} c_j(h)u_j - \frac{1}{2u}\sum_{j=1}^{n+1} u_j^2\right\} d\vec{u}_{n+1}$$

$$= \lambda^{-\frac{n}{2}}\left(\frac{\lambda}{2\pi u}\right)^{\frac{r}{2}} \exp\left\{-\frac{u}{2\lambda}\sum_{j=1}^{n}[c_j(h)]^2 - \frac{u}{2}\left(|h|^2 - \sum_{j=1}^{n}[c_j(h)]^2\right)\right\} \int_{\mathbb{R}^r} f(\vec{u}_r)$$

$$\times \exp\left\{-\frac{\lambda}{2u}\sum_{j=1}^{r} u_j^2 + i\sum_{j=1}^{r} c_j(h)u_j + \frac{u}{2\lambda}\sum_{j=1}^{r}[c_j(h)]^2\right\} d\vec{u}_r$$

$$= \lambda^{-\frac{n}{2}}\left(\frac{\lambda}{2\pi u}\right)^{\frac{r}{2}} \exp\left\{\frac{(\lambda-1)u}{2\lambda}\sum_{j=1}^{n}[c_j(h)]^2 - \frac{u}{2}|h|^2\right\} \int_{\mathbb{R}^r} f(\vec{u}_r) \exp\left\{\frac{u}{2\lambda}\sum_{j=1}^{r}\right.$$

$$\left.\left(\frac{\lambda i}{u}u_j + c_j(h)\right)^2\right\} d\vec{u}_r$$

which is the desired result. □

By Lemmas 1.1, 3.2, Theorem 2.1 and Eq. (10), we have the following lemma.

Lemma 3.3. *Let $\{e_1,\cdots,e_r,e_{r+1},\cdots,e_n\}$ be an orthonormal set in $\mathcal{H}$ with $n > r$. Further, let $h \in \mathcal{H}$ and $t_j \in \mathbb{R}$ for $j = 1,\cdots,r$. Then, for $\lambda \in \mathbb{C}_+$, we have*

$$\int_{C_0(\mathbb{B})} \exp\left\{\frac{1-\lambda}{2u}\sum_{j=1}^{n}[(e_j,x(u))^\sim]^2 + i(h,x(u))^\sim \right.$$

$$\left. +i\sum_{j=1}^{r} t_j(e_j,x(u))^\sim\right\} dm_{\mathbb{B}}(x)$$

$$= \lambda^{-\frac{n}{2}} \exp\left\{\frac{(\lambda-1)u}{2\lambda}\sum_{j=1}^{n}[c_j(h)]^2 - \frac{u}{\lambda}\sum_{j=1}^{r} c_j(h)t_j - \frac{u}{2\lambda}\sum_{j=1}^{r} t_j^2 - \frac{u}{2}|h|^2\right\},$$

where $c_j(h)$ is given by (5) with replacing r by n.

Now, we derive a relationship between Wiener integral and analytic Wiener integral over Wiener paths in abstract Wiener space.

Theorem 3.3. *Let $\{e_j : j = 1, 2, \cdots\}$ be a complete orthonormal set in $\mathcal{H}$. Moreover, let G_r be given by (8) and the assumptions be given as in Theorem 3.1. Then, for $\lambda \in \mathbb{C}_+$, we have*

$$E^{anw_\lambda}[G_r] = \lim_{n\to\infty}\left[\lambda^{\frac{n}{2}}\int_{C_0(\mathbb{B})}\exp\left\{\frac{1-\lambda}{2u}\sum_{j=1}^{n}[(e_j, x(u))^{\sim}]^2\right\}G_r(x)dm_{\mathbb{B}}(x)\right]. \quad (11)$$

Proof. Let $n \in \mathbb{N}$ with $n > r$ and let

$$\Gamma_n = \int_{C_0(\mathbb{B})}\exp\left\{\frac{1-\lambda}{2u}\sum_{j=1}^{n}[(e_j, x(u))^{\sim}]^2\right\}G_r(x)dm_{\mathbb{B}}(x).$$

By (8), Lemma 3.2 and Fubini's theorem, we have

$$\begin{aligned}
&\Gamma_n \\
&= \int_{\mathcal{H}}\int_{C_0(\mathbb{B})}\exp\left\{\frac{1-\lambda}{2u}\sum_{j=1}^{n}[(e_j, x(u))^{\sim}]^2 + i(h, x(u))^{\sim}\right\}F_r(x)dm_{\mathbb{B}}(x)d\sigma(h) \\
&= \lambda^{-\frac{n}{2}}\left(\frac{\lambda}{2\pi u}\right)^{\frac{r}{2}}\int_{\mathcal{H}}\exp\left\{\frac{(\lambda-1)u}{2\lambda}\sum_{j=1}^{n}[c_j(h)]^2 - \frac{u}{2}|h|^2\right\} \\
&\quad \int_{\mathbb{R}^r} f(\vec{u}_r)\exp\left\{\frac{u}{2\lambda}\sum_{j=1}^{r}\left(\frac{\lambda i}{u}u_j + c_j(h)\right)^2\right\}d\vec{u}_r d\sigma(h)
\end{aligned}$$

where $\vec{u}_r = (u_1, \cdots, u_r)$ and $c_j(h)$ is given by (5). By Bessel's inequality, we have, for $\lambda \in \mathbb{C}_+$,

$$\begin{aligned}
&|f(\vec{u}_r)|\left|\exp\left\{\frac{(\lambda-1)u}{2\lambda}\sum_{j=1}^{n}[c_j(h)]^2 - \frac{u}{2}|h|^2 + \frac{u}{2\lambda}\sum_{j=1}^{r}\left(\frac{\lambda i}{u}u_j + c_j(h)\right)^2\right\}\right| \\
&= |f(\vec{u}_r)|\exp\left\{-\frac{u}{2}\left(|h|^2 - \sum_{j=1}^{n}[c_j(h)]^2\right)\right. \\
&\quad \left. -\frac{u\mathrm{Re}\,\lambda}{2|\lambda|^2}\sum_{j=r+1}^{n}[c_j(h)]^2 - \frac{\mathrm{Re}\,\lambda}{2u}\sum_{j=1}^{r}u_j^2\right\} \\
&\le |f(\vec{u}_r)|\exp\left\{-\frac{\mathrm{Re}\,\lambda}{2u}\sum_{j=1}^{r}u_j^2\right\}
\end{aligned}$$

which is integrable on $\mathcal{H} \times \mathbb{R}^r$ by Hölder's inequality. By the dominated convergence theorem and Parseval's relation, we obtain

$$\begin{aligned}
&\lim_{n\to\infty} \lambda^{\frac{n}{2}}\Gamma_n \\
&= \lim_{n\to\infty}\left[\left(\frac{\lambda}{2\pi u}\right)^{\frac{r}{2}} \int_{\mathcal{H}} \exp\left\{\frac{(\lambda-1)u}{2\lambda}\sum_{j=1}^{n}[c_j(h)]^2 - \frac{u}{2}|h|^2\right\} \int_{\mathbb{R}^r} f(\vec{u}_r)\exp\left\{\frac{u}{2\lambda}\right.\right. \\
&\quad \left.\left.\times\sum_{j=1}^{r}\left(\frac{\lambda i}{u}u_j + c_j(h)\right)^2\right\} d\vec{u}_r d\sigma(h)\right] \\
&= \left(\frac{\lambda}{2\pi u}\right)^{\frac{r}{2}} \int_{\mathcal{H}} \exp\left\{\frac{(\lambda-1)u}{2\lambda}|h|^2 - \frac{u}{2}|h|^2\right\} \int_{\mathbb{R}^r} f(\vec{u}_r)\exp\left\{\frac{u}{2\lambda}\sum_{j=1}^{r}\left(\frac{\lambda i}{u}u_j\right.\right. \\
&\quad \left.\left. + c_j(h)\right)^2\right\} d\vec{u}_r d\sigma(h) \\
&= \left(\frac{\lambda}{2\pi u}\right)^{\frac{r}{2}} \int_{\mathcal{H}}\int_{\mathbb{R}^r} f(\vec{u}_r)\exp\left\{\frac{u}{2\lambda}\left[\sum_{j=1}^{r}\left(\frac{\lambda i}{u}u_j + c_j(h)\right)^2 - |h|^2\right]\right\} d\vec{u}_r d\sigma(h) \\
&= E^{anw_\lambda}[G_r]
\end{aligned}$$

where the last equality follows from (9) in Theorem 3.1. □

Now, we derive a relationship between Wiener integral and analytic Feynman integral over Wiener paths in abstract Wiener space. Using the method given as in the proof of Theorem 3.3, the following corollary immediately follows from Corollary 3.1, Parseval's relation and the dominated convergence theorem.

Corollary 3.4. *Let $\{e_j : j = 1, 2, \cdots\}$ be a complete orthonormal set in $\mathcal{H}$. Then, under the assumptions given as in Corollary 3.1, we have*

$$\begin{aligned}
&E^{anf_q}[G_r] \\
&= \lim_{n\to\infty}\left[\lambda_n^{\frac{n}{2}} \int_{C_0(\mathbb{B})} \exp\left\{\frac{1-\lambda_n}{2u}\sum_{j=1}^{n}[(e_j, x(u))^{\sim}]^2\right\} G_r(x)dm_{\mathbb{B}}(x)\right] \qquad (12)
\end{aligned}$$

for any sequence $\{\lambda_n\}_{n=1}^{\infty}$ in $\mathbb{C}_+$ with $\lambda_n \to -iq$ as $n \to \infty$.

Theorem 3.4. *Let $\{e_j : j = 1, 2, \cdots\}$ be a complete orthonormal set in $\mathcal{H}$. Let G_r be given as in Theorem 3.2. Then, for $\lambda \in \mathbb{C}_+$, Eq. (11) holds.*

Proof. Let $n \in \mathbb{N}$ with $n > r$ and let

$$\Gamma_n = \int_{C_0(\mathbb{B})} \exp\left\{\frac{1-\lambda}{2u}\sum_{j=1}^{n}[(e_j, x(u))^{\sim}]^2\right\} G_r(x)dm_{\mathbb{B}}(x).$$

By Fubini's theorem and Lemma 3.3, we have, for $\lambda \in \mathbb{C}_+$,

$$\Gamma_n = \int_{\mathcal{H}}\int_{\mathbb{R}^r}\int_{C_0(\mathbb{B})} \exp\bigg\{\frac{1-\lambda}{2u}\sum_{j=1}^{n}[(e_j, x(u))^\sim]^2$$

$$+i(h, x(u))^\sim + i\sum_{j=1}^{r} t_j(e_j, x(u))^\sim\bigg\} dm_{\mathbb{B}}(x)d\rho(\vec{t}_r)d\sigma(h)$$

$$= \lambda^{-\frac{n}{2}}\int_{\mathcal{H}}\int_{\mathbb{R}^r}\exp\bigg\{\frac{(\lambda-1)u}{2\lambda}\sum_{j=1}^{n}[c_j(h)]^2 - \frac{u}{\lambda}\sum_{j=1}^{r}c_j(h)t_j$$

$$-\frac{u}{2\lambda}\sum_{j=1}^{r}t_j^2 - \frac{u}{2}|h|^2\bigg\}d\rho(\vec{t}_r)d\sigma(h)$$

where $\vec{t}_r = (t_1, \cdots, t_r)$ and $c_j(h)$ is given by (5). Now, we have

$$\bigg|\exp\bigg\{\frac{(\lambda-1)u}{2\lambda}\sum_{j=1}^{n}[c_j(h)]^2 - \frac{u}{\lambda}\sum_{j=1}^{r}c_j(h)t_j - \frac{u}{2\lambda}\sum_{j=1}^{r}t_j^2 - \frac{u}{2}|h|^2\bigg\}\bigg|$$

$$= \exp\bigg\{-\frac{u}{2}\bigg(|h|^2 - \sum_{j=1}^{n}[c_j(h)]^2\bigg) - \frac{u\mathrm{Re}\,\lambda}{2|\lambda|^2}\bigg[\sum_{j=r+1}^{n}[c_j(h)]^2$$

$$+\sum_{j=1}^{r}[t_j + c_j(h)]^2\bigg]\bigg\} \le 1$$

by Bessel's inequality. Hence, by the dominated convergence theorem and Parseval's relation, we have the theorem from Theorem 3.2. □

Corollary 3.5. *Let q be a non-zero real number and $\{\lambda_n\}_{n=1}^{\infty}$ be a sequence in $\mathbb{C}_+$ with $\lambda_n \to -iq$ as $n \to \infty$. Then, under the assumptions given as in Theorem 3.4, Eq. (12) holds.*

Corollary 3.6. *Let $\{e_j : j = 1, 2, \cdots\}$ be a complete orthonormal set in $\mathcal{H}$. Moreover, let $G_r = F(F_r + K_r)$, where F is given by (4) and F_r, K_r are given as in Theorems 3.1, 3.2, respectively. Then, for $\lambda \in \mathbb{C}_+$, Eq. (11) holds.*

Corollary 3.7. *Let q be a non-zero real number and $\{\lambda_n\}_{n=1}^{\infty}$ be a sequence in $\mathbb{C}_+$ with $\lambda_n \to -iq$ as $n \to \infty$. Then, under the assumptions given as in Corollary 3.6 with one exception $F_r \in \mathcal{A}_{r,u}^{(1)}$, Eq. (12) holds.*

Our main result, namely, a change of scale formula for Wiener integrals over Wiener paths in abstract Wiener space, immediately follows from Corollary 3.6.

Theorem 3.5. *Let the assumptions be given as in Corollary 3.6. Then, for $\gamma > 0$, we have*

$$E[G_r(\gamma x)] = \lim_{n\to\infty}\left[\gamma^{-n}\int_{C_0(\mathbb{B})}\exp\left\{\frac{\gamma^2-1}{2u\gamma^2}\sum_{j=1}^{n}[(e_j,x(u))^{\sim}]^2\right\}G_r(x)dm_{\mathbb{B}}(x)\right]. \quad (13)$$

Proof. Letting $\lambda = \gamma^{-2}$ in (11), we have Eq. (13). □

Corollary 3.8. *Let $\{e_j : j = 1, 2, \cdots\}$ be a complete orthonormal set in $\mathcal{H}$ and let F be given by (4). Then, for $\gamma > 0$, Eq. (13) holds with replacing G_r by F.*

Proof. Let ρ be the Dirac measure concentrated at $\vec{0}$ in Eq. (10). Then $1 \in \hat{\mathrm{M}}(\mathbb{R})$ and hence, letting $F_r = 0$ and $K_r = 1$ in Theorem 3.5, we have the result. □

4. A change of scale formula on abstract Wiener space

In this section we drive a change of scale formula for Wiener integrals of possibly unbounded functions on abstract Wiener space as a special case of the formula over Wiener paths in abstract Wiener space.

The Fresnel class $\mathcal{F}(\mathbb{B})$ is defined as the space of all s-equivalence classes of functions F_1 which have the form, for $\sigma \in \mathcal{M}(\mathcal{H})$,

$$F_1(x_1) = \int_{\mathcal{H}}\exp\{i(h,x_1)^{\sim}\}d\sigma(h) \quad (14)$$

for $x_1 \in \mathbb{B}$. It is well-known that $\mathcal{F}(\mathbb{B})$ is a Banach algebra and the mapping $\sigma \to F_1$ is a Banach algebra isomorphism where σ and F_1 are related by (14) [4,9]. Moreover, it is not difficult to show that $\mathcal{F}(C_0(\mathbb{B}); u)$ is isomorphic to $\mathcal{F}(\mathbb{B})$ as Banach algebras[8].

Corollary 4.1. *(Ref. [14], Theorem 3.1). Let $G(x_1) = F_1(x_1)f(X_r(x_1))$ for s-a.e. $x_1 \in \mathbb{B}$ where X_r and F_1 are given by (2) and (14), respectively, and f is in $L_p(\mathbb{R}^r)(1 \le p \le \infty)$. Then, for $\lambda \in \mathbb{C}_+$, we have*

$$E^{anw_\lambda}[G] = \left(\frac{\lambda}{2\pi}\right)^{\frac{r}{2}}\int_{\mathcal{H}}\int_{\mathbb{R}^r} f(\vec{u}_r)\exp\left\{\frac{1}{2\lambda}\left[\sum_{j=1}^{r}[\lambda i u_j + c_j(h)]^2 - |h|^2\right]\right\}d\vec{u}_r d\sigma(h)$$

where $\vec{u}_r = (u_1, \cdots, u_r)$ and $c_j(h)$ is given by (5).

Proof. For convenience, define

$$T_1(h) = \frac{1}{\sqrt{u}}h \quad \text{and} \quad f_u(\vec{u}_r) = f\left(\frac{1}{\sqrt{u}}\vec{u}_r\right)$$

for $h \in \mathcal{H}$ and $\vec{u}_r \in \mathbb{R}^r$. Then $f \in L_p(\mathbb{R}^r)$ if and only if $f_u \in L_p(\mathbb{R}^r)$. Let $F_{r,u}$ and F_u be given by (3) and (4), respectively, with replacing f by f_u and σ by $T_1^{-1} \circ \sigma$. Moreover, let $G_{r,u} = F_u F_{r,u}$. Now, for $\lambda > 0$, we have

$$\begin{aligned}
&E[G^\lambda] \\
&= \int_{\mathbb{B}} F_1^\lambda(x_1) f(X_r^\lambda(x_1)) dm(x_1) \\
&= \int_{C_0(\mathbb{B})} \int_{\mathcal{H}} \exp\{i(T_1(h), \lambda^{-\frac{1}{2}} x(u))^\sim\} f_u(X_r^\lambda(x(u))) d\sigma(h) dm_{\mathbb{B}}(x) \\
&= \int_{C_0(\mathbb{B})} F_{r,u}^\lambda(x) \int_{\mathcal{H}} \exp\{i(h, \lambda^{-\frac{1}{2}} x(u))^\sim\} d(T_1^{-1} \circ \sigma)(h) dm_{\mathbb{B}}(x) \\
&= E[G_{r,u}^\lambda]
\end{aligned}$$

by Theorem 2.1 and the change of variable theorem. By the uniqueness of the analytic extension, we have

$$E^{anw_\lambda}[G] = E^{anw_\lambda}[G_{r,u}] \tag{15}$$

for $\lambda \in \mathbb{C}_+$ and hence we have

$$\begin{aligned}
&E^{anw_\lambda}[G] \\
&= \left(\frac{\lambda}{2\pi u}\right)^{\frac{r}{2}} \int_{\mathcal{H}} \int_{\mathbb{R}^r} f_u(\vec{v}_r) \exp\left\{\frac{u}{2\lambda}\left[\sum_{j=1}^r \left(\frac{\lambda i}{u} v_j + c_j(h)\right)^2 - |h|^2\right]\right\} \\
&\quad d\vec{v}_r d(T_1^{-1} \circ \sigma)(h) \\
&= \left(\frac{\lambda}{2\pi u}\right)^{\frac{r}{2}} \int_{\mathcal{H}} \int_{\mathbb{R}^r} f\left(\frac{1}{\sqrt{u}}\vec{v}_r\right) \exp\left\{\frac{u}{2\lambda}\left[\sum_{j=1}^r \left(\frac{\lambda i}{u} v_j + \frac{1}{\sqrt{u}} c_j(h)\right)^2 - \frac{1}{u}|h|^2\right]\right\} \\
&\quad d\vec{v}_r d\sigma(h)
\end{aligned}$$

by (9) in Theorem 3.1 where $\vec{v}_r = (v_1, \cdots, v_r)$. Let $\vec{u}_r = \frac{1}{\sqrt{u}}\vec{v}_r$. Again, by the change of variable theorem, we have

$$\begin{aligned}
&E^{anw_\lambda}[G] \\
&= \left(\frac{\lambda}{2\pi}\right)^{\frac{r}{2}} \int_{\mathcal{H}} \int_{\mathbb{R}^r} f(\vec{u}_r) \exp\left\{\frac{1}{2\lambda}\left[\sum_{j=1}^r [\lambda i u_j + c_j(h)]^2 - |h|^2\right]\right\} d\vec{u}_r d\sigma(h)
\end{aligned}$$

which is the desired result. □

Corollary 4.2. *(Ref. [14], Corollary 3.2). Let the assumptions be given as in Corollary 4.1 except for* $p = 1$*. Then, for a non-zero real* q*,* $E^{anf_q}[G]$ *exists and it is given by*

$$E^{anf_q}[G] = \left(\frac{q}{2\pi i}\right)^{\frac{r}{2}} \int_{\mathcal{H}} \int_{\mathbb{R}^r} f(\vec{u}_r) \exp\left\{\frac{i}{2q}\left[\sum_{j=1}^{r}[qu_j + c_j(h)]^2 - |h|^2\right]\right\} d\vec{u}_r d\sigma(h)$$

where $\vec{u}_r = (u_1, \cdots, u_r)$ *and* $c_j(h)$ *is given by (5).*

Proof. Using the similar method given as in the proof of Eq. (15), it is not difficult to show that

$$E^{anf_q}[G] = E^{anf_q}[G_{r,u}].$$

Now, the result follows from Corollary 3.1 and the change of variable theorem. □

Corollary 4.3. *(Ref. [14], Theorem 3.3). Let* $G(x_1) = F_1(x_1)\phi(X_r(x_1))$ *for s-a.e.* $x_1 \in \mathbb{B}$ *where* X_r*,* ϕ *and* F_1 *are given by (2), (10) and (14), respectively. Then, for* $\lambda \in \mathbb{C}_+$*, we have*

$$E^{anw_\lambda}[G] = \int_{\mathcal{H}} \int_{\mathbb{R}^r} \exp\left\{-\frac{1}{2\lambda}\left[|h|^2 + 2\sum_{j=1}^{r} t_j c_j(h) + \sum_{j=1}^{r} t_j^2\right]\right\} d\rho(\vec{t}_r) d\sigma(h)$$

where $\vec{t}_r = (t_1, \cdots, t_r)$ *and* $c_j(h)$ *is given by (5).*

Proof. For convenience, define $T_2 : \mathbb{R}^r \to \mathbb{R}^r$ by

$$T_2(\vec{t}_r) = \frac{1}{\sqrt{u}}\vec{t}_r$$

and let ϕ_ρ be given by (10) with replacing ρ by $T_2^{-1} \circ \rho$. Moreover, let $K_{r,u}(x) = \phi_\rho(X_r(x(u)))$ for s-a.e. $x \in C_0(\mathbb{B})$. Now, for $\lambda > 0$, we have

$$\begin{aligned} &E[G^\lambda] \\ &= \int_{\mathbb{B}} F_1^\lambda(x_1)\phi(X_r^\lambda(x_1))dm(x_1) \\ &= \int_{C_0(\mathbb{B})} \int_{\mathcal{H}} \exp\{i(T_1(h), \lambda^{-\frac{1}{2}}x(u))^\sim\}\phi(T_2(X_r^\lambda(x(u))))d\sigma(h)dm_{\mathbb{B}}(x) \\ &= \int_{C_0(\mathbb{B})} K_{r,u}^\lambda(x) \int_{\mathcal{H}} \exp\{i(h, \lambda^{-\frac{1}{2}}x(u))^\sim\}d(T_1^{-1} \circ \sigma)(h)dm_{\mathbb{B}}(x) \\ &= E[F_u^\lambda K_{r,u}^\lambda] \end{aligned}$$

by Theorem 2.1 and the change of variable theorem, where T_1 and F_u are given as in the proof of Corollary 4.1. By the uniqueness of the analytic extension, we have

$$E^{anw_\lambda}[G] = E^{anw_\lambda}[F_u K_{r,u}] \tag{16}$$

for $\lambda \in \mathbb{C}_+$ and hence, by the change of variable theorem and Theorem 3.2, we have

$$\begin{aligned} &E^{anw_\lambda}[G] \\ &= \int_{\mathcal{H}}\int_{\mathbb{R}^r} \exp\Big\{-\frac{u}{2\lambda}\Big[|h|^2 + 2\sum_{j=1}^{r} c_j(h)t_j + \sum_{j=1}^{r} t_j^2\Big]\Big\} \\ &\quad d(T_2^{-1}\circ\rho)(\vec{t}_r)d(T_1^{-1}\circ\sigma)(h) \\ &= \int_{\mathcal{H}}\int_{\mathbb{R}^r} \exp\Big\{-\frac{1}{2\lambda}\Big[|h|^2 + 2\sum_{j=1}^{r} c_j(h)t_j + \sum_{j=1}^{r} t_j^2\Big]\Big\}d\rho(\vec{t}_r)d\sigma(h) \end{aligned}$$

which is the desired result. □

Using the similar method given as in the proof of Eq. (16), we have the following corollary from the change of variable theorem and Corollary 3.2.

Corollary 4.4. *(Ref. [14], Corollary 3.4). Let the assumptions be given as in Corollary 4.3. Then, for a non-zero real q, $E^{anf_q}[G]$ exists and it is given by*

$$E^{anf_q}[G] = \int_{\mathcal{H}}\int_{\mathbb{R}^r} \exp\Big\{\frac{1}{2qi}\Big[|h|^2 + 2\sum_{j=1}^{r} c_j(h)t_j + \sum_{j=1}^{r} t_j^2\Big]\Big\}d\rho(\vec{t}_r)d\sigma(h)$$

where $\vec{t}_r = (t_1, \cdots, t_r)$ and $c_j(h)$ is given by (5).

Corollary 4.5. *(Ref. [14], Theorem 3.8). Let $\{e_j : j = 1, 2, \cdots\}$ be a complete orthonormal set in $\mathcal{H}$. Then, under the assumptions given as in Corollary 4.1, we have*

$$E^{anw_\lambda}[G] = \lim_{n\to\infty}\Big[\lambda^{\frac{n}{2}}\int_{\mathbb{B}} \exp\Big\{\frac{1-\lambda}{2}\sum_{j=1}^{n}[(e_j, x_1)^{\sim}]^2\Big\}G(x_1)dm(x_1)\Big] \tag{17}$$

for $\lambda \in \mathbb{C}_+$.

Proof. By (11), (15) and Theorem 2.1, we have

$$\begin{aligned}&E^{anw_\lambda}[G]\\&=E^{anw_\lambda}[G^\lambda_{r,u}]\\&=\lim_{n\to\infty}\left[\lambda^{\frac{n}{2}}\int_{C_0(\mathbb{B})}\exp\left\{\frac{1-\lambda}{2u}\sum_{j=1}^{n}[((e_j,x(u))^{\sim}]^2\right\}G_{r,u}(x)dm_{\mathbb{B}}(x)\right]\end{aligned}$$

where $G_{r,u}$ is given as in the proof of Corollary 4.1. Using the similar method given as in the proof of Corollary 4.1 with Theorem 2.1, we have

$$E^{anw_\lambda}[G]=\lim_{n\to\infty}\left[\lambda^{\frac{n}{2}}\int_{\mathbb{B}}\exp\left\{\frac{1-\lambda}{2}\sum_{j=1}^{n}[(e_j,x_1)^{\sim}]^2\right\}G(x_1)dm(x_1)\right]$$

which is the desired result. □

Now, if $p=1$ in Corollary 4.1, then we can obtain the following relationship between Wiener integral and analytic Feynman integral on abstract Wiener space using Theorem 2.1, Corollary 3.4 and the change of variable theorem, but not Corollary 4.2.

Corollary 4.6. *(Ref. [14], Theorem 3.9). Let $\{e_j : j = 1, 2, \cdots\}$ be a complete orthonormal set in $\mathcal{H}$. Then, under the assumptions given as in Corollary 4.2, we have*

$$E^{anf_q}[G]=\lim_{n\to\infty}\left[\lambda_n^{\frac{n}{2}}\int_{\mathbb{B}}\exp\left\{\frac{1-\lambda_n}{2}\sum_{j=1}^{n}[(e_j,x_1)^{\sim}]^2\right\}G(x_1)dm(x_1)\right]\quad(18)$$

for any sequence $\{\lambda_n\}_{n=1}^\infty$ in $\mathbb{C}_+$ with $\lambda_n\to -iq$ as $n\to\infty$.

By Theorems 2.1, 3.4, Corollary 3.5 and the change of variable theorem, we have the following corollary.

Corollary 4.7. *(Ref. [14], Theorem 3.10 and Corollary 3.11). Let $\{e_j : j = 1, 2, \cdots\}$ be a complete orthonormal set in $\mathcal{H}$. Then, under the assumptions given as in Corollaries 4.3 and 4.4, Eqs. (17) and (18) hold.*

As a corollary of Theorems 2.1 and 3.5, we can obtain the following change of scale formula for Wiener integrals on abstract Wiener space.

Corollary 4.8. *(Ref. [14], Theorem 3.14). Let ϕ be given by (10) and let $\{e_j : j = 1, 2, \cdots\}$ be a complete orthonormal set in $\mathcal{H}$. Moreover, let F_1,*

f and X_r be given as in Corollary 4.1. Then, for $\gamma > 0$, we have

$$\int_{\mathbb{B}} F_1(\gamma x_1)[(f+\phi)(X_r(\gamma x_1))]dm(x_1)$$
$$= \lim_{n\to\infty}\left[\gamma^{-n}\int_{\mathbb{B}}\exp\left\{\frac{\gamma^2-1}{2\gamma^2}\sum_{j=1}^{n}[(e_j,x_1)^{\sim}]^2\right\}\right.$$
$$\left. F_1(x_1)[(f+\phi)(X_r(x_1))]dm(x_1)\right].$$

Remark 4.1. Using the similar methods, we can prove Corollaries 3.12, 3.13 and 3.15 of Ref. [14] from Corollaries 3.6, 3.7 and 3.8, respectively.

Remark 4.2. With Cameron and Storvick's Banach algebra $\mathcal{S}$ in Ref. [3], the change of scale formula for Wiener integrals on classical Wiener space, say, Corollaries 4.1, 4.2, 4.3, 4.4 and 4.5 of Ref. [14] can be obtained by using the process in proof of Corollaries 4.1, 4.3 and the orthonormalization process in Corollary 4.1 of Ref. [14]. Note that, in Corollaries 4.3 and 4.4 of Ref. [14], Paley-Wiener-Zygmund integral is the stochastic inner product on classical Wiener space which is a kind of abstract Wiener space.

References

1. R. H. Cameron, The translation pathology of Wiener space, *Duke Math. J.* **21** (1954):623-628.
2. R. H. Cameron and W. T. Martin, The behavior of measure and measurability under change of scale in Wiener space, *Bull. Amer. Math. Soc.* **53** (1947):130-137.
3. R. H. Cameron and D. A. Storvick, Some Banach algebras of analytic Feynman integrable functionals, *Lecture Notes in Math.*, Springer-Verlag, N. Y., **798**(1980):18-67.
4. R. H. Cameron and D. A. Storvick, Relationships between the Wiener integral and the analytic Feynman integral, *Supplimento ai Rendiconti del Circolo Matematico di Palermo, Serie II-Numero* **17** (1987):117-133.
5. R. H. Cameron and D. A. Storvick, Change of scale formula for Wiener integral, *Supplimento ai Rendiconti del Circolo Matematico di Palermo, Serie II-Numero* **17** (1987):105-115.
6. K. S. Chang, D. H. Cho and I. Yoo, *A conditional analytic Feynman integral over Wiener paths in abstract Wiener space*, Intern. Math. J. **2** (2002), no. 9, 855-870.
7. K. S. Chang, G. W. Johnson and D. L. Skoug, *Functions in the Fresnel class*, Proc. Amer. Math. Soc. **100** (1987), 309-318.
8. D. H. Cho, *Conditional analytic Feynman integral over product space of Wiener paths in abstract Wiener space*, Rocky Mount. J. Math. (2005), to appear.

9. G. Kallianpur and C. Bromley, *Generalized Feynman integrals using analytic continuation in several complex variables, Stochastic analysis and applications*, Dekker, N. Y., 1984, 217-267.
10. H. H. Kuo, *Gaussian measures in Banach spaces*, Lecture Notes in Math. **463**, Springer-Verlag, Berlin, 1975.
11. K. S. Ryu, *The Wiener integral over paths in abstract Wiener space*, J. Korean Math. Soc. **29** (1992), no. 2, 317-331.
12. I. Yoo and D. L. Skoug, *A change of scale formula for Wiener integrals on abstract Wiener spaces*, Int. J. Math. Math. Sci. **17** (1994), 239-248.
13. I. Yoo and D. L. Skoug, *A change of scale formula for Wiener integrals on abstract Wiener spaces II*, J. Korean Math. Soc. **31** (1994), 115-129.
14. I. Yoo, T. S. Song, B. S. Kim, and K. S. Chang, *A change of scale formula for Wiener integrals of unbounded functions*, Rocky Mount. J. Math. **34** (2004), no. 1, 371-389.
15. I. Yoo and G. J. Yoon, *Change of scale formulas for Yeh-Wiener integrals*, Comm. Korean Math. Soc. **6** (1991), 19-26.

THE COMPLEX OSCILLATION OF SOLUTIONS FOR DIFFERENTIAL EQUATIONS WITH PERIODIC COEFFICIENTS*

CHEN ZONG-XUAN
Department of Mathematics
South China Normal University
Guangzhou, 510631 P. R. China
E-mail: chzx@sina.com

GAO SHI-AN
Department of Mathematics
South China Normal University
Guangzhou, 510631 P. R. China
E-mail: gaosha@scnu.edu.cn

In this paper, we reveal the properties of solutions for a class of periodic differential equations with dominant coefficients of orders more extensive than positive integers or infinity for the first time, and simultaneously discover that the dominant coefficients for this class of equations only need to possess "single-side property".

Keywords: Function in the exterior of a circle; differential polynomial; periodic differential equation; complex oscillation.

2000 Mathematics Subject Classification: 30D35

1. Introduction and Preliminaries

We use standard notations from the value distribution theory in this paper (for example, see [11, 12]): $T(r,f), m(r,f), N(r,f), \overline{N}(r,f), S(r,f)$, etc, with $f(z)$ meromorphic in the plane. In addition, we denote the order of growth of $f(z)$ by $\sigma(f)$, and denote respectively the exponent of convergence

*Project supported by the National Natural Science Foundation of China (No: 10161006) and by the Natural Science Foundation of Guangdong Province in China (No: 04010360).

of the zeros and the exponent of convergence of the distinct zeros for $f(z)$ by $\lambda(f)$ and $\overline{\lambda}(f)$.

For the periodic meromorphic functions of transcendental type, their orders are often infinity which is not comparable. We now define another new order applying to periodic functions. For a meromorphic function $f(z)$, we define

$$\sigma_e(f) = \overline{\lim_{r\to\infty}} \frac{\log T(r,f)}{r} \tag{1.1}$$

to be the e–type order of $f(z)$. In addition, we define respectively

$$\lambda_e(f) = \overline{\lim_{r\to\infty}} \frac{\log^+ N(r,1/f)}{r}, \quad \overline{\lambda}_e(f) = \overline{\lim_{r\to\infty}} \frac{\log^+ \overline{N}(r,1/f)}{r}$$

to be the e–type exponent of convergence of the zeros and the e–type exponent of convergence of the distinct zeros for $f(z)$[5].

For the periodic differential equations with dominant coefficients of orders not equal to positive integers or infinity, the researches for properties of solutions and complex oscillation have been developed by several works (for example, [1, 3, 5, 6, 7]). But for the periodic differential equations with dominant coefficients of orders equal to positive integers or infinity, the corresponding researches have not appeared. In this paper, we reveal the properties of solutions and complex oscillation for a class of periodic differential equations with dominant coefficients of orders more extensive than positive integers or infinity for the first time, and simultaneously discover that the dominant coefficients for this class of equations only need to possess "single-side property".

Assume $A(z)$ to be a periodic entire function with period $2\pi i$. Thus, we can write $A(z) = B(e^z)$, where $B(\zeta)$ is clearly analytic in $0 < |\zeta| < \infty$ (for example, see [1, p.7]). For $B(\zeta)$, it is easy to see that it has the representation (from its Laurent expansion)

$$B(\zeta) = g_1(\frac{1}{\zeta}) + g_2(\zeta),$$

where both $g_1(t)$ and $g_2(t)$ are entire functions. It is shown in [5] that

$$\sigma_e(A) = \max\{\sigma(g_1), \sigma(g_2)\}.$$

We need to make use of the functions analytic in $R_0 < |z| < \infty$ ($R_0 > 0$) in this paper. If $f(z)$ is such a function, then it has the representation (see [13, p.15])

$$f(z) = z^m \psi(z) F(z), \tag{1.2}$$

where m is an integer, $\psi(z)$ is analytic and non-vanishing on $|z| > R_0$ (including $z = \infty$), $F(z)$ is an entire function with

$$F(z) = u(z)e^{h(z)},$$

$u(z)$ is a Weierstrass product formed with the zeros of $f(z)$ in $R_0 < |z| < \infty$, $h(z)$ is an entire function.

We may regard $B(\zeta)$ to be analytic in $R_0 < |\zeta| < \infty$. By (1.2), we have the representation

$$B(\zeta) = \zeta^n \phi(\zeta) b(\zeta), \tag{1.3}$$

where n is an integer, $\phi(\zeta)$ is analytic and non-vanishing on $|\zeta| > R_0$ (including $\zeta = \infty$), $b(\zeta)$ is entire. It is shown in [5] that

$$\sigma(g_2) = \sigma(b). \tag{1.4}$$

We first can get the following two lemmas.

Lemma 1.1 (the fundamental lemma of logarithmic derivative) *If $w(z)$ is meromorphic in $R_0 < |z| < \infty$, then*

$$m_1(r, \frac{w^{(k)}}{w}) = S_1(r, w), \tag{1.5}$$

where $S_1(r, w)$ denotes any non-negative quantity satisfying $S_1(r, w) = o\{T_1(r, w)\}$ n.e. ("n.e." denotes: as $r \to \infty$ outside possibly a set of r with finite linear measure).

Lemma 1.2 *$B(\zeta)$ is analytic in $0 < |\zeta| < \infty$ if and only if it can be represented to*

$$B(\zeta) = \zeta^{n_0} h_1(\frac{1}{\zeta}) h_2(\zeta), \tag{1.6}$$

where n_0 is an integer, both $h_1(t)$ and $h_2(t)$ are entire functions with $h_1(0)h_2(0) \neq 0$.

Representation (1.6) may change as R_0 changes, so it is not unique. But from (1.6) we may get that as making from (1.3)

$$\begin{cases} \sigma_e(A) = \max\{\sigma(h_1), \sigma(h_2)\}, \\ \lambda_e(A) = \max\{\lambda(h_1), \lambda(h_2)\}, \\ \overline{\lambda}_e(A) = \max\{\overline{\lambda}(h_1), \overline{\lambda}(h_2)\}, \\ \sigma_1(B) = \sigma(h_2), \quad \sigma_1(B^*) = \sigma(h_1), \\ \lambda_1(B) = \lambda(h_2), \quad \lambda_1(B^*) = \lambda(h_1), \\ \overline{\lambda}_1(B) = \overline{\lambda}(h_2), \quad \overline{\lambda}_1(B^*) = \overline{\lambda}(h_1). \end{cases} \tag{1.7}$$

2. Lemmas for Proof of Main Results

The following Lemma 2.1 is a generalized Clunie type lemma. Its proof is similar completely to that for Lemma 2.1 in [9] and for Lemma 3.3 in [11].

Lemma 2.1 *Let $w(z)$ be a function meromorphic in $R_0 < |z| < \infty$ and satisfying $w^n P(w) = Q(w)$, where $P(w)$ and $Q(w)$ are differential polynomials in $w(z)$ with coefficients $b_j(z)$ meromorphic in $R_0 < |z| < \infty$, and the degree of $Q(w)$ is at most n. Then for $r > R_0$,*

$$m_1(r, P(w)) = O\{\sum_j m_1(r, b_j) + S_1(r, w)\},$$

or in fact,

$$m_1(r, P(w)) \leq \sum_j m_1(r, b_j) + S_1(r, w).$$

In the following Lemma 2.2, we use the "combined dominant condition", so, it is a generalization of Theorem 3.9 in [11]. Its proof is similar to that for Theorem 3.9 in [11].

Lemma 2.2 *Let $w(z)$ be a function meromorphic and transcendental in $R_0 < |z| < \infty$, and*

$$g(z) = w(z)^n + P_{n-1}(w),$$

providing that there exists a positive constant $d < \sigma_1(w)$ (if $\sigma_1(w) = 0$, set $d = 0$) such that

$$N_1(r, w) + \overline{N}_1(r, 1/g) = S_1(r, w) + o(r^d),$$

where $P_{n-1}(w)$ is a differential polynomial in $w(z)$ with degree at most $n-1$, its coefficients are $b_j(z)$ meromorphic in $R_0 < |z| < \infty$ and satisfying

$$T_1(r, b_j) = S_1(r, w) + o(r^d).$$

Then $g(z) = h(z)^n$, where $h(z) = w(z) + a(z), a(z)$ is a function meromorphic in $R_0 < |z| < \infty$ and satisfying

$$T_1(r, a) = S_1(r, w) + o(r^d),$$

and $na(z)h(z)^{n-1}$ can be obtain by the following method: it is equal to the part with degree $n-1$ in $P_{n-1}(w)$, but needing to substitute h for w, h' for w', etc., in this part.

The following Lemma 2.3 is Lemma 3.5 in [11].

Lemma 2.3 *Suppose that $F(z)$ is meromorphic in a domain D and set $F'(z)/F(z) = w(z)$. Then we have for $n \geq 1$*

$$\frac{F^{(n)}(z)}{F(z)} = w^n + \frac{n(n-1)}{2}w^{n-2}w' + \alpha_n w^{n-3}w'' + \beta_n w^{n-4}w'^2 + P_{n-3}(w),$$

where $\alpha_n = \frac{1}{6}n(n-1)(n-2), \beta_n = \frac{1}{8}n(n-1)(n-2)(n-3)$, and $P_{n-3}(w)$ is a differential polynomial in $w(z)$ with constant coefficients, which vanishes identically for $n \leq 3$ and has degree $n-3$ when $n > 3$.

Lemma 2.4 *Let $F(z)$ be meromorphic in $R_0 < |z| < \infty$ with F'/F transcendental. And let $k \geq 2$, and $D_0(z), D_1(z), \cdots, D_{k-1}(z)$ be analytic in $R_0 < |z| < \infty$. Provide again that there exists a positive constant $d < \sigma_1(F'/F)$ (if $\sigma_1(F'/F) = 0$, set $d = 0$) such that $T_1(r, D_j) = S_1(r, F'/F) + o(r^d), j = 0, 1, \cdots, k-1$, and*

$$\overline{N}_1(r, F) + \overline{N}_1(r, \frac{1}{F}) + \overline{N}_1(r, \frac{1}{L_k(F)}) = S_1(r, \frac{F'}{F}) + o(r^d),$$

where

$$L_k(F) = F^{(k)} + D_{k-1}F^{(k-1)} + \cdots + D_0 F.$$

Then $L_k(F)/F = e^{k\varphi(z)+c}$, where c is a constant, both $\varphi'(z)$ and $e^{k\varphi(z)+c}$ are analytic in $R_0 < |z| < \infty$ with $e^{k\varphi(z)+c}$ no zeros in $R_0 < |z| < \infty$, and we have

$$\frac{k(k-1)(k+1)}{24}(\varphi'^2 - 2\varphi'') - \frac{k-1}{2k}D_{k-1}^2 - \frac{k-1}{2}D'_{k-1} + D_{k-2} \equiv 0. \quad (2.1)$$

Lemma 2.5 *Let $k \geq 2$, and $D_0(z), D_1(z), \cdots, D_{k-1}(z), D(z)$ be analytic in $R_0 < |z| < \infty$ with $D(z)$ transcendental. Also assume that there exists a positive constant $d < \sigma_1(D)$ (if $\sigma_1(D) = 0$, set $d = 0$) such that*

$$\overline{N}_1(r, 1/D) = S_1(r, D) + o(r^d), \; T_1(r, D_j) = S_1(r, D) + o(r^d), \; j = 0, 1, \cdots, k-1.$$

If the equation

$$y^{(k)} + D_{k-1}y^{(k-1)} + \cdots + D_1 y' + (D_0 + D)y = 0 \quad (2.2)$$

has an analytic solution $F(z) \not\equiv 0$ in $R_0 < |z| < \infty$ satisfying that there exists a positive constant $d_1 < \sigma_1(D)$ (if $\sigma_1(D) = 0$, set $d_1 = 0$) such that

$$\overline{N}_1(r, 1/F) = S_1(r, D) + o(r^{d_1}),$$

then D has no zeros in $R_0 < |z| < \infty$ and has the form $D = e^{k\varphi + c_0}$, where φ' is analytic in $R_0 < |z| < \infty$, c_0 is a constant, and we have the relation (2.1).

The following Lemma 2.6 is Theorem 1 in [10] which generalizes Theorem 1 in [3].

Lemma 2.6 *Let $k \geq 2$, and $A_0(z), \cdots, A_{k-2}(z)$ be periodic entire functions with period $2\pi i$, for $k \geq 3$, satisfying that there exists a positive constant $d < \sigma_e(A_0)$ (if $\sigma_e(A_0) = 0$, set $d = 0$) such that*

$$T(r, A_j) = S(r, A_0) + o(e^{dr}), j = 1, \cdots, k-2.$$

If the equation

$$w^{(k)} + A_{k-2}w^{(k-2)} + \cdots + A_0 w = 0$$

has a solution $f(z) \not\equiv 0$ satisfying that there exists a positive constant $d_1 < \sigma_e(A_0)$ (if $\sigma_e(A_0) = 0$, set $d_1 = 0$) such that

$$\log^+ N(r, \frac{1}{f}) = S(r, A_0) + o(e^{d_1 r}),$$

then there exists an integer q, $1 \leq q \leq k$, such that $f(z)$ and $f(z + q2\pi i)$ are linearly dependent.

3. Main Results

Theorem 3.1 *Let $k \geq 2$, and $A(z), A_0(z), \cdots, A_{k-2}(z)$ be periodic entire functions with period $2\pi i$, i.e. $A(z) = B(e^z), A_j(z) = B_j(e^z)$ with $B(\zeta)$ and $B_j(\zeta)$ analytic in $0 < |\zeta| < \infty$, $j = 0, \cdots, k-2$, satisfying*

(i) $\overline{\lambda}_1(B) < \sigma_1(B)$;

(ii) *if $k \geq 3$, for $j \geq 0, \sigma_1(B_j) < \sigma_1(B)$;*

(iii) *if $k \geq 3$, for $j \geq 0$, $\sigma_1(B_j^*) < \max\{\sigma_1(B), \sigma_1(B^*)\}$, where $B^*(t) = B(1/t)$ and $B_j^*(t) = B_j(1/t), j = 0, \cdots, k-2$.*

If the equation

$$w^{(k)} + A_{k-2}w^{(k-2)} + \cdots + (A_0 + A)w = 0 \tag{3.1}$$

has a solution $f(z) \not\equiv 0$ satisfying

(iv) $\overline{\lambda}_e(f) < \sigma_1(B)$,

then $B(\zeta)$ has no zeros in $0 < |\zeta| < \infty$ (and thus $A(z)$ has no zeros yet), and has the form

$$B(\zeta) = \zeta^m e^{g(\zeta)}, \tag{3.2}$$

where m is an integer, $g(\zeta)$ is analytic in $0 < |\zeta| < \infty$.

Proof It follows from part 1 that $\sigma_e(A_j) < \sigma_e(A)$ for $j \geq 0$ and $\sigma_e(A_j) < \sigma_e(A_0 + A)$ for $j > 0$. Thus, there exists a positive constant $d < \sigma_e(A_0 + A)$ such that $T(r, A_j) = o(e^{dr})$ for $j > 0$, and this leads to

$$T(r, A_j) = S(r, A_0 + A) + o(e^{dr}).$$

Since

$$N(r, \frac{1}{f}) \leq (k-1)\overline{N}(r, \frac{1}{f}),$$

we have $\overline{\lambda}_e(f) = \lambda_e(f)$, thus (ii) and (iv) give

$$\lambda_e(f) < \sigma_1(B_0 + B) \leq \sigma_e(A_0 + A),$$

and this implies that there exists a positive constant $d_1 < \sigma_e(A_0 + A)$ such that $N(r, 1/f) = o(e^{d_1 r})$. It further results in $\log^+ N(r, 1/f) = S(r, A_0 + A) + o(e^{d_1 r})$. Hence, the conditions of Lemma 2.6 are satisfied. So there exists an integer $q, 1 \leq q \leq k$, such that $f(z)$ and $f(z + q2\pi i)$ are linearly dependent. From, for example, [1, p.14], $f(z)$ can be represented to $f(z) = e^{\beta z}U(z)$, where β is a constant, $U(z)$ is a periodic entire function with period $q2\pi i$, i.e. we may write $U(z) = G(e^{z/q})$ with $G(\zeta)$ analytic in $0 < |\zeta| < \infty$. Evidently, $\overline{\lambda}_e(f) = \overline{\lambda}_e(U)$. We may deduce

$$\overline{\lambda}_e(f) = \frac{1}{q}\max\{\overline{\lambda}_1(G), \overline{\lambda}_1(G^*)\}. \tag{3.3}$$

Substituting $f(z) = \zeta^{\beta q}G(\zeta)$ into equation (3.1), and noting that $\zeta = e^{z/q}$, gives

$$G^{(k)} + D_{k-1}(\zeta)G^{(k-1)} + \cdots + D_1(\zeta)G' + (D_0(\zeta) + D(\zeta))G = 0, \tag{3.4}$$

where

$$\begin{cases} D(\zeta) = \frac{q^k}{\zeta^k}B(\zeta^q), \\ D_{k-j}(\zeta) = \frac{1}{\zeta^j}[c_{k-j}^{(1)} + c_{k-j}^{(2)}B_{k-2}(\zeta^q) + \cdots + c_{k-j}^{(j)}B_{k-j}(\zeta^q)], 1 \leq j \leq k, \end{cases} \tag{3.5}$$

$c_{k-j}^{(i)}(1 \leq i \leq j)$ are constants, and for $j \geq 2$, $c_{k-j}^{(j)} = q^j$. Clearly, $D(\zeta)$ and $D_{k-j}(\zeta)(1 \leq j \leq k)$ are analytic in $0 < |\zeta| < \infty$, and $\sigma_1(D) = q\sigma_1(B)$,

$$\sigma_1(D_{k-j}) \leq q\max\{\sigma_1(B_{k-2}), \cdots, \sigma_1(B_{k-j})\} < q\sigma_1(B) = \sigma_1(D),\ 1 \leq j \leq k,$$

$$\overline{\lambda}_1(D) = q\overline{\lambda}_1(B) < \sigma_1(D).$$

Then, taking an arbitrary constant $R_0 > 0$, there exists a positive constant $d < \sigma_1(D)$ such that

$$T_1(r, D_{k-j}) = o(r^d), 1 \le j \le k, \overline{N}_1(r, 1/D) = o(r^d)$$

in $R_0 < |\zeta| < \infty$, and these lead to

$$T_1(r, D_{k-j}) = S_1(r, D) + o(r^d),\ 1 \le j \le k,$$

$$\overline{N}_1(r, 1/D) = S_1(r, D) + o(r^d).$$

And for the solution $G(\zeta)$ of the equation (3.4),

$$\overline{\lambda}_1(G) \le q\overline{\lambda}_e(U) = q\overline{\lambda}_e(f) < q\sigma_1(B) = \sigma_1(D).$$

Hence, there exists a positive constant $d_1 < \sigma_1(D)$ such that

$$\overline{N}_1(r, 1/G) = o(r^{d_1})$$

in $R_0 < |\zeta| < \infty$, and this causes $\overline{N}_1(r, 1/G) = S_1(r, D) + o(r^{d_1})$. In addition, $D(\zeta)$ is clearly transcendental in $R_0 < |\zeta| < \infty$. Therefore, the conditions of Lemma 2.5 are satisfied, and so $D(\zeta)$ has no zeros in $R_0 < |\zeta| < \infty$. By the arbitrariness of R_0, we see that $D(\zeta)$ has no zeros in $0 < |\zeta| < \infty$. In view of (3.5), $B(\zeta)$ has no zeros in $0 < |\zeta| < \infty$ yet. From Lemma 1.2, we may get (3.2). The proof is completed.

References

1. S. Bank, I. Laine, Representations of solutions of periodic second order linear differential equations, *J Reine Angew Math,* **344**(1983): 1-21.
2. S. Bank, I. Laine, On the oscillation theory of $f'' + Af = o$ where $A$is entire,*Trans. Amer. Math.Soc.* **273**(1982):351-363.
3. S. Bank, J. Langley, Oscillation theorems for higher order linear differential equations with entire periodic coefficients, *Comment Math Univ San Pau,* **41**(1992): 65~85.
4. L. Bieberbach, *Theorie der gewöhnlichen Differentialgleichungen,* Berlin-Heidelberg-New York: Springer-Verlag, 1965.
5. Y. M. Chiang, S. A. Gao, On a problem in complex oscillation theory of periodic second order linear differential equations and some related perturbation results, *Ann Acad Sci Fenn Math,* **27**(2002): 273-290.
6. S. A. Gao, A further result on the complex oscillation theory of periodic second order linear differential equations, *Proc Edinburgh Math Soc,* **33**(1990): 143-158.
7. S. A. Gao, A property of solutions and the complex oscillation for a class of higher order periodic linear differential equations, *Acta Math Appl Sinica,* **25** (4)(2002): 642-649 (in Chinese).

8. S. A. Gao, On the complex oscillation for a class of homogeneous linear differential equations, *Proc Edinburgh Math Soc,* **43**(2000): 1-13.
9. S. A. Gao, Z. X. Chen, T. W. Chen, *Complex Oscillation Theory of Linear Differential Equations,* Wuhan: Huazhong University of Science and Technology Press, 1998 (in Chinese).
10. S. A. Gao, A key property of solutions for periodic linear differential equtions, *Math. Applicata,* **13**(4) (2000):106-110 (in Chinese).
11. W. Hayman, *Meromorphic Functions,* Oxford: Clarendon Press, 1964
12. I. Laine, *Nevanlinna Theory and Complex Differential Equations,* W. de Gruyter, Berlin, 1993.
13. G. Valiron, *Lectures on the General Theory of Integral Functions,* New York: Chelsea, 1949.

Q_P-SPACES: GENERALIZATIONS TO BOUNDED SYMMETRIC DOMAINS*

MIROSLAV ENGLIŠ

MÚ AV ČR, Žitná 25, 11567 Praha 1, Czech Republic
E-mail: englis@math.cas.cz

Q_p spaces on the unit disc were introduced, and their basic properties established, in 1995 by Aulaskari, Xiao and Zhao. Later some of these results were extended also to the unit ball or even to strictly pseudoconvex domains in the complex n-space. We briefly review the theory of bounded symmetric domains, of which the disc and the ball are the simplest examples, and then discuss the Q_p spaces in this setting. It turns out that some new phenomena appear, most notably concerning the relationships of these spaces to the various kinds of Bloch spaces on symmetric domains.

2000 Mathematics Subject Classification: Primary 32M15; Secondary 32A37, 32A18

1. Introduction

The Q_p spaces on the unit disc $\mathbf{D}$ were introduced in 1995 by Aulaskari, Xiao and Zhao [4] by

$$f \in Q_p \iff \sup_{a\in\mathbf{D}} \int_{\mathbf{D}} |f'(z)|^2 g(z,a)^p \, dz < \infty,$$

the square root of the right-hand side being, by definition, the (semi)norm in Q_p. Here $g(z,a)$ stands for the Green function

$$g(z,a) = \log \left| \frac{z-a}{1-\bar{a}z} \right|,$$

and dz denotes the Lebesgue area measure. It is not difficult to see that one gets the same spaces, with equivalent seminorms, upon replacing the

*The author was supported by GA AV ČR grant A1019304.

Green function by the function $\log|\frac{z-a}{1-\overline{a}z}|$ which has (for each fixed a) the same boundary behaviour:

$$f \in Q_p \iff \sup_{a\in\mathbf{D}} \int_{\mathbf{D}} |f'(z)|^2 \left(1 - \left|\frac{z-a}{1-\overline{a}z}\right|^2\right)^p dz < \infty. \tag{1}$$

We will adhere to this latter definition throughout the sequel.

The most notable feature of the Q_p spaces is that they are Möbius invariant. Indeed, any Möbius map (i.e. a biholomorphic self-map of $\mathbf{D}$) is of the form $\phi(z) = \epsilon \dfrac{a-z}{1-\overline{a}z}$, with $|\epsilon| = 1$ and $a \in \mathbf{D}$. Thus the right-hand side of (1) can be rewritten as

$$\begin{aligned}
&\sup_{\phi\in\mathrm{Aut}(\mathbf{D})} \int_{\mathbf{D}} |f'(z)|^2 \, (1-|\phi(z)|^2)^p \, dz \\
&\quad = \sup_{\phi\in\mathrm{Aut}(\mathbf{D})} \int_{\mathbf{D}} \Delta|f|^2(z) \, (1-|\phi(z)|^2)^p \, dz \\
&\quad = \sup_{\phi\in\mathrm{Aut}(\mathbf{D})} \int_{\mathbf{D}} (\widetilde{\Delta}|f|^2)(z) \, (1-|\phi(z)|^2)^p \, d\mu(z) \\
&\quad = \sup_{\phi\in\mathrm{Aut}(\mathbf{D})} \int_{\mathbf{D}} \widetilde{\Delta}|f\circ\phi(z)|^2 \, (1-|z|^2)^p \, d\mu(z),
\end{aligned}$$

where $\widetilde{\Delta} = (1-|z|^2)^2 \dfrac{\partial^2}{\partial z \partial \overline{z}}$ and $d\mu(z) = \dfrac{dz}{(1-|z|^2)^2}$ are the invariant Laplacian and the invariant measure on $\mathbf{D}$, respectively. From the last formula it is apparent that $f \in Q_p$ implies $f\circ\phi \in Q_p$ and f and $f\circ\phi$ have the same norm in Q_p, for all $\phi \in \mathrm{Aut}(\mathbf{D})$.

It was shown in [4] that

$$\begin{aligned}
p > 1 &\implies Q_p = \mathcal{B}, \quad \text{the Bloch space,} \\
p = 1 &\implies Q_p = BMOA, \\
0 \le p_1 < p_2 \le 1 &\implies Q_{p_1} \subsetneq Q_{p_2}, \\
p = 0 &\implies Q_p = \mathcal{D}, \quad \text{the Dirichlet space,} \\
p < 0 &\implies Q_p = \{\text{const}\}.
\end{aligned}$$

Thus the Q_p spaces provide a whole range of Möbius-invariant function spaces on $\mathbf{D}$ lying strictly between the Dirichlet space on the one hand, and $BMOA$ and the Bloch space on the other.

The Q_p spaces subsequently attracted a lot of attention; see e.g. the book by Xiao [8] and the references therein. They were generalized to the unit ball $\mathbf{B}^d \subset \mathbf{C}^d$ in 1998 by Ouyang, Yang and Zhao [5]:

$$f \in Q_p \iff \sup_{a \in \mathbf{B}^d} \int_{\mathbf{B}^d} \widetilde{\Delta}|f(z)|^2 \, G(z,a)^p \, d\mu(z) < \infty$$
$$\iff \sup_{\phi \in \mathrm{Aut}(\mathbf{B}^d)} \int_{\mathbf{B}^d} \widetilde{\Delta}|f \circ \phi|^2 \, G(z,0)^p \, d\mu(z) < \infty,$$

where $\widetilde{\Delta}$, $d\mu$ and $G(z,a)$ denote the invariant Laplacian, the invariant measure and the Green function of $\widetilde{\Delta}$ on $\mathbf{B}^d$, respectively. Again, these spaces are Möbius invariant, and

$$\begin{aligned}
p \geq \frac{d}{d-1}, \quad &\implies \quad Q_p = \{\text{const}\}, \\
1 < p < \frac{d}{d-1} \quad &\implies \quad Q_p = \mathcal{B}(\mathbf{B}^d), \quad \text{the Bloch space}, \\
p = 1 \quad &\implies \quad Q_p = BMOA(\mathbf{B}^d), \\
\frac{d-1}{d} < p_1 < p_2 \leq 1 \quad &\implies \quad Q_{p_1} \subsetneq Q_{p_2}, \\
p \leq \frac{d-1}{d} \quad &\implies \quad Q_p = \{\text{const}\}.
\end{aligned}$$

The cut-off at $p = \frac{d}{d-1}$ turns out to due to the pole of $G(z,a)$ at $z = a$, and disappears if we replace (as we did for the disc) the Green function by $(1 - \|\frac{z-a}{1-\langle a,z\rangle}\|^2)^d$, i.e. upon setting

$$f \in Q_p \iff \sup_{\phi \in \mathrm{Aut}(\mathbf{B}^d)} \int_{\mathbf{B}^d} \widetilde{\Delta}|f \circ \phi|^2(z) \, (1 - \|z\|^2)^{pd} \, d\mu(z) < \infty.$$

Then $Q_p = \mathcal{B}(\mathbf{B}^d)$ $\forall p > 1$, while the other cases remain unchanged. (We again stick to this latter definition in the sequel.)

Note that, in contrast to the disc, for $d > 1$ the Dirichlet space does not turn up as one of the Q_p's, though in all other cases the situation is the same as for $\mathbf{D}$.

Other generalizations include Q_p spaces on smoothly bounded strictly pseudoconvex domains [1] or the $F(p,q,s)$ spaces of Rättyä and Zhao [6,9]. In this talk, we will consider generalization in another direction, suggested by the appearance of the invariant Laplacians $\widetilde{\Delta}$, the invariant measures $d\mu$, and the invariance of the spaces under Möbius maps — namely, the generalization to bounded symmetric domains.

2. Bounded symmetric domains

Recall that a bounded domain $\Omega \subset \mathbf{C}^d$ is called *symmetric* if $\forall x \in \Omega$ there exists $s_x \in \mathrm{Aut}(\Omega)$ such that $s_x \circ s_x = \mathrm{id}$ and x is an isolated fixed-point of s_x. One calls s_x the geodesic symmetry at x. The motivating example behind this is, of course, the complex n-space $\Omega = \mathbf{C}^n$ with $s_x(z) = 2x - z$ (except that this is not a bounded domain). Another example is the unit disc $\mathbf{D}$ with $s_0(z) = -z$ and $s_x = \phi_x \circ s_0 \circ \phi_x$, where $\phi_x(z) = \dfrac{x-z}{1-\overline{x}z}$ is the geodesic symmetry interchanging 0 and x.

A more general example is the unit ball $I_{r\times R}$ of $r \times R$ complex matrices $(R, r \geq 1)$, again with $s_0(z) = -z$ and $s_x = \phi_x \circ s_0 \circ \phi_x$, the geodesic symmetry ϕ_x interchanging 0 and x being now given by

$$s_x(z) = (I_r - xx^*)^{-1/2}(x - z)(I_R - x^*z)^{-1}(I_R - x^*x)^{1/2}.$$

Note that this includes the unit ball $\mathbf{B}^d \subset \mathbf{C}^d$ as the special case I_{1d}.

It turns out that symmetry implies homogeneity: the group $\mathrm{Aut}(\Omega) := G$ acts transitively on Ω. (In fact, already the symmetries s_x do.) It is a semisimple Lie group. A bounded symmetric domain is called irreducible if it is not biholomorphic to a Cartesian product of two other bounded symmetric domains. Irreducible bounded symmetric domains were completely classified by E. Cartan. There are four infinite series of such domains plus two exceptional domains in $\mathbf{C}^{16}$ and $\mathbf{C}^{27}$:

Domain	Description	
I_{rR}	$Z \in \mathbf{C}^{r\times R}$: $\Vert Z\Vert_{\mathbf{C}^R\to\mathbf{C}^r} < 1$	$R \geq r \geq 1$
II_r	$Z \in I_{rr}$, $Z = Z^t$	$r \geq 2$
III_m	$Z \in I_{mm}$, $Z = -Z^t$	$m \geq 5$
IV_n	$Z \in \mathbf{C}^{n\times 1}$, $\vert Z^tZ\vert < 1$, $1 + \vert Z^tZ\vert^2 - 2Z^*Z > 0$	$n \geq 5$
V	$Z \in \mathbf{O}^{1\times 2}$, $\Vert Z\Vert < 1$	
VI	$Z \in \mathbf{O}^{3\times 3}$, $Z = Z^*$, $\Vert Z\Vert < 1$	

The restrictions on R, r, m, n stem from a few isomorphisms in low dimensions:

$$IV_1 \cong III_2 \cong II_1 \cong I_{1,1}(\cong \mathbf{D}), \quad IV_3 \cong II_2, \quad IV_2 \cong \mathbf{D} \times \mathbf{D},$$

$$IV_4 \cong I_{2,2}, \quad III_3 \cong I_{1,3}, \quad III_4 \cong IV_6, \quad I_{rR} \cong I_{Rr}.$$

Up to biholomorphic equivalence, any irreducible bounded symmetric domain is uniquely determined by three integers, namely its rank r and its characteristic multiplicities a, b.

domain	r	a	b	d	p
$I_{rR}\quad (r \leq R)$	r	2	$R-r$	rR	$r+R$
II_r	r	1	0	$\frac{1}{2}r(r+1)$	$r+1$
$III_{2r+\epsilon},\ \epsilon \in \{0,1\}$	r	4	2ϵ	$r(2r+2\epsilon-1)$	$4r+2\epsilon-2$
IV_n	2	$n-2$	0	n	n
V	2	6	4	16	12
VI	3	8	0	27	18

The two other important quantities given in the table, the genus p and the dimension d are related to a, b and r by

$$p = (r-1)a + b + 2, \qquad d = \frac{r(r-1)}{2}a + rb + r.$$

The domains with $b = 0$ are in some respects "simpler" than others and are called tube domains. Thus, for instance, I_{rR} is tube $\iff r = R$. The unit balls $\mathbf{B}^d = I_{1d}$ are the only bounded symmetric domains of rank 1, and the only bounded symmetric domains with smooth boundary. The domains in the list above are called Cartan domains. Clearly, any Cartan domain is convex, contains the origin, and is circular with respect to it.

From now on, we will suppose (unless explicitly stated otherwise) that Ω is a Cartan domain, and for each $x \in \Omega$ we denote by ϕ_x the (unique) geodesic symmetry interchanging 0 and x. We further denote by K the stabilizer of the origin in G,

$$K := \{k \in G : \ k0 = 0\}.$$

It is a consequence of one of Cartan's theorems that any $k \in K$ is automatically a unitary linear map on $\mathbf{C}^d$. Note that from the definition of K it is immediate that any $\phi \in G$ is of the form $\phi = \phi_x k$, where $k \in K$, $x \in \Omega$. (In fact $x = \phi(0)$.) Having recalled the definition of bounded symmetric domains, we can turn to the Q_p spaces on them. We have seen in the Introduction that their definition involves three ingredients — namely, the invariant Laplacian $\widetilde{\Delta}$, the invariant measure $d\mu$, and powers of the function $1 - |z|^2$ (or, for the ball, $1 - \|z\|^2$). Let us now clarify what are the counterparts of these on a general Cartan domain.

3. Invariant differential operators

A differential operator L on a Cartan domain Ω is called invariant if

$$L(f \circ \phi) = (Lf) \circ \phi \qquad \forall \phi \in G = \mathrm{Aut}(\Omega).$$

It is well known that on the unit disc, invariant operators are precisely the polynomials of the invariant Laplacian $\widetilde{\Delta} = (1 - |z|^2)^2 \Delta$. The same is true for $\mathbf{B}^d$. For a general bounded symmetric domain, the situation is more complicated: namely, the algebra of all invariant differential operators consists of all polynomials in r commuting differential operators $\Delta_1, \ldots, \Delta_r$, of orders $2, 4, \ldots, 2r$, respectively, where r is the rank. In particular, the monomials $\Delta_1^{n_1} \ldots \Delta_r^{n_r}$ form a linear basis of all invariant differential operators. However, often it is much more convenient to use another basis, the construction of which we now describe.

For any invariant differential operator L, let L_0 be the (non-invariant) linear differential operator obtain upon freezing the coefficients of L at the origin, that is, $Lf(0) =: L_0 f(0)$. From the invariance of L it follows that

$$k \in G, \ k0 = 0 \implies L_0(f \circ k) = (L_0 f) \circ k$$

(i.e. L_0 is K-invariant) and

$$Lf(z) = L_0(f \circ \phi_z)(0). \tag{2}$$

Conversely, if L_0 is a K-invariant constant-coefficient differential operator, then the recipe (2) clearly defines an invariant differential operator L on Ω. Thus there is a 1-to-1 correspondence between (G-)invariant linear differential operators on Ω and K-invariant linear constant-coefficients differential operators on $\mathbf{C}^d$.

Further, any constant-coefficient linear differential operator L_0 can be written in the form $L_0 = p(\partial, \overline{\partial})$ for some polynomial p on $\mathbf{C}^d \times \mathbf{C}^d$. It is not difficult to see that such operator is K-invariant if and only if the polynomial p is K-invariant in the sense that $p(x, \overline{y}) = p(kx, \overline{ky}) \ \forall x, y \in \mathbf{C}^d \ \forall k \in K$.

Combining this with the observation in the preceding paragraph, we thus see that invariant differential operators are in 1-to-1 correspondence with K-invariant polynomials.

Example Since K consists of unitary maps, the simplest K-invariant polynomial (apart from the constants) is $p(x, \overline{y}) = \langle x, y \rangle$. The corresponding invariant differential operator is

$$Lf(a) = \Delta(f \circ \phi_a)(0).$$

This operator is called the invariant Laplacian of Ω; it coincides with the Laplace-Beltrami operator with respect to the Bergman metric on Ω. Note that for f holomorphic,

$$L|f|^2(a) = \sum_{j=1}^{d} \left| \frac{\partial (f \circ \phi_a)(0)}{\partial z_j} \right|^2 = \|\partial (f \circ \phi_a)(0)\|^2$$

is what we might call the *invariant holomorphic gradient* of f.

In a moment, we will see that there exists a very natural basis for K-invariant polynomials (which will thus yield the sought basis for invariant differential operators). Prior to that, however, we review some facts about Bergman kernels on Cartan domains.

4. Bergman spaces and kernels

The function $h(x, y) := 1 - x\overline{y}$ on $\mathbf{D}$ is noteworthy in a number of ways. First of all, h^{-2} is, up to a constant factor, the Bergman kernel $K(x,y) = \dfrac{1}{\pi(1-x\overline{y})^2}$. Second, $d\mu(z) = \dfrac{dz}{h(z,z)^2}$ is the invariant measure on $\mathbf{D}$. Finally, for any $\alpha > -1$, the Bergman kernel of $L^2_{\text{hol}}(\mathbf{D}, h(z,z)^\alpha \, dz)$ is given by

$$K_\alpha(x,y) = \frac{\text{const}}{h(x,y)^{\alpha+2}}.$$

The same properties are also possessed by the function $h(x,y) = 1 - \langle x, y \rangle$ on the ball, only in all three formulas 2 must be replaced by $d+1$.

It is a notable fact that the same phenomenon persists for a general Cartan domain. Namely, the Bergman kernel of a Cartan domain has the form

$$K(x,y) = \frac{1}{\operatorname{vol}\Omega} \, h(x,y)^{-p},$$

where $h(x,y)$ is an irreducible polynomial, analytic in x and $\overline{y}$, and such that $h(0,z) = h(z,0) = 1 \geq h(z,z) \geq 0 \; \forall z \in \Omega$. The degree of h is equal to the rank, r, and p is the genus (this is always an integer ≥ 2). Finally, the Bergman kernel of $L^2_{\text{hol}}(\Omega, h(z,z)^\alpha \, dz)$ equals

$$K_\alpha(x,y) = \frac{\text{const}}{h(x,y)^{\alpha+p}},$$

for any $\alpha > -1$, and

$$d\mu(z) := \frac{dz}{h(z,z)^p}$$

is an invariant volume element on Ω.

For the domains I and II in Cartan's list, h is given by $h(X, Y) = \det(I - XY^*)$; for domains of type III, the determinant gets replaced by the Pfaffian. Explicit formulas are known also for the types IV–VI.

Comparing all the facts above with the situations for the disc and the ball, we see that we should define the Q_p spaces on a Cartan domain for any $\nu \in \mathbf{R}$ and any invariant differential operator L as follows:

$$f \in Q_{\nu,L} \iff \sup_{\phi \in G} \int_\Omega L|f \circ \phi|^2(z)\, h(z,z)^\nu \, d\mu(z) < \infty. \tag{3}$$

(Here we have started using the subscript ν instead of p since the letter p is already reserved for the genus.) Clearly, this reduces to the original definitions for $\Omega = \mathbf{D}$ (or $\mathbf{B}^d$) and L the invariant Laplacian.

A small catch here is, however, that in order to have the square-root of the right-hand side for a seminorm, we need this right-hand side to be nonnegative for all holomorphic functions f. It is precisely at this point that the promised linear basis for invariant differential operators comes to the rescue; so let us exhibit it without further delay.

5. Peter-Weyl decomposition

Let $\mathcal{P}$ denote the vector space of all (holomorphic) polynomials on $\mathbf{C}^d$. We endow $\mathcal{P}$ with the Fock inner product

$$\begin{aligned} \langle f, g \rangle_F : &= f(\partial)\, g^*(0), \qquad \text{where} \quad g^*(z) := \overline{g(\overline{z})}, \\ &= \pi^{-d} \int_{\mathbf{C}^d} f(z)\overline{g(z)}\, e^{-\|z\|^2}\, dz. \end{aligned}$$

This makes $\mathcal{P}$ into a pre-Hilbert space, and the action

$$f \mapsto f \circ k, \qquad k \in K,$$

is a unitary representation of K on $\mathcal{P}$. It is a deep result of W. Schmidt that this representation has a multiplicity-free decomposition into irreducibles

$$\mathcal{P} = \sum_{\mathbf{m}}^{\oplus} \mathcal{P}_{\mathbf{m}}$$

where $\mathbf{m}$ ranges over all signatures, i.e. r-tuples $\mathbf{m} = (m_1, m_2, \ldots, m_r) \in \mathbf{Z}^r$ satisfying $m_1 \geq m_2 \geq \cdots \geq m_r \geq 0$. Polynomials in $\mathcal{P}_{\mathbf{m}}$ are homogeneous of degree $|\mathbf{m}| := m_1 + m_2 + \cdots + m_r$; in particular, $\mathcal{P}_{(0)}$ are the constants and $\mathcal{P}_{(1)}$ the linear polynomials. Any holomorphic function thus has a decomposition $f = \sum_{\mathbf{m}} f_{\mathbf{m}}$, $f_{\mathbf{m}} \in \mathcal{P}_{\mathbf{m}}$, which refines the usual homogeneous expansion.

Since the spaces $\mathcal{P}_{\mathbf{m}}$ are finite dimensional, they automatically possess a reproducing kernel: there exist functions $K_{\mathbf{m}}(x,y)$ on $\mathbf{C}^d \times \mathbf{C}^d$ such that for each $f \in \mathcal{P}_{\mathbf{m}}$ and $y \in \mathbf{C}^d$, $f(y) = \langle f, K(\cdot,y)\rangle_F$. Explicitly, for any orthonormal basis $\{\psi_j\}_{j=1}^{\dim \mathcal{P}_{\mathbf{m}}}$ of $\mathcal{P}_{\mathbf{m}}$, $K_{\mathbf{m}}$ is given by

$$K_{\mathbf{m}}(x,y) = \sum_{j=1}^{\dim \mathcal{P}_{\mathbf{m}}} \psi_j(x)\overline{\psi_j(y)}. \tag{4}$$

It follows from the definition of the $\mathcal{P}_{\mathbf{m}}$ spaces that the kernels $K_{\mathbf{m}}(x,y)$ are K-invariant. By the discussion in the penultimate section, we therefore know that each $K_{\mathbf{m}}$ defines an invariant differential operator

$$\Delta_{\mathbf{m}} f(a) := K_{\mathbf{m}}(\partial,\partial)(f \circ \phi_a)(0), \qquad a \in \Omega. \tag{5}$$

Further, one can show that $K_{\mathbf{m}}$ are actually a basis of all K-invariant polynomials, and, consequently, $\Delta_{\mathbf{m}}$ are a linear basis for invariant differential operators. Further, from (4) and (5) we see that for any f holomorphic,

$$\Delta_{\mathbf{m}}|f|^2(a) = \sum_j |\psi_j(\partial)(f \circ \phi_a)(0)|^2 \geq 0.$$

What makes the basis $\Delta_{\mathbf{m}}$ important for our applications to the Q_ν-spaces is the following converse to the last inequality.

Theorem *An invariant differential operator*

$$L = \sum_m l_m \Delta_m$$

satisfies $L|f|^2 \geq 0 \ \forall f$ *holomorphic if and only if*

$$l_m \geq 0 \qquad \forall m.$$

6. Bloch spaces and Q_ν spaces on bounded symmetric domains

Thus we see that the invariant differential operators L that can be used in (3) are precisely those which are linear combinations of $\Delta_{\mathbf{m}}$ with nonnegative coefficients. The most basic among such L are evidently the operators $\Delta_{\mathbf{m}}$ themselves. We are thus lead to the following definitions.

Definition *For each signature* $\mathbf{m}$*, the* $\mathbf{m}$*-Bloch space is defined by*

$$\mathcal{B}_m = \{f \text{ holomorphic on } \Omega : \ \|\Delta_m |f|^2\|_\infty < \infty\}.$$

Definition* *For each signature $\mathbf{m}$ and $\nu \in \mathbf{R}$, the space $Q_{\nu,\mathbf{m}}$ is defined by requiring that, for f holomorphic,*

$$\begin{aligned} f \in Q_{\nu,\mathbf{m}} &\iff \sup_{\phi \in G} \int_\Omega \Delta_{\mathbf{m}} |f \circ \phi|^2 \, h^\nu \, d\mu < \infty \\ &\iff \sup_{a \in \Omega} \int_\Omega \Delta_{\mathbf{m}} |f \circ \phi_a|^2 \, h^\nu \, d\mu < \infty \\ &\iff \sup_{a \in \Omega} \int_\Omega \Delta_{\mathbf{m}} |f|^2 \, (h \circ \phi_a)^\nu \, d\mu < \infty. \end{aligned}$$

Here $h(z) \equiv h(z,z)$. (All the three conditions above are equivalent owing to the invariance of $\Delta_{\mathbf{m}}$ and $d\mu$.)

Clearly, the above definitions reduce to the usual definitions for $\Omega = \mathbf{D}$ or $\mathbf{B}^d$ and $\mathbf{m} = (1)$ (so that $\Delta_{\mathbf{m}}$ is the invariant Laplacian). In particular, the case of $\mathbf{m} \neq (1)$ gives something new even for the unit disc.

Let us work out the simplest special cases of Bloch spaces.

Example* For $\mathbf{m} = (0)$ we have $\Delta_{\mathbf{m}} = I$, so $\mathcal{B}_{(0)} = H^\infty$. Also,

$$Q_{\nu,(0)} = \begin{cases} H^\infty & \nu > p - 1, \\ \{0\} & \nu \le p - 1. \end{cases}$$

Example For $\mathbf{m} = (1)$ we have $\Delta_{(1)} = \widetilde{\Delta}$, so that

$$\mathcal{B}_{(1)} = \{f : \quad \sup_a \|\partial(f \circ \phi_a)(0)\|^2 < \infty\}.$$

This is known as the <u>Timoney Bloch space</u> [7].

Example Assume that Ω is of tube type and $s := d/r \in \mathbf{Z}$. Let $\mathbf{m} = (s, s, \ldots, s) \equiv (s^r)$. It is known that in that case the space $\mathcal{P}_{(s^r)} = \mathbf{C}N^s$ is one-dimensional (for the unit disc, $N(z) = z$; for I_{rr}, $N(Z) = \det Z$), the kernel $K_{\mathbf{m}}$ is given (up to a constant factor) by $K_{\mathbf{m}}(\partial, \partial) = N(\partial)^s N(\overline{\partial})^s$, and the so-called Bol's lemma says that for any f holomorphic, $N(\partial)^s (f \circ \phi_a)(0) = \text{const} \cdot h(a)^s N(\partial)^s f(a)$. (For the disc, this reads $(f \circ \phi_a)'(0) = -(1 - |a|^2)\, f'(a)$.) Hence,

$$\Delta_{(s^r)} |f|^2 = h^p |N(\partial)^s f|^2 \tag{6}$$

and

$$\mathcal{B}_{(s^r)} = \{f \text{ holomorphic: } h^p |N(\partial)^s f|^2 \text{ is bounded}\}.$$

We might call this the <u>Arazy Bloch space</u> [2].

7. Example: the polydisc

For clarity, let us also see what is the situation for the bidisc $\Omega = \mathbf{D}^2$. This is definitely NOT a Cartan domain (it is not irreducible), but in many respects it behaves like a Cartan domain with the rank, dimension and genus $r = p = d = 2$, multiplicities $a = b = 0$, and $h(x,y) = (1 - x_1\overline{y}_1)(1 - x_2\overline{y}_2)$. Namely, the invariant measure is $h(z,z)^{-2}\,dz$; the invariant differential operators are precisely the symmetric polynomials in $\widetilde{\Delta}_1, \widetilde{\Delta}_2$, where $\widetilde{\Delta}_j := (1-|z_j|^2)^2\partial_j\overline{\partial}_j$; the Peter-Weyl spaces are given by $\mathcal{P}_{\mathbf{m}} = \mathbf{C}z_1^{m_1}z_2^{m_2} + \mathbf{C}z_1^{m_2}z_2^{m_1}$, for $\mathbf{m} = (m_1, m_2)$; and, up to a constant factor, $K_{\mathbf{m}}(x,y) = (x_1\overline{y}_1)^{m_1}(x_2\overline{y}_2)^{m_2} + (x_1\overline{y}_1)^{m_2}(x_2\overline{y}_2)^{m_1}$.

It follows that

$$\Delta_{(1,0)} = \widetilde{\Delta}_1 + \widetilde{\Delta}_2, \qquad \Delta_{(1,1)} = \widetilde{\Delta}_1\widetilde{\Delta}_2.$$

Thus the Timoney Bloch space $\mathcal{B}_{(1,0)}$ consists of all f holomorphic on $\mathbf{D}^2$ for which

$$(1-|z_1|^2)^2\Big|\frac{\partial f}{\partial z_1}\Big|^2 + (1-|z_2|^2)^2\Big|\frac{\partial f}{\partial z_2}\Big|^2 \qquad \text{is bounded,}$$

while the Arazy Bloch space $\mathcal{B}_{(1,1)}$ consists of all f holomorphic on $\mathbf{D}^2$ for which

$$(1-|z_1|^2)^2(1-|z_2|^2)^2\Big|\frac{\partial^2 f}{\partial z_1\partial z_2}\Big|^2 \qquad \text{is bounded.}$$

We thus see that the Timoney Bloch space is contained in the Arazy Bloch space, and is a proper subset thereof: any holomorphic function of the form $f(z_1, z_2) = g(z_1)$ belongs to $\mathcal{B}_{(1,1)}$, but does not belong to $\mathcal{B}_{(1,0)}$ unless g belongs to the Bloch space on the disc. We also see that the Timoney-Bloch norm vanishes precisely on the constants, while the Arazy-Bloch norm vanishes precisely on functions of the form $f(z_1) + g(z_2)$.

Similar situation can be seen to prevail for a general polydisc $\mathbf{D}^n$: there are n Bloch spaces, of which Timoney is the smallest, and Arazy the largest.

Remark The big Hnkel operator H_f is compact $\iff$ f belongs to the Timoney Bloch space.

8. Composition series

The phenomenon that we have observed for the polydiscs is connected with the existence of the <u>composition series.</u> Let us explain this concept on the example of the unit disc $\mathbf{D}$. There the following assertion holds.

Claim Let E be any topological vector space of holomorphic functions on $\mathbf{D}$ which is Möbius invariant and on which the group of rotations acts strongly-continuously. Then either $E = \{0\}$, or $E = \{\text{constants}\}$, or E contains all polynomials.

Proof. Let $E \ni f = \sum_{k=0}^{\infty} f_k z^k$; then by rotation invariance,

$$\int_0^{2\pi} f(e^{i\theta}z)\, e^{-mi\theta}\, \frac{d\theta}{2\pi} = f_m z^m \in E. \tag{7}$$

Thus if $f_m \neq 0$ for some m, then $z^m \in E$; hence, by invariance, $(\frac{a-z}{1-\overline{a}z})^m \in E\ \forall a \in \mathbf{D}$. Taking this for the f in (7) and 0 for the m in (7), we get $a^m \mathbf{1} \in E$; thus the constants are in E.

If even $f_m \neq 0$ for some $m \geq 1$, then, applying (7) to the same function again but this time taking 1 for the m in (7) and noting that $(\frac{a-z}{1-\overline{a}z})^m = a^m - m(1-|a|^2)z + O(z^2)$, we see that $z \in E$; thus by invariance $\frac{a-z}{1-\overline{a}z} = a - (1-|a|^2)z\sum_{j=1}^{\infty} \overline{a}^j z^j$ belongs to E, for any $a \in \mathbf{D}$. Taking the last function for the f in (7) shows that $z^j \in E\ \forall j$, i.e. all polynomials are in E.

This completes the proof. □

The last theorem admits the following reformulation. Denote $\mathcal{M}_1 = \{\text{all holomorphic functions}\}$, $\mathcal{M}_0 = \{\text{constants}\}$, $\mathcal{M}_{-1} = \{0\}$. Then

$$E \setminus \mathcal{M}_{j-1} \neq \emptyset \implies \mathcal{P} \cap \mathcal{M}_j \subset E.$$

It turns out that, in a sense, precisely the same thing holds for a general Cartan domain. Namely, let

$$(x)_k := x(x+1)\cdots(x+k-1)$$

denote the familiar Pochhammer symbol, and for a signature $\mathbf{m} = (m_1, m_2, \ldots, m_r)$, consider the function

$$x \mapsto (x)_{m_1}(x - \frac{a}{2})_{m_2}(x-a)_{m_3}\cdots(x - \frac{r-1}{2}a)_{m_r}, \qquad x \in \mathbf{C}.$$

Let $q(\mathbf{m})$ be the multiplicity of zero of this function at $x = 0$:

$$q(\mathbf{m}) := \operatorname{card}\{j : m_j > \frac{j-1}{2}a \in \mathbf{Z}\}.$$

Also denote by q the maximum possible value of $q(\mathbf{m})$, i.e.

$$q = \begin{cases} r & a \text{ even}, \\ \left[\dfrac{r+1}{2}\right] & a \text{ odd}. \end{cases}$$

For $-1 \le j \le q$, let

$$\mathcal{M}_j = \{f = \sum_{\mathbf{m}} f_{\mathbf{m}} \text{ holomorphic} : f_{\mathbf{m}} = 0 \text{ if } q(\mathbf{m}) > j\}.$$

Thus, in particular,

$$\mathcal{M}_{-1} \subset \mathcal{M}_0 \subset \mathcal{M}_1 \subset \cdots \subset \mathcal{M}_q,$$
$$\mathcal{M}_{-1} = \{0\}, \quad \mathcal{M}_0 = \{\text{constants}\}, \quad \mathcal{M}_q = \{\text{all holomorphic}\}.$$

The following deep result is due to Orsted, Faraut and Koranyi.

Theorem (1) *Each $\mathcal{M}_j$ is G-invariant;* (2) *for any G-invariant space E of holomorphic functions on which the action of K is strongly continuous,*

$$E \setminus \mathcal{M}_{j-1} \neq \emptyset \implies \mathcal{P} \cap \mathcal{M}_j \subset E.$$

Example* For the bidisc $\mathbf{D}^2$, $q = 2$ and

$$\begin{aligned} &\mathcal{M}_{-1} = \{0\}, \quad \mathcal{M}_0 = \{\text{constants}\}, \\ &\mathcal{M}_1 = f(z_1) + g(z_2), \qquad f, g \text{ holomorphic on } \mathbf{D}, \\ &\mathcal{M}_2 = \{\text{all holomorphic}\}. \end{aligned}$$

9. Results

After all the preparations, we can finally state our results.

Theorem 9.1. *If $j < q(\boldsymbol{m})$, then the $Q_{\nu,\boldsymbol{m}}$-norm vanishes on $\mathcal{M}_j$; thus $\mathcal{M}_j$ is contained in $Q_{\nu,\boldsymbol{m}}$ in a trivial way. The same is true also for the Bloch space $\mathcal{B}_{\boldsymbol{m}}$.*

Theorem 9.2. *If $\nu > p - 1$, then $\mathcal{B}_{\boldsymbol{m}} \subset Q_{\nu,\boldsymbol{m}}$ continuously.*

Theorem 9.3. *If $q(\boldsymbol{m}) \le q(\boldsymbol{n})$, then $Q_{\nu,\boldsymbol{m}} \subset \mathcal{B}_{\boldsymbol{n}}$ continuously.*

Corollary$\nu > p - 1 \implies Q_{\nu,\boldsymbol{m}} = \mathcal{B}_{\boldsymbol{m}}$, *with equivalent norms.* $q(\boldsymbol{m}) \le q(\boldsymbol{n}) \implies \mathcal{B}_{\boldsymbol{m}} \subset \mathcal{B}_{\boldsymbol{n}}$ *continuously.* $q(\boldsymbol{m}) = q(\boldsymbol{n}) \implies \mathcal{B}_{\boldsymbol{m}} = \mathcal{B}_{\boldsymbol{n}}$, *with equivalent norms.* $q(\boldsymbol{m}) = q(\boldsymbol{n})$, $\nu > p - 1 \implies Q_{\nu,\boldsymbol{m}} = Q_{\nu,\boldsymbol{n}}$, *with equivalent norms. The last corollary exhausts the case $\nu > p-1$ completely. What about $\nu \le p - 1$?*

Theorem 9.4. *If $\nu < 0$, then $Q_{\nu,\boldsymbol{m}} = \mathcal{M}_{q(\boldsymbol{m})-1}$.*

Theorem 9.5. *For $r > 1$ and $m = (1, 0, \dots, 0)$, that is,*

$$f \in Q_{\nu,(1)} \iff \sup_{a\in\Omega} \int_\Omega \widetilde{\Delta}|f \circ \phi_a|^2 \, h^\nu \, d\mu,$$

we have

$$Q_{\nu,(1)} = \begin{cases} \mathcal{B}_{(1)}, \text{ the Timoney Bloch space} & \nu > p - 1, \\ \{\text{constants}\} & \nu \le p - 1. \end{cases}$$

Note that the last theorem means that the situation for $r > 1$ differs radically from the one for $r = 1$ (disc, ball): there Q_ν is nontrivial also for $p - 2 < \nu \le p - 1$ (for the disc, even for $p - 2 \le \nu \le p - 1$).

Theorem 9.6. *For a tube domain with $s = \frac{d}{r} \in \mathbf{Z}$, and $m = (s^r)$,*

$$Q_{\nu,(s^r)} = \begin{cases} \mathcal{B}_{(s^r)}, \text{ the Arazy Bloch space} & \nu > p - 1 \\ \mathcal{D} & \nu = 0 \\ \mathcal{M}_{q-1} & \nu < 0. \end{cases}$$

Here $\mathcal{D}$ is the Dirichlet space

$$\mathcal{D} = \{f \text{ holomorphic: } N(\partial)^s f \in L^2(\Omega, dz)\}.$$

At the moment, we do not know what are the spaces $Q_{\nu,\mathbf{m}}$ for ν between 0 and $p-1$ and $|\mathbf{m}| > 1$ — for instance, whether they are properly increasing with ν. We can offer somewhat more complete information only for the polydisc:

Theorem 9.7. *For the polydisc $\mathbf{D}^r$ (so that $p = 2$, $q(m) = \#\{j : m_j > 0\}$, and $q = r$),*

$$q(m) < r \implies Q_{\nu,(m)} = \begin{cases} \mathcal{B}_m & \nu > 1, \\ \mathcal{M}_{q(m)-1} & \nu \le 1; \end{cases}$$

$$q(m) = r \implies Q_{\nu,(m)} = \begin{cases} \text{Arazy-Bloch} & \nu > 1, \\ \mathcal{D} & \nu = 0, \\ \mathcal{M}_{q(m)-1} & \nu < 0. \end{cases}$$

All this suggests the following conjecture about the nontriviality of the spaces $Q_{\nu,\mathbf{m}}$.

Conjecture *For tube domains with $s = \frac{d}{r} \in \mathbf{Z}$ and $q(m) = q$,*

$$\mathcal{M}_{q(m)-1} \subsetneq Q_{\nu,m} \iff \nu \ge 0;$$

in all other cases,

$$\mathcal{M}_{q(m)-1} \subsetneq Q_{\nu,m} \iff \nu > p-1.$$

10. Sketches of proofs

We proceed to give some hints about the proofs of the theorems. Details will appear in [3].

Theorem 10.1. $j < q(m) \implies \mathcal{M}_j \subset Q_{\nu,m}$ *and the norm vanishes. Similarly for* $\mathcal{B}_{\mathbf{m}}$.

Proof. Recall that

$$f \in Q_{\nu,\mathbf{m}} \iff \sup_a \int_\Omega \Delta_{\mathbf{m}}|f|^2 \,(h \circ \phi_a)^\nu \,d\mu < \infty,$$
$$f \in \mathcal{B}_{\mathbf{m}} \iff \Delta_{\mathbf{m}}|f|^2 \in L^\infty.$$

Now

$$\begin{aligned}\Delta_{\mathbf{m}}|f|^2(z) &= K_{\mathbf{m}}(\partial,\partial)|f \circ \phi_z|^2(0)\\ &= \sum_j \psi_j(\partial)\overline{\psi_j(\partial)}|f \circ \phi_z|^2(0)\\ &= \sum_j |\psi_j(\partial)(f \circ \phi_z)(0)|^2\\ &= \sum_j |\langle f \circ \phi_z, \psi_j\rangle_F|^2\\ &= \|P_{\mathbf{m}}(f \circ \phi_z)\|_F^2,\end{aligned}$$

where $P_{\mathbf{m}}$ denotes the projection onto $\mathcal{P}_{\mathbf{m}}$.

Thus $f \in \mathcal{M}_j \implies f \circ \phi_z \in \mathcal{M}_j \implies P_{\mathbf{m}}(f \circ \phi_z) = 0 \implies \Delta_{\mathbf{m}}|f|^2 = 0 \implies f \in \mathcal{B}_{\mathbf{m}}$ and $f \in Q_{\nu,\mathbf{m}}$. □

Theorem 10.2. $\nu > p-1 \implies \mathcal{B}_m \subset Q_{\nu,m}$ *continuously.*

Proof. It is known that for $\nu > p-1$, the measure $h^\nu\,d\mu$ is finite. Thus $\forall a \in \Omega$,

$$\begin{aligned}\int_\Omega (\Delta_{\mathbf{m}}|f|^2) \circ \phi_a \, h^\nu \,d\mu &\le c_\nu \,\|\Delta_{\mathbf{m}}|f|^2\|_\infty\\ &= c_\nu \|f\|^2_{\mathcal{B}_{\mathbf{m}}}.\end{aligned}$$

□

Theorem 10.3. $q(m) \le q(n) \implies Q_{\nu,m} \subset \mathcal{B}_n$ *continuously.*

Proof. By the K-invariance of $\Delta_{\mathbf{m}}$ and h, the integral

$$\int_\Omega \Delta_{\mathbf{m}}(f\overline{g})\, h^\nu\, d\mu$$

is a K-invariant bilinear form of $f, g \in \mathcal{P}$. It follows from the Peter-Weyl decomposition of $\mathcal{P}$ into the $\mathcal{P}_{\mathbf{m}}$ that any such bilinear functional must be of the form

$$\sum_{\mathbf{k}} c_{\mathbf{mk}} \langle f_{\mathbf{k}}, g_{\mathbf{k}} \rangle_F,$$

for some coefficients $c_{\mathbf{mk}} \geq 0$. Suppose we can show that

$$c_{\mathbf{mn}} > 0. \tag{8}$$

Since $\Delta_{\mathbf{n}}|f|^2(0) = \|P_{\mathbf{n}} f\|_F^2 = \|f_{\mathbf{n}}\|_F^2$, it will follow that

$$\Delta_{\mathbf{n}}|f|^2(0) \leq \frac{1}{c_{\mathbf{mn}}} \int_\Omega \Delta_{\mathbf{m}}|f|^2\, h^\nu\, d\mu.$$

Replacing f by $f \circ \phi_a$, this becomes

$$\Delta_{\mathbf{n}}|f|^2(a) \leq \frac{1}{c_{\mathbf{mn}}} \int_\Omega \Delta_{\mathbf{m}}|f \circ \phi_a|^2\, h^\nu\, d\mu.$$

Taking suprema over all $a \in \Omega$ gives the assertion.

It remains to prove (8). But by the properties of the composition series,

$$\begin{aligned} c_{\mathbf{mn}} = 0 \iff & \int_\Omega \Delta_{\mathbf{m}}|f_{\mathbf{n}}|^2\, h^\nu\, d\mu = 0 \qquad \forall f_{\mathbf{n}} \in \mathcal{P}_{\mathbf{n}} \\ \iff & \Delta_{\mathbf{m}}|f_{\mathbf{n}}|^2(z) = 0 \quad \forall z\, \forall f_{\mathbf{n}} \\ \iff & \|P_{\mathbf{m}}(f_{\mathbf{n}} \circ \phi_z)\|_F^2 = 0 \quad \forall z\, \forall f_{\mathbf{n}} \\ \iff & P_{\mathbf{m}}(f_{\mathbf{n}} \circ \phi_z) = 0 \quad \forall z\, \forall f_{\mathbf{n}} \\ \iff & P_{\mathbf{m}} \mathcal{M}_{q(\mathbf{n})} = 0 \\ \iff & q(\mathbf{m}) > q(\mathbf{n}). \end{aligned}$$

□

Theorem 10.4. $\nu < 0 \implies Q_{\nu,m} = \mathcal{M}_{q(m)-1}$.

Proof. From the composition series we know that

$$\begin{aligned} \mathcal{M}_{q(\mathbf{m})-1} \subsetneq Q_{\nu,\mathbf{m}} & \implies \mathcal{P} \cap \mathcal{M}_{q(\mathbf{m})} \subset Q_{\nu,\mathbf{m}} \\ & \implies \mathcal{P}_{\mathbf{m}} \subset Q_{\nu,\mathbf{m}} \\ \implies \sup_a \int_\Omega \Delta_{\mathbf{m}}|f|^2\, (h \circ \phi_a)^\nu\, d\mu < \infty & \qquad \forall f \in \mathcal{P}_{\mathbf{m}}. \end{aligned}$$

Since $K_{\mathbf{m}}(z,z) = \sum_j |\psi_j(z)|^2$ for any basis $\{\psi_j\}$ of $\mathcal{P}_{\mathbf{m}}$, we can continue by

$$\Longrightarrow \sup_a \int_\Omega \Delta_{\mathbf{m}} K_{\mathbf{m}} \cdot (h \circ \phi_a)^\nu \, d\mu < \infty$$

where $K_{\mathbf{m}} = K_{\mathbf{m}}(z,z)$. It can be shown that

$$\exists m \gg 0 : \quad \Delta_{\mathbf{m}} K_{\mathbf{m}} \geq c\, h^m.$$

Thus we can continue by

$$\Longrightarrow \sup_a \int_\Omega h^m \, (h \circ \phi_a)^\nu \, d\mu < \infty. \tag{9}$$

Forelli-Rudin inequalities show that this happens iff $\nu \geq 0$. □

Theorem 10.5. *For rank $r > 1$ and $\boldsymbol{m} = (1, 0, \ldots, 0)$,*

$$Q_{\nu,(1)} = \begin{cases} \mathcal{B}_{(1)}, \text{ the Timoney Bloch space} & \nu > p-1, \\ \{\text{constants}\} & \nu \leq p-1. \end{cases}$$

Proof. As above,

$$\{\text{constants}\} \subsetneq Q_{\nu,(1)} \iff \sup_a \int_\Omega \widetilde{\Delta}\|z\|^2 \, (h \circ \phi_a)^\nu \, d\mu < \infty.$$

The fact that

$$\widetilde{\Delta}\|z\|^2 \approx \begin{cases} h^2 & \Omega = \mathbf{D}, \\ h & \Omega = \mathbf{B}^d, \\ 1 & r > 1 \end{cases}$$

and (9) again yield the conclusion. □

Theorem 10.6. *For a tube domain with $s = \frac{d}{r} \in \mathbf{Z}$, and $\boldsymbol{m} = (s^r)$,*

$$Q_{\nu,(s^r)} = \begin{cases} \mathcal{B}_{(s^r)}, \text{ the Arazy Bloch space} & \nu > p-1 \\ \mathcal{D} & \nu = 0 \\ \mathcal{M}_{q-1} & \nu < 0. \end{cases}$$

Proof. As mentioned before, in this case

$$\Delta_{\mathbf{m}} = h^p N(\partial)^s N(\overline{\partial})^s$$

for a certain polynomial N (the Jordan norm). Hence

$$f \in Q_{\nu,\mathbf{m}} \iff \sup_a \int_\Omega h^p \, |N(\partial)^s f|^2 \, (h \circ \phi_a)^\nu \, d\mu < \infty$$
$$\iff \sup_a \int_\Omega |N(\partial)^s f|^2 \, (h \circ \phi_a)^\nu \, dz < \infty.$$

Thus for $\nu \geq 0$, all polynomials belong to $Q_{\nu,\mathbf{m}}$. For $\nu = 0$, this coincides with the definition of the Dirichlet space. The case $\nu < 0$ was settled by the previous theorem. □

Theorem 10.7. *For the polydisc $\mathbf{D}^r$ (so that $p = 2$, $q(\boldsymbol{m}) = \#\{j : m_j > 0\}$, and $q = r$),*

$$q(\boldsymbol{m}) < r \implies Q_{\nu,(\boldsymbol{m})} = \begin{cases} \mathcal{B}_{\boldsymbol{m}} & \nu > 1, \\ \mathcal{M}_{q(\boldsymbol{m})-1} & \nu \leq 1; \end{cases}$$

$$q(\boldsymbol{m}) = r \implies Q_{\nu,(\boldsymbol{m})} = \begin{cases} \textit{Arazy-Bloch} & \nu > 1, \\ \mathcal{D} & \nu = 0, \\ \mathcal{M}_{q(\boldsymbol{m})-1} & \nu < 0. \end{cases}$$

Proof. Using explicit formulas for $K_{\mathbf{m}}$, $\Delta_{\mathbf{m}}$ etc. given in one of the preceding sections, this is easily reduced to explicit calculations on the disc. □

11. Open problems

We conclude the paper by a list of open problems. **(1)** The first of them is, of course, to determine when $Q_{\nu,\mathbf{m}}$ is nontrivial — we repeat here the conjecture stated above:

Conjecture *$Q_{\nu,\boldsymbol{m}}$ is nontrivial iff $\nu \geq 0$ (for tube domain with $\frac{d}{r} \in \mathbf{Z}$ and $q(\boldsymbol{m}) = q$) $\nu > p - 1$ (otherwise).*

(2) If $q(\boldsymbol{m}) = q(\boldsymbol{n})$, is $Q_{\nu,\boldsymbol{m}} = Q_{\nu,\boldsymbol{n}}$? (We have seen that this holds for the Bloch spaces, hence also for $\nu > p - 1$; the case of $\nu \leq p - 1$ remains unresolved.)

(3) If $\nu_1 < \nu_2$ and $Q_{\nu_1,\boldsymbol{m}}, Q_{\nu_2,\boldsymbol{m}}$ are nontrivial, is $Q_{\nu_1,\boldsymbol{m}} \subsetneq Q_{\nu_2,\boldsymbol{m}}$? (For **D***, this was proved in* [4]*, and for $\mathbf{B}^d$ in* [1]*.)*

(4) In principle, one can define $Q_{\nu,L}$ and $\mathcal{B}_L$ for any invariant differen-

tial operator L, even when the right-hand side in (3) is not nonnegative, by

$$f \in \mathcal{B}_L \iff L|f|^2 \text{ is bounded,}$$

$$f \in Q_{\nu,L} \iff \sup_a \int_\Omega \left| L|f \circ \phi_a|^2 \right| h^\nu \, d\mu < \infty.$$

If L is such that $L|f|^2 \geq 0$ *for all holomorphic f, i.e. if*

$$L = \sum_m l_m \, \Delta_m, \qquad l_m \geq 0, \tag{10}$$

then it is easy to see that

$$Q_{\nu,L} = \bigcap_{m:\, l_m > 0} Q_{\nu,m}$$

and

$$\mathcal{B}_L = \mathcal{B}_m \qquad \text{where } q(m) = \min\{q(k) : l_k > 0\}.$$

What happens for operators L not satisfying (10)?

For instance, does the space of all holomorphic f on **D** *for which*

$$\sup_{a \in \mathbf{D}} \int_{\mathbf{D}} \left| \widetilde{\Delta}^2 |f(\phi_a(z))|^2 \right| (1 - |z|^2)^{\nu - 2} \, dz < \infty$$

coincide with the Bloch space for $\nu > 1$*?*

References

1. M. Andersson, H. Carlsson, Q_p spaces in strictly pseudoconvex domains, *J. Anal. Math.* **84** (2001):335–359.
2. J. Arazy, Boundedness and compactness of generalized Hankel operators on bounded symmetric domains, *J. Funct. Anal.* **137** (1996):97–151.
3. J. Arazy, M. Engliš, *Q_p-spaces on bounded symmetric domains,* forthcoming.
4. R. Aulaskari, J. Xiao, R. Zhao, On subspaces and subsets of BMOA and UBC, *Analysis,* **15** (1995):101–121.
5. C. Ouyang, W. Yang, R. Zhao, Möbius invariant Q_p spaces associated with the Green's function on the unit ball of C^n, *Pacific J. Math.* **182** (1998):69–99.
6. J. Rättyä, On some complex function spaces and classes,*Ann. Acad. Sci. Fenn. Math. Diss.* **124** (2001): 73.
7. R.M. Timoney, Bloch functions in several complex variables I, *Bull. London Math. Soc.* **12** (1980):241–267; *II, J. reine angew. Math.* **319** (1980):1–22.
8. J. Xiao, *Holomorphic Q classes,* Lecture Notes in Mathematics **1767**, Springer-Verlag, Berlin, 2001.
9. R. Zhao,On a general family of function spaces, *Ann. Acad. Sci. Fenn. Math. Diss.* **105** (1996), 56 pp.

ORDER OF GROWTH OF PAINLEVÉ TRANSCENDENTS*

AIMO HINKKANEN

Department of Mathematics
University of Illinois
Urbana, IL 61801, U.S.A.
E-mail: aimo@math.uiuc.edu

ILPO LAINE[†]

Department of Mathematics
University of Joensuu
P.O. Box 111, FI–80101 Joensuu, Finland
E-mail: ilpo.laine@joensuu.fi[‡]

1. Introduction

The growth of transcendental solutions of the Painlevé differential equations have been recently considered in several papers by using complex analytic reasoning. In particular, this concerns the first Painlevé differential equation

$$f'' = 6f^2 + z, \tag{1}$$

as well as the second and fourth Painlevé equations

$$f'' = 2f^3 + zf + \alpha, \qquad \alpha \in \mathbb{C} \tag{2}$$

and

$$2ff'' = (f')^2 + 3f^4 + 8zf^3 + 4(z^2 - \alpha)f^2 + 2\beta, \qquad \alpha, \beta \in \mathbb{C}. \tag{3}$$

It is well known that all solutions of these differential equations are meromorphic functions in the complex plane, see, e.g., [2]. One of the recent

*This material is based upon work supported by the National Science Foundation under Grants No. 0200752 and 0457291.

†The second author has been partially supported by the Academy of Finland grant 210245.

‡Both authors were also supported by the Väisälä Fund of the Finnish Academy of Science and Letters.

results about their growth proved by complex analytic methods is the following.

Theorem 1.1. *All transcendental solutions of (2), resp. of (3), are of order of growth at least* 3/2, *resp.* 2.

Three independent proofs of this theorem, by different methods, appeared recently, see [6], [4] and [7]. This paper aims to show how the proof offered in [4] may be substantially shortened by making use of an idea originally proposed by J. Langley [5] for the case of (2).

2. Short proof of Theorem 1.1

A key tool in the proof offered in [4] is a lemma due to G. Gundersen in [3]. We recall this lemma for the convenience of the reader:

Lemma 2.1. *Let F be a transcendental meromorphic function in the complex plane, of finite order ρ, and suppose that $\varepsilon > 0$. Then there exists a set $E \subset [1,\infty)$ of finite logarithmic measure such that whenever $|z| \in [1,\infty)$, then*

$$\left|\frac{F^{(j)}(z)}{F(z)}\right| \leq |z|^{j(\rho-1)+\varepsilon}$$

for $j = 1,2$. Moreover, the same estimates hold on all radii $z = re^{i\theta}$, $r \geq R(\theta) > 1$ outside of a set $E_1 \subset [0,2\pi)$ of zero linear measure.

Case of (2). Suppose that f is a transcendental solution of (2) of order $\rho < 3/2$. Since the case $\alpha = 0$ results in an immediate contradiction, see [4], p. 647, we may assume that $\alpha \neq 0$. We now write (2) in the form

$$\frac{f''}{f} = 2f^2 + z + \frac{\alpha}{f}. \tag{4}$$

By Lemma 2.1, we have $|f''/f| = O(|z|^{1-\delta})$ for some $\delta > 0$ on almost all circles of radius r, centered at the origin. If there is now a sequence of such permitted circles, with their radii tending to infinity, such that each circle in this sequence admits at least one point z such that

$$|z|^{-1+\varepsilon} \leq |f(z)| \leq |z|^{\frac{1}{2}-\varepsilon}$$

for some $\varepsilon > 0$, an immediate contradiction to (4) follows. Therefore, we must have a sequence of circles such that either $|f(z)| > |z|^{\frac{1}{2}-\varepsilon}$ or

$|f(z)| < |z|^{-1+\varepsilon}$ holds on all of these circles. In the first of these two possibilities, (4) implies that

$$2f^2 = (-1 + o(1))z$$

and we obtain a contradiction exactly as in [4]. In the remaining final case, see [4] again, we may write (2) in the form

$$\frac{\alpha}{f} + z = \frac{f''}{f} - 2f^2 =: \frac{z}{f} y(z). \tag{5}$$

But then

$$|y(z)| \le \left|\frac{f}{z}\right| \left|\frac{f''}{f}\right| + \frac{2}{|z|}|f|^3 = O(|z|^{-1-\gamma}) \tag{6}$$

for some $\gamma > 0$, provided that $\varepsilon > 0$ is small enough. Writing now

$$f(z) = -\frac{\alpha}{z} + y(z)$$

and integrating we obtain

$$-\alpha = \sum_{|z_j|<r} \mathrm{Res}(f, z_j) - \frac{1}{2\pi i} \int_{|\zeta|=r} y(\zeta)\, d\zeta,$$

where z_j stands for the poles of f in $|z| < r$. By (6), the last integral tends to zero as $r \to \infty$. Since the residues of f are integers, and actually ± 1, we conclude that α has to be an integer. To arrive at the final contradiction, see [6], we now apply the Bäcklund transformation

$$f_{\alpha-1} := -f + \frac{\alpha - \frac{1}{2}}{f' - f^2 - \frac{z}{2}}$$

or its inverse

$$f_{\alpha+1} := -f - \frac{\alpha + \frac{1}{2}}{f' + f^2 + \frac{z}{2}}$$

to obtain a solution $f_{\alpha-1}$ of (2) with parameter $\alpha - 1$, resp. $f_{\alpha+1}$ with parameter $\alpha + 1$. Clearly, the Bäcklund transformations are order preserving. Hence, applying one of these transformations finitely many times, we obtain a transcendental solution f_0 of (2) with parameter $\alpha = 0$ and of order $< 3/2$, a contradiction.

Case of (3). We first remark that all fourth Painlevé transcendents possess infinitely many poles. This immediately follows by applying the standard Clunie lemma [1] to conclude that $m(r, f) = O(\log r)$ from (3). Here $m(r, f) = \frac{1}{2\pi} \int_0^{2\pi} \log^+ |f(re^{i\theta})|\, d\theta$ is Nevanlinna's proximity function.

Moreover, all residues of f are integers and in fact ± 1. Suppose now that f is a solution of (3) of order $\rho = \rho(f) < 2$. As shown in [4], p. 650, we may construct a sequence of circles, permitted by Lemma 2.1, with their radii tending to infinity so that just one of the following three possibilities holds on all of these circles for some $\delta > 0$:

$$f(z) = -2z + O(|z|^{1-\delta}), \tag{7}$$

$$f(z) = -\frac{2}{3}z + O(|z|^{1-\delta}), \tag{8}$$

$$f(z) = \frac{A}{z} + O(|z|^{-1-\delta}), \qquad A^2 = -\frac{\beta}{2} \neq 0. \tag{9}$$

Actually, just one of these alternatives holds in a connected set E tending to infinity, for $|z|$ large enough, see [4], p. 651. Moreover, by [4], pp. 651– 652, it suffices to consider the cases (8) and (9) only. Defining now a meromorphic function W as

$$W := -\frac{(f')^2 + 2\beta}{4f} + \frac{1}{4}f^3 + zf^2 + (z^2 - \alpha)f, \tag{10}$$

we know that $W' = f^2 + 2zf$ and that there exist entire functions u, v such that $u' = -uW$, $v = uf$, see, e.g., [2], pp. 121–122.

Consider first the case (9). Supposing first that $\alpha \neq 2A$, a simple consideration implies that

$$f(z) = \frac{A}{z} - \frac{\beta(\alpha - 2A)}{4Az^3} + y(z), \tag{11}$$

where $|y(z)| = O(|z|^{-3-\delta})$, see [4], p. 652. Substituting this expression into $W' = f^2 + 2zf$ we obtain

$$W'(z) = 2A - \frac{A(A-\alpha)}{z^2} + s(z),$$

where $|s(z)| = O(|z|^{-2-\delta})$. By integration,

$$W(z) = 2Az + \gamma + \frac{A(A-\alpha)}{z} + t(z),$$

where γ is the constant of integration and $|t(z)| = O(|z|^{-1-\delta})$. Integrating around a permitted circle similarly as in the case of (2), we get

$$\sum_{|z_j|<r} \operatorname{Res}(W, z_j) = A(A-\alpha) + \frac{1}{2\pi i}\int_{|\zeta|=r} t(\zeta)\, d\zeta.$$

Since $W = -\frac{u'}{u}$, we have $\text{Res}(W, z_j) \in -\mathbb{N}$. Since the last integral tends to zero as r tends to infinity, there are at most finitely many poles of W, hence of f in the complex plane, contradicting the Clunie argument above. Similarly, if $\alpha = 2A$, we have

$$f(z) = \frac{A}{z} + \frac{3\beta(2+\beta)}{32Az^5} + O(|z|^{-5-\delta}), \tag{12}$$

see [4], p. 652. Substituting (12) into $W' = f^2 + 2zf$ and integrating we get

$$W(z) = 2Az + \gamma + \frac{1}{2}\frac{\beta}{z} + O(|z|^{-1-\delta}).$$

Integrating around a permitted circle as in the preceding case we now obtain

$$\sum_{|z_j|<r} \text{Res}(W, z_j) = \frac{\beta}{2} + \frac{1}{2\pi i}\int_{|z|=r} O(|z|^{-1-\delta})\, dz,$$

resulting in a contradiction as before. As for the remaining case (8), we first recall that

$$f(z) = -\frac{2}{3}z + \frac{\alpha}{z} - \frac{1 + 3\alpha^2 + \frac{9}{2}\beta}{4z^3} + O(|z|^{-3-\delta}),$$

see [4], p. 653. Proceeding as in the preceding case (9), we conclude that

$$W(z) = -\frac{8}{27}z^3 + \frac{2}{3}\alpha z + \gamma + \left(\frac{1}{6} - \frac{1}{2}\alpha^2 + \frac{3}{4}\beta\right)\frac{1}{z} + O(|z|^{-1-\delta})$$

and

$$\sum_{|z_j|<r} \text{Res}(W, z_j) = \frac{1}{6} - \frac{1}{2}\alpha^2 + \frac{3}{4}\beta + \frac{1}{2\pi i}\int_{|z|=r} O(|z|^{-1-\delta})\, dz,$$

and a similar contradiction again follows.

References

1. J.G. Clunie, On integral and meromorphic functions, *J. London Math. Soc.* **37** (1962): 17–27.
2. V. Gromak, I. Laine, and S. Shimomura, Painlevé Differential Equations in the Complex Plane, *Studies in Mathematics*, W. de Gruyter, Berlin–New York **28**(2002).
3. G. Gundersen, Estimates for the logarithmic derivative of a meromorphic function, plus similar estimates, *J. London Math. Soc.* **37**(2) (1988): 88–104.
4. A. Hinkkanen, and I. Laine, Growth results for Painlevé transcendents, *Math. Proc. Camb. Phil. Soc.* **137** (2004):645–655.
5. J. Langley, *private communication.*

6. S. Shimomura, Lower estimates for the growth the fourth and the second Painlevé transcendents, *Proc. Edinb. Math. Soc.* **47** (2004):231–249.
7. N. Steinmetz, Boutroux's method vs. rescaling lower estimates for the orders of growth of the second and fourth Painlevé transcendents, *Port. Math. (N.S.)* **61** (2004):369–374.

UNIQUENESS OF MEROMORPHIC FUNCTIONS THAT SHARE FOUR VALUES*

BIN HUANG

Chansha University of Science and Technology
Changsha 410076, P. R. China
E-mail: huangbincscu@163.com

1. Problem "1CM+3IM=4CM" and Gundersen's Question

Let f and g are two meromorphic functions. That "$f(z)$ and $g(z)$ share a value c" means f takes value c if and only if g takes c. Usually, we say "$f(z)$ and $g(z)$ share a value c IM" or "$f(z)$ and $g(z)$ share a value c IM", here, IM refers to ignoring multiplicities, and CM refers to counting multiplicities.

Theorem A.[3] (4CM Theorem) *If f and g are distinct nonconstant meromorphic functions that share four distinct values a_1, a_2, a_3, a_4 CM, then f is a Möbius transformation of g; two of the values, say, a_1 and a_2, are Picard values, and the cross ratio $(a_1, a_2, a_3, a_4) = -1$.*

Rubel's question: Whether CM can be replaced by IM in Theorem A with the same conclusion or not? i.e., 4IM=4CM?

Theorem B.[5] (2CM+2IM=4CM) *If two nonconstant meromorphic functions share two values IM, and share two other values CM, then f and g share all four values CM.*

1CM+3IM Question: If two nonconstant meromorphic functions share three values IM and a fourth value CM, then do the functions necessarily share all four values CM?

*Project supported by NSFHN (No. 05jj30013).

Gundersen's Question: If two nonconstant entire functions share three finite values IM, then do the functions necessarily share all three values CM?

2. Some Auxiliary Functions

(I) The case when $f(z)$ and $g(z)$ are distinct meromorphic functions and share four values $0, 1, c, \infty$ IM. Set

$$\varphi = \frac{f'g'(f-g)^2}{f(f-1)(f-c)g(g-1)(g-c)}. \tag{2.1}$$

Then φ is an entire function satisfying $T(r, \varphi) = S(r, f)$.

(II) The case when $f(z)$ and $g(z)$ are distinct meromorphic functions sharing three values $0, 1, c$ IM and a fourth value ∞ CM. Set

$$\psi = \frac{f''}{f'} - \frac{f'}{f} - \frac{f'}{f-1} - \frac{f'}{f-c} - \{\frac{g''}{g'} - \frac{g'}{g} - \frac{g'}{g-1} - \frac{g'}{g-c}\}. \tag{2.2}$$

Then ψ satisfies $T(r, \psi) = S(r, f)$.

(III) The case when $f(z)$ and $g(z)$ are distinct meromorphic functions sharing four values $0, 1, c, \infty$ IM with property $\overline{N}(r, f) = S(r, f)$. Set

$$\Delta_0 = (\frac{f'f}{(f-1)(f-c)} - \frac{g'g}{(g-1)(g-c)})^2$$
$$\frac{(f-1)(f-c)(g-1)(g-c)}{(f-g)^2}, \tag{2.3}$$

$$\Delta_1 = (\frac{f'(f-1)}{f(f-c)} - \frac{g'(g-1)}{g(g-c)})^2 \frac{f(f-c)g(g-c)}{(f-g)^2}, \tag{2.4}$$

$$\Delta_c = (\frac{f'(f-c)}{f(f-1)} - \frac{g'(g-c)}{g(g-1)})^2 \frac{f(f-1)g(g-1)}{(f-g)^2} \tag{2.5}$$

and

$$\nabla_c = (1-c)^2\Delta_0 + c^2\Delta_1 - \Delta_c. \tag{2.6}$$

Let $f(z)$ and $g(z)$ be two distinct meromorphic functions sharing four values $0, 1, c, \infty$ IM with property $\overline{N}(r, f) = S(r, f)$, and let

$$H_c = 4\{(1-c)^2\Delta_0 - \frac{1}{2}c^2(1-c)^2\varphi\}\{c^2\Delta_1 - \frac{1}{2}c^2(1-c^2)\varphi\}$$

$$-\{\nabla_c - \frac{1}{2}c^2(1-c)^2\varphi\}^2. \tag{2.7}$$

Then $T(r, H_c) = S(r, f)$.

3. A Branch of $\sqrt{\varphi}$ and its Characteristic Function

Let $f(z)$ and $g(z)$ share values $0, 1, c, \infty$ IM ($c \neq 0, 1, \infty$). To the constant c, choose a branch of $\sqrt{z}$, still denoted by $\sqrt{z}$, such that $\sqrt{c^2} = c$. Set

$$\sigma(z) = (f-g)\sqrt{\frac{f'g'}{f(f-1)(f-c)g(g-1)(g-c)}}. \tag{3.8}$$

Then $\sigma^2 = \varphi$. Link all branch points of $\sqrt{\varphi}$ with a smooth curve L. Then σ is meromorphic in $\mathbb{C} \setminus L$. Let

$$F(\sigma) = a_m(z)\sigma^m + a_{m-1}(z)\sigma^{m-1} + \cdots + a_1(z)\sigma + a_0(z) \tag{3.9}$$

where $a_0(z), a_1(z), \cdots, a_m(z)$ are meromorphic functions. Notice that $\sigma^2 = \varphi$, we see that actually $F(\sigma) = a(z)\sigma + b(z)$, where $a(z)$ and $b(z)$ are meromorphic functions. We may chose L does not pass by any of zeros and poles of $F(\sigma(z))$ except branch points of $\sqrt{\varphi}$.

Let $z_0 \in \mathbb{C} \setminus L$. If in a neighborhood of z_0, $F(\sigma(z))$ has the following expansion

$$F(\sigma(z)) = c_{-\tau}(z-z_0)^{-\tau} + c_{-\tau+1}(z-z_0)^{-\tau+1} + \cdots, (c_{-\tau} \neq 0)$$

or

$$F(\sigma(z)) = c_\tau(z-z_0)^\tau + c_{\tau+1}(z-z_0)^{\tau+1} + \cdots, (c_\tau \neq 0)$$

then define, respectively,

$$\tau_*(z_0, F(\sigma(z)) = \tau,$$

or

$$\tau_*(z_0, \frac{1}{F(\sigma(z))}) = \tau.$$

Let $r > 0$. Define

$$n_*(r, F(\sigma(z))) = \sum_{|z|<r} \tau_*(z, F(\sigma(z))),$$

$$n_*(r, \frac{1}{F(\sigma(z))}) = \sum_{|z|<r} \tau_*(z, \frac{1}{F(\sigma(z))}).$$

Let $z_0 \in L$. If in a neighborhood (or a sector region) of z_0, $F(\sigma(z))$ has the following expansion

$$F(\sigma(z)) = c_{-\tau}(z - z_0)^{\frac{-\tau}{2}} + c_{-\tau+1}(z - z_0)^{\frac{-\tau+1}{2}} + \cdots, (c_{-\tau} \neq 0)$$

or

$$F(\sigma(z)) = c_\tau(z - z_0)^{\frac{\tau}{2}} + c_{\tau+1}(z - z_0)^{\frac{\tau+1}{2}} + \cdots (c_\tau \neq 0)$$

then define, respectively,

$$\tau_L(z_0, F(\sigma(z)) = \frac{\tau}{2},$$

or

$$\tau_L(z_0, \frac{1}{F(\sigma(z))}) = \frac{\tau}{2}.$$

Let $r > 0$. Define

$$n_L(r, F(\sigma(z))) = \sum_{|z|<r} \tau_*(z, F(\sigma(z))),$$

$$n_L(r, \frac{1}{F(\sigma(z))}) = \sum_{|z|<r} \tau_*(z, \frac{1}{F(\sigma(z))}).$$

Define

$$m(r, F(\sigma)) = \frac{1}{2\pi} \int_{0}^{2\pi} \log^{+} |F(\sigma(re^{r\theta}))| d\theta,$$

$$N_*(r, F(\sigma)) = \int_{0}^{r} \frac{n_*(t, F(\sigma)) - n_*(0, F(\sigma))}{t} dt + n_*(0, F(\sigma)) \log r,$$

$$N_L(r, F(\sigma)) = \int_{0}^{r} \frac{n_L(t, F(\sigma)) - n_L(0, F(\sigma))}{t} dt + n_L(0, F(\sigma)) \log r,$$

$$N(r, F(\sigma)) = N_*(r, F(\sigma)) + N_L(r, F(\sigma)),$$

$$T(r, F(\sigma)) = m(r, F(\sigma)) + N(r, F(\sigma)),$$

where
$n_*(0, F(\sigma(z)) = \tau_*(0, F(\sigma(z)), n_L(0, F(\sigma(z)) = \tau_L(0, F(\sigma(z)),$ and $n_*(0, \frac{1}{F(\sigma(z))}) = \tau_*(0, \frac{1}{F(\sigma(z))}), n_L(0, \frac{1}{F(\sigma(z))}) = \tau_L(0, \frac{1}{F(\sigma(z))})$.

Since $m(r, \sigma) = S(r, f), N(r, \sigma) = 0$, it is clear that

$$T(r, \sigma) = S(r, f),$$

$$T(r, \frac{1}{F(\sigma(z))}) = T(r, F(\sigma(z)) + O(1). \tag{3.10}$$

4. The Power Expansions of σ at Zeros, 1-Points, and c-points of f

We may suppose the chosen branch of $\sqrt{z}$ above is $\sqrt{|z|}e^{\frac{i \arg z}{2}}$, where $-\pi < \arg z < \pi$. To the constant c, there are the following four cases.

Case I:

$$\sqrt{c^2} = c, \quad \sqrt{(1-c)^2} = (c-1), \quad \sqrt{c^2(1-c)^2} = c(c-1); \tag{A1}$$

Case II:

$$\sqrt{c^2} = c, \quad \sqrt{(1-c)^2} = (c-1), \quad \sqrt{c^2(1-c)^2} = -c(c-1); \tag{A2}$$

Case III:

$$\sqrt{c^2} = c, \quad \sqrt{(1-c)^2} = -(c-1), \quad \sqrt{c^2(1-c)^2} = c(c-1); \tag{A3}$$

Case IV:

$$\sqrt{c^2} = c, \quad \sqrt{(1-c)^2} = -(c-1), \quad \sqrt{c^2(1-c)^2} = c(c-1). \qquad (A4)$$

We merely consider the basic case I. The other cases can be managed in the same way. Let $a \in \{0, 1, c, \infty\}$, and $S(a) = \{z : f(z) = a = g(z)\}$. Then it is easily seen that any point in $S(a)$ is not a branch point of σ. Define

$$S_{(p,q)}(a) = \{z : \tau(z, \frac{1}{f-a}) = p, \tau(z, \frac{1}{g-a}) = q, z \in S(a)\}.$$

Let $\arg c = \theta_0$, $\arg(c-1) = \theta_1$, $\arg c(c-1) = \theta_2$. Case I implies

$$-\frac{\pi}{2} < \theta_j \leq \frac{\pi}{2}, \quad j = 0, 1, 2.$$

We can find a number $\beta : 0 < \beta < \frac{\pi}{2}$ such that

$$|\theta_i - \theta_j| < \pi - \beta, \quad i, j = 0, 1, 2.$$

Set

$$\underline{\theta} = \max_{j=0,1,2}\{-\theta_j - \frac{\pi - \beta}{2}\}, \quad \overline{\theta} = \min_{j=0,1,2}\{-\theta_j + \frac{\pi-\beta}{2}\}.$$

Then $\overline{\theta} > \underline{\theta}$, and $(\underline{\theta}, \overline{\theta}) \cap (-\frac{\pi}{2}, \frac{\pi}{2}) \neq \emptyset$. Assume an interval $[\underline{\lambda}, \overline{\lambda}] \subset (\underline{\theta}, \overline{\theta}) \cap (-\frac{\pi}{2}, \frac{\pi}{2})$.

Let $z_0 \in S_{(p,q)}(f = a = g), a \in \{0, 1, c\}$, and

$$f(z) = a + (z - z_0)^p h_1(z), \quad g(z) = a + (z - z_0)^q h_2(z),$$

where $h_i(z) (i = 1, 2)$ are analytic at z_0, and $h_i(z_0) \neq 0, \infty$. Thus

$$\begin{aligned}\eta(z) &= \frac{f'g'}{f(f-1)(f-c)g(g-1)(g-c)} \\ &= \frac{pq}{A^2(z-z_0)^2}(1 + h_3(z)),\end{aligned}$$

where the constant A is c,or $c - 1$, or $c(c - 1)$ accordingly as $a = 0$, or 1, or c; $h_3(z)$ is analytic at z_0, and $h_3(z_0) = 0$. When z is near enough to z_0 in the angular region $\underline{\lambda} \le \arg(z - z_0) \le \overline{\lambda}$, we have

$$|\arg(1 + h_3(z))| < \beta, \quad \underline{\lambda} \le \arg(z - z_0) \le \overline{\lambda}, \tag{4.11}$$

Notice that $-\arg A + \frac{\pi-\beta}{2} \ge \overline{\theta}, -\arg A - \frac{\pi-\beta}{2} \le \underline{\theta}$, we see

$$-\pi < 2\arg A + 2\arg(z - z_0) + \arg(1 + h_3(z)) < \pi.$$

Hence

$$\begin{aligned}\sigma(z) &= (f(z) - g(z))\sqrt{\eta(z)} \\ &= (f(z) - g(z))\frac{\sqrt{pq}}{\sqrt{A^2}\sqrt{(z - z_0)^2}}\sqrt{1 + h_3(z)}.\end{aligned}$$

Since (A1) and (4.11), it follows that

$$\sigma(z) = \frac{\sqrt{pq}}{A(z - z_0)}\{(z - z_0)^p h_1(z) - (z - z_0)^q h_2(z)\} + h(z), \tag{4.12}$$

where $h(z)$ is analytic at z_0, satisfying $h(z_0) = 0$. By the uniqueness of Taylor expansion of σ at z_0, (4.12) holds in a sufficiently small neighborhood of z_0.

5. Jensen's Formula for $F(\sigma)$

Let $G_R = \{z : |z| < R\}$, $L_R = L \cap G_R$, and $D = G_R \setminus L_R$. Suppose L_R^+ is one side of L_R, a part of the counter-clockwise boundary of the region D, starting at $\zeta_0(|\zeta| < R)$, and ending at $\zeta_R(|\zeta| = R)$. Another side of L_R from ζ_R to ζ_0 is L_R^-. To the point $\zeta \in L_R$, we use ζ^+ and ζ_- indicating ,respectively the point on L_R^+ and the point L_R^-. Suppose $F(\sigma(z))$ has zeros $a_1, a_2, \cdots, a_\mu$ and poles $b_1, b_2, \cdots, b_\nu$ in D. If $F(\sigma(0)) \ne 0, \infty$, Let

$$g(\zeta) = \log\left\{F(\sigma(\zeta))\frac{\prod\limits_{j=1}^{\mu}\frac{R^2 - \bar{a}_j z}{R(z - a_j)}}{\prod\limits_{k=1}^{\nu}\frac{R^2 - \bar{b}_k z}{R(z - a_k)}}\right\},$$

then there holds

$$\log|F(\sigma(0))| = \frac{1}{2\pi}\int_0^{2\pi}\log\left|F(\sigma(Re^{i\varphi}))\right| d\varphi$$

$$-\sum_{j=1}^{\mu}\log\frac{R}{|a_\mu|} + \sum_{k=1}^{\nu}\log\frac{R}{|b_\nu|} + \Delta, \tag{5.13}$$

where

$$\Delta = Re\{\frac{1}{2\pi i}\int_{L_r^+}\frac{g_+(\zeta)}{\zeta}d\zeta + \frac{1}{2\pi i}\int_{L_r^-}\frac{g_-(\zeta)}{\zeta}d\zeta\}. \tag{5.14}$$

Here

$$g_-(\zeta) = g(\zeta^+) + \{g(\zeta)\}_{\gamma_\zeta}, \tag{5.15}$$

$\{g(\zeta)\}_{\gamma_\zeta}$ is the continue change of $g(\zeta)$ with respect to the initial value $g(\zeta^+)$ when ζ goes around L_R from ζ^+ to ζ^- alone the closed line $\gamma_\zeta \subset G_R$.

We point out that Theorem 3 is actually the characteristic form of the following theorem, which is Jensen's formula for $F(\sigma)$.

Let $F(\sigma)$ be defined by (3.9),and let L be a smooth line liking all the branch points of σ, on which $F(\sigma)$ takes neither zero nor ∞ except the branch points. Suppose $F(\sigma(z))$ has zeros $a_1, a_2, \cdots, a_\mu$ and poles $b_1, b_2, \cdots, b_\nu$ in $G_R \setminus L$, where $G_R = \{z : |z| < R\}$. If $F(\sigma(z)) \neq 0, \infty$, then

$$\log|F(\sigma(0))| = \frac{1}{2\pi}\int_0^{2\pi}\log\left|F(\sigma(Re^{i\varphi}))\right| d\varphi$$

$$-\sum_{j=1}^{\mu}\log\frac{R}{|a_\mu|} + \sum_{k=1}^{\nu}\log\frac{R}{|b_\nu|} + \Delta, \tag{5.16}$$

where

$$\Delta = N_L(R, F(\sigma(z)) - N_L(R, \frac{1}{F(\sigma(z)}). \tag{5.17}$$

6. Some Results on the Uniqueness of Meromorphic Functions that Share Four values

In this section we state some results on the uniqueness of meromorphic functions that share four values.

Let f be a meromorphic function. Define

$$\tau(z, f) = \begin{cases} p, & z \text{ is a p-fold pole of } f(z); \\ 0, & f(z) \neq \infty, f(z) \neq 0; \\ -p, & z \text{ is a p-fold zero of } f(z). \end{cases} \tag{6.18}$$

Let a be a complex number or ∞ shared by f and g. Define the set $S(a) = S(f = a = g) = \{z : f(z) = a, g(z) = a\}$ (if $a = \infty$, replace $\frac{1}{f-a}$ by f), and define some subsets of $S(a)$ below

$$S_E(a) = \{z : \tau(z, \frac{1}{f-a}) = \tau(z, \frac{1}{g-a}), z \in S(a)\}$$

$$S_D(a) = \{z : \tau(z, \frac{1}{f-a}) \neq \tau(z, \frac{1}{g-a}), z \in S(a)\}$$

$$S_{(p,q)}(a) = \{z : \tau(z, \frac{1}{f-a}) = p, \tau(z, \frac{1}{g-a}) = q, z \in S(a)\}.$$

We define symbols $N_{(p,q)}(r, \frac{1}{f-z})$ $N_D(r, \frac{1}{f-a})$, and $N_E(r, \frac{1}{f-a})$ to denote the counting functions of a-points of $f(z)$ in, respectively, $S_{(p,q)}(a)$, $S_D(a)$, and $S_E(a)$. Moreover, denote their correspond reduced counting functions by $\overline{N}_{(p,q)}(r, a)$, $\overline{N}_D(r, a)$, and $\overline{N}_E(r, a)$.

Let f and g be nonconstant meromorphic functions sharing the value a IM. Define

$$\tau(a) = \begin{cases} \liminf\limits_{r\to\infty} \frac{\overline{N}_E(r,a)}{\overline{N}(r,a)} & \text{if } \overline{N}(r, a) \not\equiv 0, \\ 1 & \text{if } \overline{N}(r, a) \equiv 0. \end{cases}$$

The following two theorems are respectively due to E. Muse and G. G. Gundersen(see ref. [7] and ref. [6])

Theorem A *Let f and g be nonconstant meromorphic functions that share four distinct values a_1, a_2, a_3, a_4. If a_4 is shared CM and $\tau(a_1) > 2/3$, then f and g share all four values CM.*

Theorem B *Let f and g be nonconstant meromorphic functions that share a_1, a_2, a_3 IM and a_4 CM. Suppose that there exist some real constant $\lambda > 4/5$ and some set $I \subset (0, \infty)$ that has infinite linear measure such that*

$$\frac{N(r, a_4, f)}{T(r, f)} \geq \lambda \tag{6.19}$$

for all $r \in I$. Then f and g share all four values CM.

Using some auxiliary functions like $F(\sigma)$ and applying Theorem 3, we obtain the following results.

The same consequence still holds if replace the condition $\tau(a_1) > 2/3$ by $\tau(a_1) > 1/2$ in Theorem A.

Let f and g be nonconstant meromorphic functions that share three distinct values a_1, a_2, a_3 IM and a fourth value a_4 CM. Suppose that there exist some real constant $\lambda > \frac{2}{1+\tau}$, where $\tau = \sum_{j=1}^{3} \frac{1}{1-\tau(a_j)}(\tau = \infty$ if $\tau(a_j) = 1$ for some $j \in \{1, 2, 3\})$, and some set $I \subset (0, \infty)$ which has infinite linear measure such that

$$\frac{N(r, a_4, f)}{T(r, f)} \geq \lambda \tag{6.20}$$

for all $r \in I$.Then f and g share all four values CM.

The same conclusion holds if replace the condition $\lambda > 4/5$ by $\lambda > 1/2$ in Theorem B.

To the Gundersen's question, we have the following theorem.

Let two meromorphic functions $f(z)$ and $g(z)$ share four values a_1, a_2, a_3, a_4. If $\overline{N}(r, \frac{1}{f-a_1}) = S(r, f)$, then $f(z)$ and $g(z)$ share a_1, a_2, a_3, a_4 CM.

If two nonconstant entire functions share three finite values IM, then the functions necessarily share all three values CM.

References

1. W.K. Hayman, Meromorphic Functions, *Clarendon Press,* Oxford, 1964.
2. R. Nevanlinna, Einige Eindeutigkeitssätze in der Theorie der meromorphen Funktionen, *Acta Math.* **48**(1926):367-391.
3. R. Nevanlinna, Le théorème de Picard-Borel et la théorie des fonctions méromorphes, *Gauthier-Villars,* Paris, 1929.
4. G.G. Gundersen, Meromorphic functions that share three or four values, *J. London Math. Soc.* **20**(2)(1979):457-466.
5. G.G. Gundersen, Meromorphic functions that share four values,*Trans. Am. Math. Soc.,* **277**(1983):545-567. *Correction:***304**(1987):847-850.
6. G.G. Gundersen, Meromorphic functions that share three values IM and a fourth value CM, *Complex Variables,* **20**(1992):99-106.
7. E. Mues, Meromorphic functions sharing four values, *Complex Variables,* **12**(1989):169-179.
8. S.P. Wang, On meromophic functions that share four values, *J. Math. Anal. Appl.,* **173**(1993):359-369.
9. H.X. Yi, C.D. Zhou, Meromorphic functions that share three values, *J. Shandong University (Natural Science),* **31**(1996):121-128.(in Chinese)

10. H.X. Yi, C.C. Yang, Unicity theory of meromorphic functions, *Science Press,* Beijing (1995):248-250. (in Chinese)
11. B. Huang, On the Unicity of Meromorphic Functions That Share Four Values, *Indian J. Pure Appl. Math.,* **35**(3)(2004):359-372.

ON TWO PERTURBATION RESULTS OF SECOND ORDER LINEAR DIFFERENTIAL EQUATIONS WITH PERIODIC COEFFICIENTS*

ZHIBO HUANG

School of Mathematical Sciences
South China Normal University
Guangzhou, 510631 P. R. China.
E-mail: hzbo20019@sina.com

ZONGXUAN CHEN

School of Mathematical Sciences
South China Normal University
Guangzhou, 510631 P. R. China.
E-mail: chzx@sina.com

In this paper, we obtain two perturbation results on the linear differential function

$$f'' + \Pi(z)A(z)f = 0,$$

where $\Pi(z)$ and A(z) are periodic entire functions with period $2\pi i$ and $\sigma_e(\Pi) < \sigma_e(A)$.

Keywords: Complex oscillation, differential function, period, linearly dependent, linearly independent.

2000 Mathematics Subjects Classification: 30D35; 34A20.

1. Introduction

We use standard notation from Nevanlinna theory in this paper(see [1],[5],[11]). In addition, we use the notation $\sigma(f)$ and $\lambda(f)$ respectively to denote the

*The work was supported by the National Natural Science Foundation of China (10161006) and the Natural Science Foundation of Guangdong Province (04010360).

order of growth and exponent of convergence of zeros of a meromorphic function f.
The following results were proved in [12].

Theorem A. *Let $A(z) = B(e^z)$, where $B(\zeta) = g_1(1/\zeta) + g_2(\zeta)$, g_1 and g_2 are entire functions with g_2 transcendental and $\sigma(g_2)$ not equal to a positive integer or infinity, and g_1 arbitrary.*
(i) Suppose $\sigma(g_2) > 1$.
(a) If f is a non-trivial solution of

$$f'' + A(z)f = 0, \tag{1.1}$$

with $\lambda_e(f) < \sigma(g_2)$, then $f(z)$ and $f(z + 2\pi i)$ are linearly dependent.
(b) If f_1 and f_2 are any two linearly independent solutions of (1.1), then $\lambda_e(f_1 f_2) \geq \sigma(g_2)$.
(ii) Suppose $\sigma(g_2) < 1$.
(a) If f is a non-trivial solution of (1.1) with $\lambda_e(f) < 1$, then $f(z)$ and $f(z + 2\pi i)$ are linearly dependent.
(b) If f_1 and f_2 are any two linearly independent solutions of (1.1), then $\lambda_e(f_1 f_2) \geq 1$.

We remark that the conclusion of Theorem A yet valid if we assume $\sigma(g_1)$ is not equal to an integer or infinity, and g_2 arbitrary and still assume $B(\zeta) = g_1(1/\zeta)+g_2(\zeta)$. In this case when g_1 is transcendental with its order not equal to an integer or infinity and g_2 is arbitrary, we need only consider $B^*(\eta) = B(1/\eta) = g_1(\eta) + g_2(1/\eta)$, g_1 in $0 < |\eta| < +\infty, \eta = 1/\zeta$.

Theorem B. *Let $g(\zeta)$ be a transcendental entire function and its order be not a positive and infinity. Let $A(z) = B(e^z)$, where $B(\zeta) = g(1/\zeta) + \Sigma_{j=1}^{p} b_j\zeta^j$ and p is an odd positive integer, then $\lambda(f) = +\infty$ for each non-trivial solution f of (1.1). In fact, the stronger conclusion*

$$\log^+ N(r, f) \neq o(r) \tag{1.2}$$

holds.

2. Main results

Suppose (1.1) admits a non-trivial solution that has a finite e-type exponent of convergence of zeros. If $\Pi(z)$ is a periodic entire function with period $2\pi i$ and $\sigma_e(\Pi) < \sigma_e(A)$, what can we say about the e-type exponent of convergence of zeros of any two linearly independent solutions of the equation

$$f'' + \Pi(z)A(z)f = 0? \tag{2.1}$$

A perturbation result for differential function

$$f'' + (A(z) + \Pi(z))f = 0, \tag{2.2}$$

where the coefficients for (2.2) are not necessarily periodic is given in [7], and a perturbation result for (2.2) where the coefficients are periodic is given in [12]. We answer the above problem based on the method used in [7,12] coupled with the special properties of periodic coefficients.

Theorem 1. Let $B(\zeta) = g_1(1/\zeta) + g_2(\zeta)$, $C(\zeta) = g_3$, where g_1, g_2 and g_3 are entire functions of finite order such that $\sigma(g_2)$ is a positive integer, and $\sigma(g_3) < \sigma(g_2)$. Suppose $A(z) = B(e^z), \Pi(z) = C(e^z)$ and furthermore that (1.1) admits a non-trivial solution f with $\lambda_e(f) < \sigma(g_2)$ and that $f(z)$ and $f(z + 2\pi i)$ are linearly independent. Then

(i) for any non-trivial solution h of equation (2.1) with $\lambda_e(h) < \sigma(g_2)$, $h(z)$ and $h(z + 2\pi i)$ are linearly dependent.

(ii) for any two linearly independent solutions h_1 and h_2 of (2.1), we must have $\lambda_e(h_1 h_2) \geq \sigma(g_2)$.

We easily deduce the following result from Theorem 1.

Corollary. *Under the assumption in Theorem 1, any two linearly independent solutions h_1 and h_2 of (2.1) must have $\lambda(h_1 h_2) = +\infty$.*

Theorem B investigates the oscillation properties of any non-trivial solution of (1.1) with $B(\zeta) = g(1/\zeta) + \Sigma_{j=1}^{p} b_j \zeta^j$ and p is an odd positive integer and $\sigma(g)$ is not a positive and infinity. We now investigate perturbation problem for (2.1) precisely when $\sigma(g)$ is a positive integer.

Theorem 2. *Let $g(\zeta)$ be a transcendental entire function of an integer order $\sigma(g)$, and $C(\zeta) \not\equiv 0$ be an entire function with $\sigma(C) < \sigma(g)$. Let $A(z) = B(e^z)$, where $B(\zeta) = \Sigma_{j=1}^{p} b_{-j} \zeta^{-j} + g(\zeta)$, p is an odd positive integer and $\Pi(z) = C(e^z)$. Suppose (1.1) admits a non-trivial solution f with $\lambda(f) < +\infty$. Then any non-trivial solution h of (2.1) must have $\lambda(h) = +\infty$. In fact, the stronger conclusion (1.2) holds.*

3. Preliminaries for the proofs of the main results

The main tools that we use in this paper are Nevanlinna theory in C_0 (see [5,8]) where $C_0 = C \setminus \{z : |z| \leq R_0\}$ and Valiron representation for functions analytic in C_0 (see [4]), and the fact that if f_1 and f_2 are two non-trivial, linearly independent solutions of (1.1), then the product $E(z) = f_1(z)f_2(z)$ satisfies the differential equation

$$-4A(z) = \frac{c^2}{E^2(z)} - \left(\frac{E'(z)}{E(z)}\right)^2 + 2\frac{E''(z)}{E(z)}, \tag{3.1}$$

where $c \neq 0$ is the Wronskian of f_1 and f_2 (see [1]), and

$$T(r, E) = N(r, 1/E) + \frac{1}{2}T(r, A) + O(\log rT(r, E)) \qquad n.e. \tag{3.2}$$

In general, we use "n.e." to denote that relation holds except possibly outsides a set of finite measure. Our argument actually depends on an analogue formula of (3.2) on the Valiron representation of periodic functions.

In this paper, we use the following definitions (see [2,12]).

Let $A(z)$ be an entire function. We define

$$\sigma_e(A) = \overline{\lim_{r \to +\infty}} \frac{\log T(r, A)}{r}, \tag{3.3}$$

to be the e-type order of $A(z)$, also define

$$\lambda_e(A) = \overline{\lim_{r \to +\infty}} \frac{\log^+ N(r, 1/A)}{r}, \tag{3.4}$$

to be the e-type exponent of convergence of zeros of $A(z)$. We shall have occasions to consider the zeros of $A(z)$ in the right-half plane only, in that case, we define the upper limit in (3.4) by $\lambda_{eR}(A)$ when we only count the zeros of $A(z)$ in the right-half plane, similarly, we define $\lambda_{eL}(A)$ to be the upper limit in (3.4) when we only count the zeros of $A(z)$ in the left-half plane.

Let $B(\zeta)$ be analytic in $0 < |\zeta| < +\infty$, hence we have a representation $B(\zeta) = g_1(1/\zeta) + g_2(\zeta)$, where both $g_1(\zeta)$ and $g_2(\zeta)$ are entire functions. Let $A(z) = B(e^z) = A_1(z) + A_2(z)$, where $A_1(z) = g_1(e^{-z})$ and $A_2(z) = g_2(e^z)$. Observe that the transformation $\zeta = e^z$ is a one-to-one correspondence between the sets $\{z : -\log \rho \leq Rez \leq \log \rho, -\pi < Imz \leq \pi\}$ and $\{\zeta : \rho^{-1} \leq |\zeta| \leq \rho\}$. By the periodicity of e^z, we have

$$\begin{aligned}
\max_{\rho^{-1} \leq |\zeta| \leq \rho} |B(\zeta)| &= \max_{\substack{-\log \rho \leq Rez \leq \log \rho, \\ -\pi < Imz \leq \pi}} |A(z)| \leq \max_{|z| \leq \log \rho + \pi} |A(z)| \\
&\leq \max_{\substack{-(\log \rho + \pi) \leq Rez \leq \log \rho + \pi, \\ -\pi < Imz \leq \pi}} |A(z)| \\
&= \max_{(e^\pi \rho)^{-1} \leq |\zeta| \leq (e^\pi \rho)} |B(\zeta)|,
\end{aligned}$$

so deduce that

$$\sigma_e(A) = \overline{\lim_{\rho\to+\infty}} \frac{\log\log \max\limits_{\rho^{-1}\le|\zeta|\le\rho} |B(\zeta)|}{\log\rho}. \tag{3.5}$$

From

$$\max_{\rho^{-1}\le|\zeta|\le\rho} |B(\zeta)| = \max\{\max_{|\zeta|=\rho^{-1}} |B(\zeta)|, \max_{|\zeta|=\rho} |B(\zeta)|\},$$

this together with (3.5) yields

$$\sigma_e(A) = \overline{\lim_{\rho\to+\infty}} \frac{\log\log \max\limits_{\rho^{-1}\le|\zeta|\le\rho} |B(\zeta)|}{\log\rho} = \max\{\sigma(g_1), \sigma(g_2)\}. \tag{3.6}$$

In particular,

$$\sigma_e(A_1) = \sigma(g_1), \sigma_e(A_2) = \sigma(g_2).$$

Let us now turn to the discussion of zeros. Let $n(D, 1/F)$ be the number of zeros of $F(z)$ in the set D, then we deduce

$$\begin{aligned}
& n\left(\rho^{-1} \le |\zeta| \le \rho, 1/B(\zeta)\right) = n\left(\left\{\begin{matrix} -\log\rho \le Rez \le \log\rho \\ -\pi < Imz \le \pi \end{matrix}\right\}, 1/A(z)\right) \\
& \le n\left(\{|z| \le \log\rho + \pi\}, 1/A(z)\right) \\
& \le 2\left(\frac{(\log\rho+\pi)-\pi}{2\pi}+1\right) n\left(\left\{\begin{matrix} -(\log\rho+\pi) \le Rez \le \log\rho+\pi \\ -\pi < Imz \le \pi \end{matrix}\right\}, 1/A(z)\right) \\
& = \left(\frac{\log\rho}{\pi}+2\right) n\left(\{(e^\pi\rho)^{-1} \le |\zeta| \le e^\pi\rho\}, 1/B(\zeta)\right).
\end{aligned}$$

Thus

$$\lambda_e(A) = \overline{\lim_{\rho\to+\infty}} \frac{\log n\left(\{\rho^{-1} \le |\zeta| \le \rho\}, 1/B(\zeta)\right)}{\log\rho}, \tag{3.7}$$

similarly, we have

$$\begin{aligned}
\lambda_{eR}(A) &= \overline{\lim_{\rho\to+\infty}} \frac{\log n(\{1 < |\zeta| \le \rho\}, 1/B(\zeta))}{\log\rho}; \\
\lambda_{eL}(A) &= \overline{\lim_{\rho\to+\infty}} \frac{\log n(\{\rho^{-1} \le |\zeta| \le 1\}, 1/B(\zeta))}{\log\rho}.
\end{aligned} \tag{3.8}$$

Thus

$$\lambda_e(A) = \max\{\lambda_{eL}(A), \lambda_{eR}(A)\}.$$

If f is analytic in C_0, then [4] implies that

$$f(z) = z^n\Theta(z)u(z), \tag{3.9}$$

where n is an integer, $\Theta(z)$ is analytic and non-vanishing on $C_0 \cup \{\infty\}$, and $u(z)$ is an entire function with

$$u(z) = \pi(z)e^{h(z)},$$

the function $\pi(z)$ is a Weierstrass product formed from the zeros of f in C_0, and $h(z)$ is an entire function.

Letting $R_0 = 1$, we may regard $B(\zeta)$ to ba analytic in C^*, where $C^* := C \setminus \{z : |z| \leq 1\}$. By (3.9) we have a similar representation

$$B(\zeta) = \zeta^n R(\zeta)b(\zeta), \tag{3.10}$$

where n is an integer, $R(\zeta)$ is analytic and non-vanishing on $C^* \cup \{\infty\}$, and $b(\zeta)$ is an entire function. From (3.10) we obtain

$$\sigma(g_2) = \overline{\lim_{\rho\to+\infty}} \frac{\log\log \max_{|\zeta|=\rho} |B(\zeta)|}{\log \rho} = \overline{\lim_{\rho\to+\infty}} \frac{\log\log M(\rho, b(\zeta))}{\log \rho} = \sigma(b(\zeta)). \tag{3.11}$$

Since the zeros of $B(\zeta)$ and $b(\zeta)$ coincide in $1 < |\zeta| < +\infty$, we deduce from (3.8) that

$$\lambda_{eR}(A) = \lambda(b(\zeta)). \tag{3.12}$$

Suppose $w(z)$ is meromorphic in $C_0 := \{z : R_0 < |z| < +\infty\}$. By a similar argument as in Valiron [4], $w(z)$ has a representation

$$w(z) = z^n \Theta(z) f(z), \tag{3.13}$$

where n is an integer, $\Theta(z)$ is analytic and non-vanishing on $C_0 \cup \{\infty\}$, f is a meromorphic function in C. Let $T_1(r, w)$ denote the Nevanlinna characteristic function (see [1,8]) for $w(z)$ in C_0, which is defined by $T_1(r, w) = m_1(r, w) + N_1(r, w)$, where

$$m_1(r, w) = \frac{1}{2\pi} \int_0^{2\pi} \log^+ |w(re^{i\varphi})| d\varphi,$$

and $N_1(r, w)$ is the counting function for the poles of w in C_0.

From (3.13) we deduce that

$$m_1(r, w) = m(r, f) + O(\log r), \tag{3.14}$$

and $N_1(r, w) = N(r, f)$. Thus

$$T_1(r, w) = T(r, f) + O(\log r). \tag{3.15}$$

Since $T(r, f) = T(r, 1/f) + O(1)$, so

$$\begin{aligned} T_1(r, 1/w) &= T(r, 1/f) + O(\log r) \\ &= T(r, f) + O(\log r) \\ &= T_1(r, w) + O(\log r), \end{aligned} \tag{3.16}$$

it is

$$T_1(r, 1/w) = T_1(r, w) + O(\log r).$$

As in [1,8], we define the order of w in C_0 by

$$\sigma_1(w) = \overline{\lim_{r\to+\infty}} \frac{\log T_1(r, w)}{\log r}.$$

4. Proof of Theorem 1

Lemma 4.1[1]. *Let $F(r)$ and $G(r)$ are non-decreasing function on $(0, \infty)$. If*

(i) $F(r) \le G(r)$ n.e.;

or

(ii) $F(r) \le G(r)$, where $r \notin H \cup (0, 1]$, $H \subset (0, \infty)$ is a set that has finite logarithmic measure.

Then for any constant $\alpha (\alpha > 1)$, there exists a constant $r_0 (r_0 > 0)$ such that $F(r) \le G(\alpha r)$ when $r > r_0$.

Lemma 4.2[10]. *Let $A_0(z), A_1(z), \cdots, A_{k-2}(z)$ be periodic entire functions with period $2\pi i$ and function in e^z when $k \ge 2$, and $A_j(z)$ satisfy*

$$T(r, A_j) = o\{T(r, A_0)\} \quad n.e., \quad j = 1, 2, \cdots, k-2.$$

Assume $f \not\equiv 0$ is a solution of equation

$$w^{(k)} + A_{k-2}w^{k-2} + \cdots + A_0 w = 0.$$

If

(i) A_0 is national in e^z and satisfies

$$\log^+ N(r, 1/f) = o(r),$$

or

(ii) A_0 is transcendental in e^z and satisfies

$$\log^+ N(r, 1/f) = O(r),$$

Then there exists positive number q and $1 \leq q \leq k$ such that $f(z)$ and $f(z+q2\pi i)$ are linearly dependent.

Proof of Theorem 1. (i) Let f be a non-trivial solution of (1.1) with $\lambda_e(f) < \sigma(g_2)$, and $f(z)$ and $f(z+2\pi i)$ are linearly independent. Since $\lambda_e(f) < \sigma(g_2) < +\infty$, Lemma 4.2 implies that $f(z)$ and $f(z+4\pi i)$ must be linearly dependent. Let $E(z) = f(z)f(z+2\pi i)$, then

$$E(z+2\pi i) = f(z+2\pi i)f(z+4\pi i) = c_1 E(z),$$

for some non-zero constant c_1. Clearly $E'(z)/E(z)$ and $E''(z)/E(z)$ are both periodic functions with period $2\pi i$, while $A(z)$ is periodic by definition. Since (3.1) shows that $E^2(z)$ is also a periodic function with period $2\pi i$. Thus we can find an analytic function $\Phi(\zeta)$ in $0 < |\zeta| < +\infty$ such that $E^2(z) = \Phi(e^z)$. Substituting this representation into (3.1) yields

$$-4B(\zeta) = \frac{c^2}{\Phi} + \zeta\frac{\Phi'}{\Phi} - \frac{3}{4}\zeta^2\left(\frac{\Phi'}{\Phi}\right)^2 + \zeta^2\frac{\Phi''}{\Phi}. \tag{4.1}$$

Since both $B(\zeta)$ and $\Phi(\zeta)$ are analytic in $C^* := \{\zeta : 1 < |\zeta| < +\infty\}$, the Valiron theory gives their representation as

$$B(\zeta) = \zeta^n R(\zeta) b(\zeta), \quad \Phi(\zeta) = \zeta^{n_1} K_1(\zeta)\phi(\zeta) \tag{4.2}$$

where n and n_1 are some integers, $R(\zeta)$ and $K_1(\zeta)$ are functions that are analytic and non-vanishing on $C^* \cup \{\infty\}$ and $b(\zeta)$ and $\phi(\zeta)$ are entire functions. By (3.14) and the lemma on logarithmic derivative in [6], we deduce from (4.1) that

$$m_1(\rho, 1/\Phi) = m_1(\rho, B) + O\{\log(\rho \log T(\rho, \phi))\} \quad n.e. \tag{4.3}$$

Since $N_1(\rho, B) \equiv 0$, (4.3) implies

$$T_1(\rho, 1/\Phi) = N_1(\rho, 1/\Phi) + T_1(\rho, B) + O\{\log(\rho \log T(\rho, \phi))\}. \tag{4.4}$$

Applying (3.15) and (3.16) to (4.4) and using the fact that $N_1(\rho, 1/\Phi) = N(\rho, 1/\phi)$, we deduce

$$T(\rho, 1/\phi) = N(\rho, 1/\phi) + T(\rho, b) + O\{\log(\rho \log T(\rho, \phi))\}. \tag{4.5}$$

It is easy to see that $\lambda_e(f) = \lambda_e(E) = \lambda_e(E^2)$. Since $\lambda_e(f) < \sigma(g_2)$, so $\lambda_{eR}(E^2) \leq \lambda_e(E^2) = \lambda_e(f) < \sigma(g_2)$. As in (3.12), $\lambda(\phi) = \lambda_{eR}(\Phi) = \lambda_{eR}(E^2)$. But $\sigma(g_2) = \sigma(b(\zeta))$ by (3.11). Hence $\lambda(\phi) < \sigma(b(\zeta))$. Notice that (4.5) satisfied by ϕ is an analogous formula to (3.2) satisfied by E. It follows from (4.5) and Lemma 4.1 that $\sigma(\phi) = \sigma(b(\zeta))$, thus $\lambda(\phi) < \sigma(\phi)$. We know $\sigma(g_2)$ is a positive integer, so are $\sigma(b(\zeta))$ and $\sigma(\phi(\zeta))$. Thus we may rewrite

$\phi(\zeta) = \pi_1(\zeta)e^{P(\zeta)}$, where $\pi_1(\zeta)$ is an entire function, $\sigma(\pi_1) < \sigma = \sigma(g_2)$, $P(\zeta) = \alpha\zeta^\sigma$, and α is a non-zero constant.

Let us now suppose (2.1) possesses a non-trivial solution $h(z)$ such that $\lambda_e(h) < \sigma(g_2)$ but $h(z)$ and $h(z+2\pi i)$ are linearly independent. Let $F(z) = h(z)h(z+2\pi i)$. By a similar argument that we have applied to $E(z)$ above, we conclude that there exists an analytic function $\Psi(\zeta)$ in $0 < |\zeta| < +\infty$ such that $F^2(z) = \Psi(e^z)$. Similarly, $\Psi(\zeta)$ has a Valiron representation

$$\Psi(\zeta) = \zeta^{n_2} K_2(\zeta)\psi(\zeta) \tag{4.6}$$

in C^*, where n_2 is an integer, $K_2(\zeta)$ is analytic and non-vanishing on $C^* \cup \{\infty\}$, and $\psi(\zeta)$ is an entire function in C.

We now substitute $F^2(z) = \Psi(\zeta)$, with $\zeta = e^z$, into (3.1) with $A(z)$ replaced by $\Pi(z)A(z)$. This yields

$$-4C(\zeta)B(\zeta) = \frac{c_2^2}{\Psi} + \zeta\frac{\Psi'}{\Psi} - \frac{3}{4}\zeta^2\left(\frac{\Psi'}{\Psi}\right)^2 + \zeta^2\frac{\Psi''}{\Psi}, \tag{4.7}$$

where c_2 is a the Wronskian of $h(z)$ and $h(z+2\pi i)$.

In a similar fashion of $\phi(\zeta)$, we have

$$\lambda(\psi) = \lambda_{eR}(F^2) \le \lambda_e(F^2) = \lambda_e(h) < \sigma(g_2) = \sigma(b(\zeta)).$$

We then apply a similar argument to (4.7) to obtain

$$T(\rho,\psi) = N(\rho, 1/\psi) + T(\rho, d) + O\{\log\rho\log T(\rho,\psi)\} \tag{4.8}$$

as to (4.1) for (4.5), where $d(\zeta)$ is an entire function appearing in the Valiron representation of

$$C(\zeta)B(\zeta) = \zeta^{n_3} R_d(\zeta)d(\zeta),$$

where the functions $R_d(\zeta)$ and $d(\zeta)$ play the same roles of the corresponding functions in (4.2), and it is easy to check that $\sigma(d) = \sigma(b(\zeta))$. It follows from (4.8) and Lemma 4.1 that $\sigma(\psi) = \sigma(b(\zeta))$. Thus $\lambda(\psi) < \sigma(\psi)$. Since $\sigma(g_2)$ is a positive integer, so are $\sigma(b(\zeta))$ and $\sigma(\psi)$. Thus we may rewrite $\psi(\zeta) = \pi_2(\zeta)e^{Q(\zeta)}$, where $\pi_2(\zeta)$ is an entire function, $Q(\zeta) = \beta\zeta^\sigma$, $\sigma(\pi_2) < \sigma = \sigma(g_2)$, and β is a non-zero constant.

Let $\Phi(\zeta) = R_1(\zeta)e^{P(\zeta)}$ and $\Psi(\zeta) = R_2(\zeta)e^{Q(\zeta)}$, where $R_1(\zeta) = \zeta^{n_1}K_1(\zeta)\pi_1(\zeta)$ and $R_2(\zeta) = \zeta^{n_2}K_2(\zeta)\pi_2(\zeta)$, $\max\{\sigma_1(R_1), \sigma_1(R_2)\} < \sigma =$

$\sigma(g_2)$, and substitute them into (4.1) and (4.7) respectively yielding

$$\begin{aligned} -4B(\zeta) = & \frac{c^2}{R_1 e^P} + \zeta\left(\frac{R_1'}{R_1} + P'\right) \\ & - \frac{3}{4}\left\{\left(\zeta\frac{R_1'}{R_1}\right)^2 + 2\zeta^2\frac{R_1'}{R_1}P' + \zeta^2 P'^2\right\} \\ & + \zeta^2\left(\frac{R_1''}{R_1} + 2\frac{R_1'}{R_1}P' + P'^2 + P''\right), \end{aligned} \tag{4.9}$$

and

$$\begin{aligned} -4C(\zeta)B(\zeta) = & \frac{c_2^2}{R_2 e^Q} + \zeta\left(\frac{R_2'}{R_2} + Q'\right) \\ & - \frac{3}{4}\left\{\left(\zeta\frac{R_2'}{R_2}\right)^2 + 2\zeta^2\frac{R_2'}{R_2}Q' + \zeta^2 Q'^2\right\} \\ & + \zeta^2\left(\frac{R_2''}{R_2} + 2\frac{R_2'}{R_2}Q' + Q'^2 + Q''\right). \end{aligned} \tag{4.10}$$

To multiple (4.9) with $C(\zeta)$ yields

$$\begin{aligned} -4C(\zeta)B(\zeta) = & \frac{c^2}{R_1 e^P}C(\zeta) + \zeta\left(\frac{R_1'}{R_1} + P'\right)C(\zeta) \\ & - \frac{3}{4}\left\{\left(\zeta\frac{R_1'}{R_1}\right)^2 + 2\zeta^2\frac{R_1'}{R_1}P' + \zeta^2 P'^2\right\}C(\zeta) \\ & + \zeta^2\left(\frac{R_1''}{R_1} + 2\frac{R_1'}{R_1}P' + P'^2 + P''\right)C(\zeta). \end{aligned} \tag{4.11}$$

Substituting (4.10) from (4.11) yields

$$0 \equiv \frac{c^2}{R_1 e^P}C(\zeta) - \frac{c_2^2}{R_2 e^Q} + H(\zeta), \tag{4.12}$$

where $H(\zeta)$ is meromorphic function in C^*. In fact, $H(\zeta)$ is a differential polynomial in $\frac{R_1'}{R_1}, \frac{R_2'}{R_2}, C(\zeta), P', Q'$ and their derivatives. We can deduce from the definitions of $R_1, R_2, C(\zeta), P, Q$ that $\sigma_1(H) < \sigma = \sigma(g_2)$. Rewriting (4.12) as

$$e^{-P} + H_1 e^{-Q} = H_2, \tag{4.13}$$

where H_1 and H_2 are meromorphic functions in C^*, with

$$H_1 = -\frac{c_2^2}{c^2}\frac{R_1}{R_2}\frac{1}{C(\zeta)}, \quad H_2 = -\frac{R_1}{c^2}\frac{H(\zeta)}{C(\zeta)},$$

and $\max\{\sigma_1(H_1), \sigma_1(H_2)\} < \sigma(g_2)$.
Differentiating (4.13) yields

$$e^{-P} + \frac{H_1 Q' - H_1'}{P'} e^{-Q} = -\frac{H_2'}{P'}. \tag{4.14}$$

Using (4.14) to eliminate e^{-P} from (4.13) yields

$$H_3 e^{-Q} = H_4, \tag{4.15}$$

where $H_3 = H_1' + (P' - Q')H_1$ and $H_4 = H_2 P' + H_2'$ are meromorphic functions in C^* with $\sigma_1(H_3), \sigma_1(H_4) < \sigma(g_2)$. Thus $H_4 \equiv 0$, i.e. $H_2 = c_3 e^{-P}$, where c_3 is a non-zero constant, from a simple order consideration in (4.15). We can obtain $\sigma_1(H_2) = \sigma = \sigma(g_2)$. This is a contradiction to $\sigma_1(H_2) < \sigma(g_2)$. Hence $h(z)$ and $h(z + 2\pi i)$ must be linearly dependent.

(ii) Suppose (2.1) possesses two non-trivial solutions h_1 and h_2 that are linearly independent and $\lambda_e(h_1 h_2) < \sigma(g_2)$. Then $\lambda_e(h_j) < \sigma(g_2)$ for $j = 1, 2$. Part (i) implies that $h_j(z)$ and $h_j(z + 2\pi i)$ are linearly dependent for $j = 1, 2$. Let $E(z) = f(z)f(z + 2\pi i)$ and $F(z) = h_1(z)h_2(z)$. Then $F(z + 2\pi i) = c_4 F(z)$ for some non-zero constant c_4. Applying a similar argument to $E(z)$ and $F(z)$ as in part (i) yields $\sigma_1(H_2) = \sigma(g_2)$. This is a contradiction. Hence $\lambda_e(h_1 h_2) \geq \sigma(g_2)$. This completes the proof of Theorem 1.

5. Proof of Theorem 2

Lemma 5.1 [3]. *Let $A(z) = B(e^{\alpha z})$ ba a periodic entire function with period $\omega = 2\pi i/\alpha$, and be transcendent in $e^{\alpha z}$, i.e. $B(\zeta)$ is transcendent and analytic on $0 < |\zeta| < +\infty$. If $B(\zeta)$ has a pole of odd order at $\zeta = \infty$ or $\zeta = 0$ (including those which can be changed into this case by varying the period of $A(z)$), and equation (1.1) has a solution $f \not\equiv 0$ which satisfies condition*

$$\log^+ N(r, 1/f) = o(r) \quad as \quad r \to +\infty, \tag{5.1}$$

then $f(z)$ and $f(z + \omega)$ are linearly independent.

Proof of Theorem 2. Suppose (1.1) possess a non-trivial solution f with $\lambda(f) < +\infty$. Hence $\lambda_e(f) = 0 < \sigma(g)$ by definition (3.4). Thus Lemma 5.1 implies that $f(z)$ and $f(z + 2\pi i)$ are linearly independent solutions of (1.1). Thus $f(z)$ and $f(z + 2\pi i)$ satisfy the hypotheses of Theorem 1. Suppose (2.1) admits a non-trivial solution $h(z)$ with $\lambda_e(h) < \sigma(g)$. Then

Lemma 5.1 again implies that $h(z)$ and $h(z+2\pi i)$ are linearly independent. Part (ii) of Theorem 1 shows that $\lambda_e(h(z)h(z+2\pi i)) \geq \sigma(g) > 0$. So $\lambda_e(h) = \lambda_e(h(z)h(z+2\pi i)) > 0$. Thus (1.2) holds. This complete the proof.

References

1. S. A Gao, Z. X. Chen and T. W. Chen, *Oscillation Theory of Linear Differential Equations,* Huazhong University of Science and Technology Press, 1998(Chinese).
2. Gao Shi-An. A result concerning the complex oscillation for periodic second order linear differential equations of transcendental type. *Math. App.,* **15**(2)(2002):85-88.
3. S. A. Gao, A further result on the complex oscillation theory of periodic second order linear differential equations. *Proc. Edinburgh Math. Soc.,* **33**(1990):143-158.
4. G. Valiron, *Lectures on the genaral theory of integer functions,* Chelsea, New York,1949.
5. I. Laine, *Nevanlinna Theory and Complex Differential Equations,* Walter de Gruyter, Berlin,1993.
6. L. Yang. *Value distribution theory and it's new researches,* Peking: Science Press, 1982.
7. S. B. Bank, I. Laine and J. Langley, Oscillation results for solutions of linear differential equations in the complex domain. *Resulate Math.,* **16**(1989):3-15.
8. S. B. Bank, I. Laine, Representations of solutions of periodic second order linear differential equations, *J.Reine Angew.Math.,* **344**(1983):1-21.
9. S. B. Bank, I. Lanie, On the oscillation theory of $f'' + Af = 0$ where A is entire, *Trans. Amer. Math. Sco.,* **344**(1982):1-22.
10. S. B. Bank and J. Langley, Oscinallation theorems for higher order linear differential equations with entire periodic coefficients. *Comment. Math. Univ. St. Paul.,* **41**(1992):65-85.
11. W. K. Hayman. *Meromorphic Functions*, Clarendon Press,Oxford,1975.
12. Y. M. Chiang, S. A. Gao, mOn a problem in complex oscillation theory of periodic second order linear differential equations and some related perturbation results, *Annales Academic Sciedtiarum Fennice,* **27**(2002).

A REMARK ON HOLOMORPHIC SECTIONS OF CERTAIN HOLOMORPHIC FAMILIES OF RIEMANN SURFACES

YOICHI IMAYOSHI

Department of Mathematics
Osaka City University
Sugimoto, Sumiyoshi-ku, Osaka 558-8585, Japan
E-mail: imayoshi@sci.osaka-cu.ac.jp

TOSHIHIRO NOGI

Department of Mathematics
Osaka City University
Sugimoto, Sumiyoshi-ku, Osaka 558-8585, Japan
E-mail: nogicchi@ezweb.ne.jp

In this paper, we study a special holomorphic family (M, π, R) of closed Riemann surfaces of genus two over a fourth punctured torus R, which is a kind of a Kodaira surface and is constructed by Riera. We give two explicit defining equations for (M, π, R) by using elliptic functions, and determine all the holomorphic sections of (M, π, R). Proofs will appear elsewhere.

Keywords: Holomorphic section, holomorphic family of Riemann surfaces, function field, Kodaira surface, moduli space, Teichmüller space, Diophantine equation.

2000 Mathematics Subject Classification: Primary 11G30, Secondary 14H15, 32G15, 32J25

1. Introduction

Let us consider a Diophantine equation

$$\sum_{i+j+k=N} A_{ijk} X^i Y^j Z^k = 0 \tag{1}$$

over the function field K of a closed Riemann surgace $\hat{R}$. Here K is the field of all meromorphic functions on $\hat{R}$ and the coefficients A_{ijk} are elements of K. The problem is to find the solutions $(X, Y, K) \in P^2(K)$ of the function equation (1) over $\hat{R}$.

This problem is reformulated geometrically in the following way. It is assumed that we find a Zariski open subset R of $\hat{R}$ and a Zariski open subset M of the algebraic surface $\hat{M}$ defined by

$$\hat{M} = \{([x,y,z],t) \in P^2(\mathbf{C}) \times \hat{R} \mid \sum_{i+j+k=N} A_{ijk}(t)x^i y^j z^k = 0\}$$

such that the holomorphic map $\pi\colon M \to R$ given by $\pi([x,y,z],t) = t$ satisfies the two conditions:

(1) π is of maximal rank at every point of M, and
(2) for every $t \in R$, the fiber $S_t = \pi^{-1}(t)$ of M over t is a Riemann surface of fixed finite type (g,n), where g is the genus of S_t and n is the number of punctures of S_t.

We call such a triplet (M,π,R) is a *holomorphic family of Riemann surfaces* of type (g,n) over R.

We assume throughout this paper that a holomorphic family (M,π,R) of Riemann surfaces is of type (g,n) with $2g-2+n>0$ and the base surface R is of finte type, i.e., a Riemann surface obtained by removing at most a finite number of points from a closed Riemann surface.

Two holomorphic families (M_1,π_1,R) and (M_2,π_2,R) of Riemann surfaces are called *isomorphic* if there exists a biholomorphic map $f\colon M_1 \to M_2$ with $\pi_1 = \pi_2 \circ f$. A holomorphic family (M,π,R) of Riemann surfaces is *locally trivial* if for every point $p_0 \in R$ there exsits a neighborhood U_0 of p_0 in R such that $(\pi^{-1}(U_0), \pi|\pi^{-1}(U_0), U_0)$ is isomorphic to the trivial holomorphic family $(U_0 \times S_{p_0}, \pi_0, U_0)$, where $\pi_0\colon U_0 \times S_{p_0} \to U_0$ is the canonical projection. It is known that a holomorphic family (M,π,R) of Riemann surfaces is locally trivial if and only if the fibers S_t are all isomorphic.

A holomorphic map $s\colon R \to M$ is said to be a *holomorphic section* of a holomorphic family (M,π,R) of Riemann surfaces if s stisfies $\pi \circ s = \mathrm{id}$ on R. Note that if $(X,Y,K) \in P^2(K)$ satisfies the function equation (1), then $s(t) = ([X(t), Y[t], Z[t], t)$ gives rise to a holomorphic section of (M,π,R).

By using theory of Teichmüller space, Imayoshi and Shiga in [6] gave a proof for a finiteness theorem of sections (Mordell conjecture) and a finiteness theorem of families (Shafarevich conjecture). See also Arakelov [1], Faltings [3], Grauert [4], Jost and Yau [7], Manin [9], McMullen [10], and Parshin [11].

In this paper we deal with a special holomorphic family (M,π,R) of closed Riemann surfaces of genus two over a fourth punctured torus R, which is a kind of a Kodaira surface as [8] and is constructed by Riera in [12]. We give two explicit defining equations for (M,π,R) by using elliptic functions, and determine all the holomorphic sections of (M,π,R).

2. Construction of a certain holomorphic family (M, π, R) of Riemann surfaces

Now we explain briefly a construction of our holomorphic family (M, π, R) of Riemann surfaces, which is due to Riera [12].

Take a point τ in the upper half-plane $\mathbf{H}$ in the complex plane $\mathbf{C}$. Let $\Gamma_{1,\tau}$ be the discrete subgroup of $\mathrm{Aut}(\mathbf{C})$ generated by two translations $z \mapsto z+1$ and $z \mapsto z+\tau$. Denote by $\hat{T}$ a torus defiend by the quotient space $\mathbf{C}/\Gamma_{1,\tau} = \{[z] \mid z \in \mathbf{C}\}$. We set $p_0 = [0] \in \hat{T}$ and $T = \hat{T} \setminus \{p_0\}$.

For a point $p \in T$ we take two replicas of the torus $\hat{T}$ cut along a simple arc from p to p_0, and call them sheet I and sheet II. The cut on each sheet has two edges, labeled + edge and − edge. To construct a Riemann surface X_p, we attach the + edge on sheet I and the − edge on sheet II, and then attach the + edge on sheet II and the − edge on sheet I. Then we obtain a closed Riemann surface X_p of genus two and the two-sheeted covering $X_p \to \hat{T}$ which is branched over p_0 and p with branch order 2. It should be noted that the above procedure depends, of course, not only on the choice of the point p but also on the choice of the "cut" from p to p_0. Essentially we can take four different "cuts" $\alpha_1, \alpha_2, \alpha_3$, and α_4 between p and p_0 (see Fig.1).

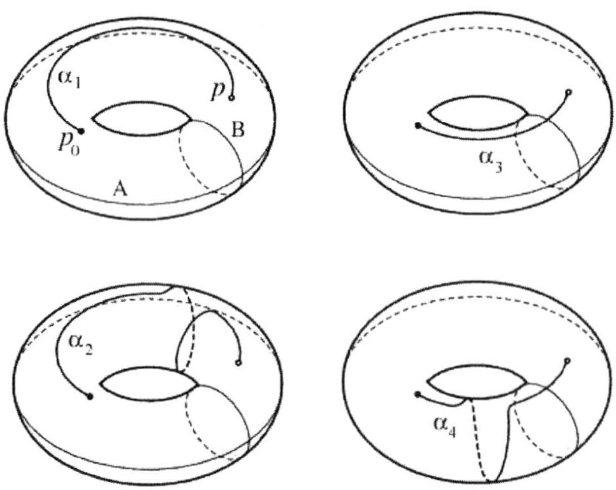

Figure 1. For each $p \in T$, there are precisely four two-sheeted branched coverings $S_{t_i} \to \hat{T}$, where $t_i \in R$ with $\hat{\rho}(t_i) = p$ and the cut α_i is given by the projection $\hat{\rho}(\beta_i)$ of a simple arc β_i between [0] and t_i on $\hat{R}$.

To specify the "cut" we construct a four-sheeted unbranched covering

$$\rho\colon R \to T \tag{2}$$

of T such that R is a torus with four punctures as follows: Let $\Gamma_{2,2\tau}$ be the discrete subgroup of $\mathrm{Aut}(\mathbf{C})$ generated by two translations $z \mapsto z+2$ and $z \mapsto z+2\tau$. Denote by $\hat{R}$ a torus defiend by the quotient space $\mathbf{C}/\Gamma_{2,2\tau} = \{[z] \mid z \in \mathbf{C}\}$. Let $\hat{\rho}\colon \hat{R} \to \hat{T}$ be the canonical projection given by $\hat{\rho}([z]) = [z]$. We set $R = \hat{\rho}^{-1}(T)$ and $\rho = \hat{\rho}|R$. The good thing is that a point $t = [z] \in R$ determines a point $p = \rho([z]) \in T$ and a "cut" $\alpha = \hat{\rho}(\beta)$ from p to $p_0 = [0]$, where β is a simple arc on $\hat{R}$ from $[0]$ to t. Denote by S_t the closed Riemann surface of genus two which is a two-sheeted branched covering surface of $\hat{T}$ constructed by a "cut" $\alpha = \hat{\rho}(\beta)$. Note that the two-sheeted branched covering $\Pi_t\colon S_t \to \hat{T}$ is uniquely determined by the choice of $t \in R$ and does not depend on β.

Now we have the following theorem (see Riera [12]):

Theorem 2.1. *In the above situation, let*

$$M = \bigsqcup_{t\in R} \{t\} \times S_t,$$

$$\pi\colon M \to R,\ \pi(t,q) = t.$$

Then (M, π, R) is a holomorphic family of closed Riemann surfaces of genus two over a fourth punctured torus R.

3. Defining equations for (M, π, R)

In this section we find defining equations for (M, π, R). For any point $t = [\tilde{t}] \in R$, Abel's theorem shows there exists a meromorphic function f_t on $\hat{T}$ which has two zeros $[0]$ and $\rho(t)$ of order one, and a pole $q_t = \rho(t)/2$ of order two. Moreover in order to determine f_t uniquely, we assume that $(df_t/dz)([0]) = 1$. This function f_t is given explicitly as follows (see Clemens [2]):

$$\begin{aligned} f_t([z]) = {} & \frac{1}{\theta'(1/2+\tau/2)} \times \frac{\theta\left(-\tilde{t}/2+1/2+\tau/2\right)^2}{\theta\left(-\tilde{t}+1/2+\tau/2\right)} \times \\ & \frac{\theta\left(z+1/2+\tau/2\right)\theta\left(z-\tilde{t}+1/2+\tau/2\right)}{\theta\left(z-\tilde{t}/2+1/2+\tau/2\right)^2}. \end{aligned} \tag{3}$$

Here the theta function $\theta(z,\tau)$ is defined by

$$\theta(z,\tau) = \sum_{k=-\infty}^{\infty} e^{\pi i(k^2\tau+2kz)}, \quad z \in \mathbf{C}.$$

Then we have the following assertion:

Theorem 3.1. *In the above situation, let*

$$M_E = \{(t,p,w) \in R \times \hat{T} \times \hat{\mathbf{C}} \mid w^2 = f_t(p)\},$$
$$\pi_E : M_E \to R, \ \pi_E(t,p,w) = t.$$

Then the triplet (M_E, π_E, R) is a holomorphic family of closed Riemann surfaces of genus two, and it is isomorphic to $(\mathcal{M}, \pi, R)$ in Theorem 1.

We find another defining equation for (M, π, R). The holomorophic map $f_t \colon \hat{T} \to \hat{\mathbf{C}}$ has four branch points q_t (pole), $a(t), b(t)$, and $c(t)$, where

$$a(t) = f_t([(\tilde{t}+1)/2]),$$
$$b(t) = f_t([(\tilde{t}+\tau)/2]),$$
$$c(t) = f_t([(\tilde{t}+1+\tau)/2]).$$

Let g_t be the meromorphic function on $\hat{T}$ of degree 3 satisfying

(1) g_t has simple zeros $[(\tilde{t}+1)/2], [(\tilde{t}+\tau)/2], [(\tilde{t}+1+\tau)/2]$,
(2) g_t has a pole $[\tilde{t}]$ of order 3, and
(3) $g_t([0]) = i$.

This function g_t is given by

$$g_t(z) = ie^{-2\pi i z} \times \frac{\theta\left(-\tilde{t}/2+1/2+\tau/2\right)^3}{\theta\left(-\tilde{t}/2\right)\theta\left(-\tilde{t}/2+1/2\right)\theta\left(-\tilde{t}/2+\tau/2\right)} \times \frac{\theta\left(z-\tilde{t}/2\right)\theta\left(z-\tilde{t}/2+1/2\right)\theta\left(z-\tilde{t}/2+\tau/2\right)}{\theta\left(z-\tilde{t}/2+1/2+\tau/2\right)^3}.$$

Setting $x = f_t, y = g_t$, we have a functional relation

$$y^2 = \frac{1}{a(t)b(t)c(t)}\left(x-a(t)\right)\left(x-b(t)\right)\left(x-c(t)\right) \tag{4}$$

on $\hat{T}$.

Now we have the following theorem:

Theorem 3.2. *In the above situation, let*

$$P_t(x) = \frac{1}{a(t)b(t)c(t)}(x^2-a(t))(x^2-b(t))(x^2-c(t)),$$
$$M_{HE} = \{(t,x,y) \in R \times \hat{\mathbf{C}} \times \hat{\mathbf{C}} \,|\, y^2 = P_t(x)\},$$
$$\pi_{HE} \colon M_{HE} \to R, \ \pi_{HE}(t,x,y) = t.$$

Then the triplet (M_{HE}, π_{HE}, R) is a holomorphic family of closed Riemann surfaces of genus two, and it is isomorphic to (M, π, R) in Theorem 1.

4. Holomorphic sections (M, π, R)

Let us study the holomorphic sections of (M, π, R) in Theorem 1. The following is our main theorem in this paper.

Theorem 4.1. *The holomorphic family (M, π, R) of closed Riemann surfaces of genus two in Theorem 1 has exactly two holomorphic sections s_1, s_2. These sections are given by $s_1(t) = (t, p_0)$ and $s_2(t) = (t, \rho(t))$ for every $t \in R$.*

In order to prove this theorem, we need the following two theorems (cf. Imayoshi [5], Theorem 4 and Theorem 5):

Theorem 4.2. *The holomorphic family (M, π, R) in Theorem 1 has a canonical completion $(\hat{M}, \hat{\pi}, \hat{R})$, where $\hat{M}$ is a compact two dimensional normal complex analytic space and $\hat{\pi}: \hat{M} \to \hat{R}$ is holomorphic. Moreover every holomorphic section $s: R \to M$ has a holomorphic extension $\hat{s}: \hat{R} \to \hat{M}$.*

Theorem 4.3. *The holomorphic map $\Pi: M = \bigsqcup_{t \in R}\{t\} \times S_t \to \hat{T}$ defined by $\Pi(t, q) = \Pi_t(q)$ has a holomorphic extension $\hat{\Pi}: \hat{M} \to \hat{T}$.*

Theorem 6 is proved by Theorems 2, 3, and 5.

Now we can prove Theorem 4 as follows: Let $s: R \to M$ be an arbitrary holomorphic section of (M, π, R). Theorem 5 and 6 imply that the holomorphic map $\varphi = \Pi \circ s: R \to \hat{T}$ has a holomorphic extention $\hat{\varphi} = \hat{\Pi} \circ \hat{s}: \hat{R} \to \hat{T}$. Let $\tilde{\varphi}: \mathbf{C} \to \mathbf{C}$ is a lift of $\hat{\varphi}: \hat{R} \to \hat{T}$. Then $\tilde{\varphi}(z) = Az + B, z \in \mathbf{C}$ for some constants $A, B \in \mathbf{C}$. Since $\varphi = \Pi \circ s$, we can show that we may assume that $A = 0, B = 0$, or $A = 1, B = 0$. In the case $A = 0, B = 0$, we have the section $s_1(t) = (t, p_0)$, and in the case $A = 1, B = 0$, we have the section $s_2(t) = (t, \rho[t])$.

5. Injectivity of the moduli map J of (M, π, R)

Let M_2 be the moduli space of all biholomorphic equivalence classes $[S]$ of closed Riemann surfaces S of genus two. Then we have the following assertion:

Theorem 5.1. *For the holomorphic family of closed Riemann surfaces of genus two in Theorem 1, the holomorphic map $J: R \to M_2$, $J(t) = [S_t]$, is injective.*

This assertion is proved by Theorem 3 and the well-known fact:Two closed Riemann surfaces S and S' of genus two defined by algebraic equations $y^2 = (x - a_1)(x - a_2) \cdots (x - a_6)$ and $y^2 = (x - a'_1)(x - a'_2) \cdots (x - a'_6)$ are biholomorphically equivalent if and only if there exists a Möbius transformation $\gamma: \mathbf{C} \to \mathbf{C}$ such that $\gamma(\{a_1, a_2, \cdots, a_6\}) = \{a'_1, a'_2, \cdots, a'_6\}$.

Let $\hat{M}_2$ be the Deligne-Mumford compctification of M_2, that is, $\hat{M}_2$ is the compact normal complex analytic space of closed Riemann surfaces of genus two with or without nodes. Then the holomorphic map $J: R \to M_2$ has a holomorphic extension $\hat{J}: \hat{R} \to \hat{M}_2$ (see Imayoshi [5], Lemma 1).

Theorem 5.2. *Let* $t_1 = [0], t_2 = [1], t_3 = [\tau], t_4 = [1+\tau] \in \hat{R} \setminus R$. *Then the images* $\hat{J}(t_j), j = 1, 2, \ldots, 4$, *are represened by the Riemann surfaces* S_{t_j} *with one node:*

(1) S_{t_1} *has two components isomorphic to the torus* T,

(2) S_{t_2} *is the algebraic curve defiend by*

$$y^2 = x^2\left(x^2 - \frac{\theta(0,\tau)^2}{\theta(1/2,\tau^2)}\right)\left(x^2 - \frac{\theta(1/2,\tau)^2}{\theta(0,\tau)^2}\right),$$

(3) S_{t_3} *is the algebraic curve defiend by*

$$y^2 = x^2\left(x^2 - e^{-\pi i\tau}\frac{\theta(0,\tau)^2}{\theta(\tau/2,\tau)^2}\right)\left(x^2 - \frac{\theta(\tau/2,\tau)^2}{\theta(0,\tau)^2}\right),$$

(4) S_{t_4} *is the algebraic curve defiend by*

$$y^2 = x^2\left(x^2 + e^{-\pi i\tau}\frac{\theta(1/2,\tau)^2}{\theta(\tau/2,\tau)^2}\right)\left(x^2 - \frac{\theta(\tau/2,\tau)^2}{\theta(1/2,\tau)^2}\right).$$

Moreover, if $\tau \in \mathbf{H}$ *satisfies* $-1/2 < \operatorname{Re}\tau < 0,\ |\tau| > 1$, *or* $0 \leq \operatorname{Re}\tau \leq 1/2,\ |\tau| \geq 1$, *then it follows that*

(1) $\hat{J}(t_2) = \hat{J}(t_3)$ *if and only if* $\tau = i, e^{\pi i/3}$, *or* $(1+\sqrt{15}i)/4$,

(2) $\hat{J}(t_2) = \hat{J}(t_4)$ *if and only if* $\tau = e^{\pi i/3}$, *and*

(3) $\hat{J}(t_3) = \hat{J}(t_4)$ *if and only if* $\tau = e^{\pi i/3}, (1+\sqrt{15}i)/4$, *or* $(1+\sqrt{7}i)/2$.

The first part of Theorem 8 is prove by the explict representations for f_t, S_t as (3), (4). In order to show the second part, we recall the universal covering $\lambda\colon \mathbf{H} \to \hat{\mathbf{C}} \setminus \{0, 1, \infty\}$ definded by

$$\lambda(\tau) = -e^{\pi i\tau}\left(\frac{\theta(\tau/2,\tau)}{\theta(1/2,\tau)}\right)^4.$$

Then the second part is proved by using the following facts (cf. Clemens [2]):

(1) by the blowing up at $(x, y) = (0, 0)$, the normaliztions of S_{t_2}, S_{t_3},and S_{t_4} are isomorphic to elliptic curves defined by $y^2 = (x-1)(x-\lambda(2\tau))$, $y^2 = x(x-1)(x-\lambda(\tau/2))$, and $y^2 = (x-1)(x-\lambda(1/2+\tau/2))$, respectively,

(2) the universal covering transformation group of the universal covering $\lambda\colon \mathbf{H} \to \hat{\mathbf{C}} \setminus \{0, 1, \infty\}$ is a Fuchsian gropup generated by two elements $\tau \mapsto \tau + 2$ and $\tau \mapsto \tau/(2\tau+1)$, which is a subgroup of $\mathrm{PSL}(2, \mathbf{Z})$ of index 6, and

(3) for any two points $\tau_1, \tau_2 \in \mathbf{H}$, elliptic curves defined by $y^2 = x(x-1)(x-\lambda(\tau_1))$ and $y^2 = x(x-1)(x-\lambda(\tau_2))$ are isomorphic if and only if there exists an element $\gamma \in \mathrm{PSL}(2, \mathbf{Z})$ satisfying $\tau_2 = \gamma(\tau_1)$.

References

1. S. Arakelov, Families of algebraic curves with fixed degeneracies, *Math. USSR Izv.* **35** (1971):1269–1293.
2. C.H. Clemens, *A Scrapbook of Complex Curve Theory*, Plenum Press, New York and London, 1980.
3. G. Faltings, Endlichkeitssätze für abelsche Varietäten über Zahlkörpern, *Invent. Math.* **73**(1983):349–366.
4. H. Grauert, Mordells Vermutung über rationale Punkte auf algebraischen Kurven und Funktionenkörper, *IHES Publ. Math.* **25**(1965):131–149.
5. Y. Imayoshi, Holomorphic families of Riemann surfaces and Teichmüller spaces, in Riemann Surfaces and Related Topics, Stony Brook Conference, *Ann. Math. Studies,* **97**(1978):277-300, edited by I. Kra and B. Maskit, Princeton University Press, Princeton, New Jersey, 1981.
6. Y. Imayoshi and H. Shiga, A finiteness theorem of holomorphic families of Riemann surfaces, in *Holomorphic Functions and Moduli*, Vol. II, pp.207-219, edited by D. Drain et al, Springer-Verlag:MSRI Publications Vol.10, 1988.
7. J. Jost and S-T. Yau, Harmonic mappings and algebraic varieties over function fields, *Amer. J. of Math.* **115** (1993):1197–1227.
8. K. Kodaira, A certain type of irregular algebraic surfaces, *J. d'Analyse Math.* **19** (1967):207-215.
9. Y. Manin, A proof of the analog of the Mordell conjecture for algebraic curves over function fields, *Soviet Math. Dokl.* **152**(1963):1061–1063.
10. C. McMullen, From dynamics on surfaces to rational points on curves, *Bull. Amer. Math. Soc.* **37** (2000):119–140.
11. A.N. Parshin, Albegraic curves over function fields, *Soviet Math. Dokl.* **183**(1968):524–526.
12. G. Riera, Semi-direct products of Fuchsian groups and uniformization, *Duke Math. J.* **44** (1977):291-304.

α-ASYMPTOTICALLY CONFORMAL FIXED POINTS AND HOLOMORPHIC MOTIONS*

YUNPING JIANG

Department of Mathematics
Queens College of the City University of New York
Flushing, NY 11367-1597
and
Department of Mathematics
Graduate School of the City University of New York
365 Fifth Avenue, New York, NY 10016
and
Academy of Mathematics and System Sciences
Chinese Academy of Sciences
Beijing 100080 P. R. China
E-mail: Yunping.Jiang@qc.cuny.edu

An α-asymptotically conformal fixed point for a quasiconformal map defined on a domain is defined for any $0 \leq \alpha \leq 1$. Furthermore, we prove that for $\alpha > 0$, there is a normal form for this kind attracting or repelling or super-attracting fixed point under quasiconformal changes of coordinate which is also asymptotically conformal at this point. These results generalize König's Theorem and Böttcher's Theorem in classical complex analysis. The idea in proofs is new and uses holomorphic motion theory and provides a new understanding of the inside mechanism of these two famous theorems.

Keywords: α-asymptotically conformal fixed point, holomorphic motion, normal form.

2000 Mathematics Subject Classification: Primary 37F99, Secondary 32H02

*The research is partially supported by NSF grants and PSC-CUNY awards and Hundred Talents Program from Academia Sinica.

1. Introduction

Two of the fundamental theorems in complex dynamical systems are König's Theorem and Bötthcher's Theorem in classical complex analysis which were proved back to 1884 [19] and 1904 [8], respectively, by using some well-known methods in complex analysis. These theorems say that an attracting or repelling or super-attracting fixed point of an analytic map can be written into a normal form under suitable conformal changes of coordinate. These theorems become two fundamental results in the recent study of the dynamics of a polynomial or a rational map.

However, it becomes more and more clear in recent years that only conformal change of coordinate is not enough in the study of many problems in dynamics and in geometry, for examples, in the study of monotonicity of the entropy function for the family $|x|^3 + t$ [29] and in the study of quasiconformal structures on a 4-manifold [10]–in these studies, quasiconformal changes of coordinate are appealed. The quasiconformal changes of coordinate may still have asymptotical conformality property just at one point but definitely not conformal. (It is a big difference between asymptotically conformal and conformal, see definition in Section 2.)

During the study of complex dynamical systems, a subject called holomorphic motions becomes more and more interesting and useful. The subject of holomorphic motions over the open unit disk shows some interesting connections between classical complex analysis and problems on moduli. This subject even becomes an interesting branch in complex analysis [4,6,14,27,31,35].

In this paper, we will use holomorphic motions over the open unit disk to study the quasiconformal changes of coordinate which are aymptotically conformal at one point. The paper is organized as follows. In Section 2, we give an overview about holomorphic motions and quasiconformal theory. In Section 3, we define an asymptotically conformal fixed point. We then definite an attracting or repelling α-asymptotically conformal fixed point. In Section 4, we prove one of our main theorems in this paper:

Theorem 1.1. *Let f be a quasiconformal homeomorphism defined on a neighborhood about 0. Suppose 0 is an attracting or repelling α-asymptotically conformal fixed point of f for some $0 < \alpha \leq 1$. Then there is a quasiconformal homeomorphism $\phi : \Delta_\delta \to \phi(\Delta_\delta) \subset U$ from an open disk of radius $\delta > 0$ centered at 0 into U which is asymptotically conformal at 0 such that*

$$\phi^{-1} \circ f \circ \phi(z) = \lambda z, \quad z \in \Delta_\delta.$$

The conjugacy ϕ^{-1} is unique up to multiplication of constants.

To present our idea clearly, we first use the same idea in the proof of above theorem to give a new proof of König's Theorem in classical complex analysis in Section 3. Then we prove Theorem 1.1 in the same section.

We definite an asymptotically conformal super-attracting fixed point in Section 3. In Section 5, we prove the other main theorem in this paper:

Theorem 1.2. *Let $g = f(z^n)$ be a quasiregular map defined on a neighborhood about 0. Suppose 0 is a super-attracting α-asymptotically conformal fixed point of g for some $0 < \alpha \leq 1$. Then there is a quasiconformal homeomorphism $\phi : \Delta_\delta \to \phi(\Delta_\delta) \subset U$ from an open disk of radius $\delta > 0$ centered at 0 into U which is asymptotically conformal at 0 such that*

$$\phi^{-1} \circ g \circ \phi(z) = z^n, \quad z \in \Delta_\delta.$$

The conjugacy ϕ^{-1} is unique up to multiplication by $(n-1)^{th}$-roots of the unit.

Again, we will first give a new proof of Böttecher's Theorem in classical complex analysis in Section 5. Then we prove Theorem 1.2 in the same section.

2. Holomorphic Motions and Quasiconformal Maps

In the study of complex analysis, the measurable Riemann mapping theorem plays an important role. Consider the Riemann sphere $\hat{\mathbb{C}}$. A measurable function μ on $\hat{\mathbb{C}}$ is called a Beltrami coefficient if there is a constant $0 \leq k < 1$ such that $\|\mu\|_\infty \leq k$, where $\|\cdot\|_\infty$ means the L^∞-norm of μ on $\hat{\mathbb{C}}$. The equation

$$H_{\overline{z}} = \mu H_z$$

is called the Beltrami equation with the given Beltrami coefficient μ. The measurable Riemann mapping theorem says that the Beltrami equation has a solution H which is a quasiconformal homeomorphism of $\hat{\mathbb{C}}$ whose quasiconformal dilatation is less than or equal to $K = (1+k)/(1-k)$. The study of the measurable Riemann mapping theorem has a long history since Gauss considered in 1820's the connection with the problem of finding isothermal coordinates for a given surface. As early as 1938, Morrey [32] systematically studied homeomorphic L^2-solutions of the Beltrami equation (see [24,25]). But it took almost twenty years until in 1957 Bers [5] observed that these solutions are quasiconformal (refer to [20]). Finally the existence of a solution to the Beltrami equation under the most general possible circumstance, namely, for measurable μ with $\|\mu\|_\infty < 1$, was shown by Bojarski [7]. In this generality the existence theorem is sometimes called the measurable Riemann mapping theorem (refer to [16]).

If one only considers a normalized solution in the Beltrami equation (a solution fixes 0, 1, and ∞), then H is unique, which is denoted as H^μ. The solution H^μ is expressed as a power series made up of compositions of singular integral operators applied to the Beltrami equation on the Riemann sphere. In this expression, if one considers μ as a variable, then the solution H^μ depends on μ analytically. This analytic dependence was emphasized by Ahlfors and Bers in their 1960 paper [2] and is essential in determining a complex structure for Teichmüller space (refer to [1,16,20,26,33]). Note that when $\mu \equiv 0$, H^0 is the identity map. A 1-quasi-conformal map is conformal. Twenty years later, due to the development of complex dynamics, this analytic dependence presents an even more interesting phenomenon called holomorphic motions as follows.

Let $\Delta_r = \{c \in \mathbb{C} \mid |c| < r\}$ be the disk centered at 0 and of radius $r > 0$. In particular, we use Δ to denote the unit disk. Given a Beltrami coefficient $\mu \not\equiv 0$, consider a family of Beltrami coefficients $c\mu/\|\mu\|_\infty$ for $c \in \Delta$ and the family of normalized solutions $H^{\frac{c\mu}{\|\mu\|_\infty}}$. Note that $H^{\frac{c\mu}{\|\mu\|_\infty}}$ is a quasiconformal homeomorphism whose quasiconformal dilatation is less than or equal to $(1+|c|)/(1-|c|)$. Moreover, $H^{\frac{c\mu}{\|\mu\|_\infty}}$ is a family which is holomorphic on c. Consider a subset E of $\hat{\mathbb{C}}$ and its image $E_c = H^{\frac{c\mu}{\|\mu\|_\infty}}(E)$. One can see that E_c moves holomorphically in $\hat{\mathbb{C}}$ when c moves in Δ. That is, for any point $z \in E$, $z(c) = H^{\frac{c\mu}{\|\mu\|_\infty}}(z)$ traces a holomorphic path starting from z as c moves in the unit disk. Although E may start out as smooth as a circle and although the points of E move holomorphically, E_c can be an interesting fractal with fractional Hausdorff dimension for every $c \neq 0$ (see [15]).

Surprisingly, the converse of the above fact is true too. This starts from the famous λ-lemma of Mañé, Sad, and Sullivan [28] in complex dynamical systems. Let us start to understand this fact by first defining holomorphic motions.

Definition 2.1. *[Holomorphic Motions] Let E be a subset of $\hat{\mathbb{C}}$. Let*

$$h(c,z) : \Delta_r \times E \to \hat{\mathbb{C}}$$

be a map. Then h is called a holomorphic motion of E parametrized by Δ_r if

(1) $h(0, z) = z$ *for* $z \in E$;
(2) for any fixed $c \in \Delta_r$, $h(c, \cdot) : E \to \hat{\mathbb{C}}$ *is injective;*
(3) for any fixed z, $h(\cdot, z) : \Delta_r \to \hat{\mathbb{C}}$ *is holomorphic.*

For example, for a given Beltrami coefficient μ,

$$H(c, z) = H^{\frac{c\mu}{\|\mu\|_\infty}}(z) : \Delta \times \hat{\mathbb{C}} \to \hat{\mathbb{C}}$$

is a holomorphic motion of $\hat{\mathbb{C}}$ parametrized by Δ.

Note that even continuity does not directly enter into the definition; the only restriction is in the c direction. However, continuity is a consequence of the hypotheses from the proof of the λ-lemma of Mañé, Sad, and Sullivan [28]. Moreover, Mañé, Sad, and Sullivan prove in [28] that

Lemma 2.1. *[λ-Lemma] A holomorphic motion of a set* $E \subset \hat{\mathbb{C}}$ *parametrized by* Δ *can be extended to a holomorphic motion of the closure of* E *parametrized by the same* Δ.

Furthermore, Mañé, Sad, and Sullivan show in [28] that $f(c, \cdot)$ satisfies the Pesin property. In particular, when the closure of E is a domain, this property can be described as the quasiconformal property. A further study of this quasiconformal property is given by Sullivan and Thurston [35] and Bers and Royden [6]. In [35], Sullivan and Thurston prove that there is a universal constant $a > 0$ such that any holomorphic motion of any set $E \subset \hat{\mathbb{C}}$ parametrized by the open unit disk Δ can be extended to a holomorphic motion of $\hat{\mathbb{C}}$ parametrized by Δ_a. In [6], Bers and Royden show, by using classical Teichmüller theory, that this constant actually can be taken to be $1/3$. Moreover, in the same paper, Bers and Royden show that in any holomorphic motion $H(c, z) : \Delta \times \hat{\mathbb{C}} \to \hat{\mathbb{C}}$, $H(c, \cdot) : \hat{\mathbb{C}} \to \hat{\mathbb{C}}$ is a quasiconformal homeomorphism whose quasiconformal dilatation less than or equal to $(1 + |c|)/(1 - |c|)$ for $c \in \Delta$. In the both papers [35,6], they expect $a = 1$. This was eventually proved by Slodkowski in [34].

Theorem 2.1. *[Slodkowski's Theorem] Suppose*

$$h(c, z) : \Delta \times E \to \hat{\mathbb{C}}$$

is a holomorphic motion of a set $E \subset \hat{\mathbb{C}}$ *parametrized by* Δ *. Then* h *can be extended to a holomorphic motion*

$$H(c, z) : \Delta \times \hat{\mathbb{C}} \to \hat{\mathbb{C}}$$

of $\hat{\mathbb{C}}$ parametrized by also Δ. Moreover, following [6]*, for every $c \in \Delta$, $H(c,\cdot) : \hat{\mathbb{C}} \to \hat{\mathbb{C}}$ is a quasiconformal homeomorphism whose quasiconformal dilatation*

$$K(H(c,\cdot)) \leq \frac{1+|c|}{1-|c|}.$$

Holomorphic motions of a set $E \subset \hat{\mathbb{C}}$ parametrized by a connected complex manifold with a base point can be also defined. They have many interesting relationships with the Teichmüller space $T(E)$ of a closed set E (refer to [31]).

In addition to the references we mentioned above, there is a partial list of references [3,4,11,12,13,14,27,21,22] about holomorphic motions and Teichmüller theory. The reader who is interested in holomorphic motions may refer to those papers and books.

3. α-Asymptotically conformal fixed points

We define a class of quasiconformal maps and quasiregular maps fixing a point. Let f be a quasiconformal homeomorphism defined on a neighborhood U about 0 and fixing 0. Let $\mu(z) = f_{\overline{z}}/f_z$ be the complex dilatation of f on U. For any $t > 0$ such that $\Delta_t \subset U$, let $\omega(t) = \|\mu|\Delta_t\|_\infty$, where $\|\cdot\|_\infty$ means the L^∞ norm.

Definition 3.1. We call f asymptotically conformal at 0 if

$$\omega(t) \to 0 \quad \text{as} \quad t \to 0^+.$$

Furthermore, for a real number $0 \leq \alpha \leq 1$, we call f α-asymptotically conformal at 0 if

$$\lim_{t\to 0^+} t^\alpha \int_0^t \frac{\omega(s)}{s^{1+\alpha}} ds = 0,$$

If f is asymptotically conformal at 0, then f maps a tiny circle centered at 0 to an ellipse and, moreover, the ratio of the long axis and the short axis tends to 1 as the radius of the tiny circle tends to 0. But the map still can fail to be differentiable at 0 (refer to [17]). However, following Reshetnyak's 1978 paper [23]), if f is 0-asymptotically conformal at 0, then f is differentiable and conformal at 0, i.e., the limit of $f(z)/z$ exists as z goes to 0. Let

$$\lambda = \lim_{|z|\to 0} \frac{f(z)}{z}$$

and call it the multiplier of f at 0.

It is clear that for $0 \le \alpha' < \alpha \le 1$, if f is α-asymptotically conformal at 0, then it must be α'-asymptotically conformal at 0. Therefore, if f is α-asymptotically conformal at 0, then one can define its multiplier λ at 0. We call 0

i) attracting if $0 < |\lambda| < 1$;
ii) repelling if $|\lambda| > 1$;
iii) neutral if $|\lambda| = 1$.

Correspondingly, we call 0 an attracting, repelling, or neutral α-asymptotically conformal fixed point.

Let g be a quasiregular map defined on a neighborhood U about 0 and fixing 0. Assume $g = f \circ q_n$ where $q_n(z) = z^n$ and f is a quasiconformal homeomorphism. We say g is α-asymptotically conformal at 0 if f is α-asymptotically conformal at 0 with nonzero multiplier $\lambda = \lim_{|z|\to 0} f(z)/z$. In this case 0 is called a super-attracting α-asymptotically conformal fixed point of g.

The following lemma will be useful in our proofs of Theorems 1 and 2.

Lemma 3.1. *Suppose $\omega(t)$ is an increasing function of $t > 0$. Suppose, for some $0 \le \alpha \le 1$,*

$$\lim_{t\to 0^+} t^\alpha \int_0^t \frac{\omega(s)}{s^{1+\alpha}} ds = 0.$$

Suppose $0 < \sigma < 1$. Let

$$\tilde{\omega}(t) = \sum_{j=n}^{\infty} \sigma^{-\alpha n} \omega(\sigma^n t).$$

Then

$$\tilde{\omega}(t) \le \frac{t^\alpha}{-\log \sigma} \int_0^{\sigma t} \frac{\omega(s)}{s^{1+\alpha}} ds.$$

Moreover, $\tilde{\omega}(t) \to 0$ as $t \to 0^+$.

Proof. Since $\omega(t)$ is increasing for $t > 0$, we have

$$\tilde{\omega}(t) \le \int_1^\infty \sigma^{-\alpha x} \omega(\sigma^x t) dx = \frac{t^\alpha}{-\log \sigma} \int_0^{\sigma t} \frac{\omega(s)}{s^{1+\alpha}} ds. \qquad \square$$

4. Linearization for α-asymptotically conformal attracting or repelling fixed points

One of the main results in this article is Theorem 1, which says that if f is a quasiconformal homeomorphism and 0 is an attracting or repelling α-asymptotically conformal fixed point with the multiplier λ, $0 < |\lambda| < 1$ or $|\lambda| > 1$, and with $\alpha > 0$, then f can be written as a linear map $z \to \lambda z$ under some quasiconformal change of coordinate which is also asymptotically conformal at 0. The result generalizes the famous König's Theorem in classical analysis. Therefore, to present a clear idea about how we get Theorem 1, we first use the same idea to give another proof of König's Theorem. The idea of the new proof follows the viewpoint of holomorphic motions. For the classical proof of König's Theorem, the reader may refer to König's original paper [19] or most recent books [9,30]. Actually from the technical point of views, our proof is more complicate and uses a sophistical result. But from the conceptual point of views, our proof gives some inside mechanism for the linearization of an attracting or a repelling fixed point.

Theorem 4.1. *[König's Theorem] Let $f(z) = \lambda z + \sum_{j=2}^{\infty} a_j z^j$ be an analytic function defined on Δ_{r_0}, $r_0 > 0$. Suppose $0 < |\lambda| < 1$ or $|\lambda| > 1$. Then there is a conformal map $\phi : \Delta_\delta \to \phi(\Delta_\delta)$ for some $0 < \delta < r_0$ such that*

$$\phi^{-1} \circ f \circ \phi(z) = \lambda z.$$

The conjugacy ϕ^{-1} is unique up to multiplication of constants.

Proof. [A new proof of Theorem 4.1] We only need to prove it for $1 < |\lambda| < 1$. In the case of $|\lambda| > 1$, we can consider f^{-1}.

First, we can find a $0 < \delta < r_0$ such that

$$|f(z)| < |z|, \quad z \in \overline{\Delta_\delta}$$

and f is injective on $\overline{\Delta_\delta}$. For every $0 < r \leq \delta$, let

$$S_r = \{z \in \mathbb{C} \mid |z| = r\}$$

and

$$T_r = |\lambda| S_r = \{z \in \mathbb{C} \mid |z| = |\lambda| r\}.$$

Denote $E = S_r \cup T_r$. Define

$$\phi_r(z) = \begin{cases} z & z \in S_r \\ f(\frac{z}{\lambda}), & z \in T_r. \end{cases}$$

It is clear that

$$\phi_r^{-1} \circ f \circ \phi_r(z) = \lambda z$$

for $z \in S_r$.

Now write $\phi_r(z) = z\psi_r(z)$ for $z \in T_r$, where

$$\psi_r(z) = 1 + \sum_{j=1}^{\infty} \frac{a_{j+1}}{\lambda^{j+1}} z^j.$$

Define

$$h_r(c, z) = \begin{cases} z, & z \in S_r \\ z\psi_r(\frac{\delta cz}{r}), & z \in T_r \end{cases} : \Delta \times E \to \hat{\mathbb{C}}.$$

Note that

$$h(c, z) = z\psi_r\Big(\frac{cz\delta}{r}\Big) = \frac{r}{c\delta}\phi\Big(\frac{cz\delta}{r}\Big) = \frac{r}{c\delta}f\Big(\frac{cz\delta}{r\lambda}\Big), \quad z \in T_r, \quad c \neq 0.$$

For each fixed $z \in E$, it is clear that $h(c, z)$ is a holomorphic function of $c \in \Delta$. For each fixed $c \in \Delta$, the restriction $h(c, \cdot)$ to S_r and T_r, respectively, are injective. Now we claim that their images do not cross either. That is because for any $z \in T_r$, $|z| = |\lambda| r$ and $|cz\delta|/|r\lambda| \leq \delta$, so

$$|h(c, z)| = \Big|\frac{r}{c\delta}\Big|\Big|f\Big(\frac{cz\delta}{r\lambda}\Big)\Big| < \Big|\frac{r}{c\delta}\Big|\Big|\frac{cz\delta}{r\lambda}\Big| = r.$$

Therefore, $h(c, z) : \Delta \times E \to \hat{\mathbb{C}}$ is a holomorphic motion because we also have $h(0, z) = z$ for all $z \in E$. From Theorem 3, h can be extended to a holomorphic motion $H(c, z) : \Delta \times \hat{\mathbb{C}} \to \hat{\mathbb{C}}$, and moreover, for each fixed $c \in \Delta$, $H_c = h(c, \cdot) : \hat{\mathbb{C}} \to \hat{\mathbb{C}}$ is a quasiconformal homeomorphism whose quasiconformal dilatation is less than or equal to $(1 + |c|)/(1 - |c|)$. Now take $c_r = r/\delta$ and consider $H(c_r, \cdot)$. We have $H(c_r, \cdot)|E = \phi_r$. Let

$$A_{r,j} = \{z \in \mathbb{C} \mid |\lambda|^{j+1} r \leq |z| \leq |\lambda|^j r\}.$$

We still use ϕ_r to denote $H(c_r, \cdot)|A_{r,0}$.

For an integer $k > 0$, take $r = r_k = \delta|\lambda|^k$. Then

$$\overline{\Delta}_\delta = \cup_{j=-k}^{\infty} A_{r,j} \cup \{0\}.$$

Extend ϕ_r to $\overline{\Delta}_\delta$, which we still denote as ϕ_r, as follows.

$$\phi_r(z) = f^{-j}(\phi_r(\lambda^n z)), \quad z \in A_{r,j}, \quad j = -k, \cdots, -1, 0, 1, \cdots,$$

and $\phi_r(0) = 0$. Since $\phi_r|E$ is a conjugacy from f to λz, ϕ_r is continuous on Δ_δ. Since f is conformal, ϕ_r is quasiconformal whose quasiconformal dilatation is the same as that of $H(c_r, \cdot)$ on $A_{r,0}$. So the quasiconformal dilatation of ϕ_r on Δ_δ is less than or equal to $(1+r)/(1-r)$. Furthermore,

$$f(\phi_r(z)) = \phi_r(\lambda z), \quad z \in \Delta_\delta.$$

Since $f(z) = \lambda z(1 + O(z))$, $f^k(z) = \lambda^k z \prod_{i=0}^{k-1}(1 + O(\lambda^i z))$. Because $|\lambda|^k r_k = \delta$, the range of ϕ_{r_k} on Δ_δ is a Jordan domain bounded above from ∞ and below from 0 uniformly on k. In addition, 0 is fixed by ϕ_k and the quasiconformal dilatations of the ϕ_k are uniformly bounded. Therefore, the sequence $\{\phi_{r_k}\}_{k=1}^{\infty}$ is a compact family (see [1]). Let ϕ be a limiting map of this family. Then we have

$$f(\phi(z)) = \phi(\lambda z), \quad z \in \Delta_\delta.$$

The quasiconformal dilatation of ϕ is less than or equal to $(1+r_k)/(1-r_k)$ for all $k > 0$. So ϕ is a 1-quasiconformal map, and thus is conformal. This is the proof of the existence.

For the sake of completeness, we also provide the proof of uniqueness but this is not new and the reader can find it on [9,30]. Suppose ϕ_1 and ϕ_2 are two conjugacies such that

$$\phi_1^{-1} \circ f \circ \phi_1(z) = \lambda z \quad \text{and} \quad \phi_2^{-1} \circ f \circ \phi_2(z) = \lambda z, \quad z \in \Delta_\delta.$$

Then for $\Phi = \phi_2^{-1} \circ \phi_1$, we have $\Phi(\lambda z) = \lambda \Phi(z)$. This implies that $\Phi'(\lambda z) = \Phi'(z)$ for any $z \in \Delta_\delta$. Thus $\Phi'(z) = \Phi'(\lambda^n z) = \Phi(0)$. So $\Phi(z) = \text{const.}$ and $\phi_2^{-1} = \text{const.} \cdot \phi_1^{-1}$. □

Now let us prove Theorem 1.1. After understanding the proof of Theorem 4.1, the following proof is relatively easier to follow.

Proof. [Proof of Theorem 1.1] We need only to prove this theorem for attracting α-asymptotically conformal germs. In the case of a repelling asymptotically conformal germ, we can consider f^{-1}.

Let $\sigma = |\lambda|$. First, we can find a $\delta > 0$ such that

$$|f(z)| < |z|, \quad z \in \overline{\Delta}_\delta \subset U$$

and f is injective on $\overline{\Delta}_\delta$. For every $0 < r \le \delta$, let

$$S_r = \{z \in \mathbb{C} \mid |z| = r\}$$

and

$$T_r = \sigma S_r = \{z \in \mathbb{C} \mid |z| = \sigma r\}.$$

Denote $E = S_r \cup T_r$. Define

$$\phi_r(z) = \begin{cases} z & z \in S_r \\ f(\frac{z}{\lambda}), & z \in T_r. \end{cases}$$

It is clear that

$$\phi_r^{-1} \circ f \circ \phi_r(z) = \lambda z$$

for $z \in S_r$.

Now write $\phi(z) = f(\frac{z}{\lambda})$ defined on Δ_r. Suppose $\phi(r) = \tau_r$. Extend ϕ to $\hat{\mathbb{C}}$ by quasiconformal reflection with respect to S_r and $\phi(S_r)$ (see [1]). We still denote this extended map as ϕ. Let $\nu = \phi_{\bar{z}}/\phi_z$ be the complex dilatation of the extended ϕ. Then

$$a(r) = \|\nu\|_\infty = O(\|\mu|\Delta_{\sigma r}\|_\infty) = O(\omega(\sigma^{-1}r)).$$

Consider $\nu_c = ca_0 a(r)^{-1}\nu$ and the unique solution $\phi_c = \phi^{\nu_c}$ that maps 0, r, and ∞ to 0, τ_r, and ∞, respectively. Here a_0 is a constant independent of r such that $|\phi_c(z)| < r$ for all $|z| \leq \sigma r$ and $|c| < 1$. (Since ϕ_c can be written as a power series in c and $\|\nu_c\| \to 0$ uniformly as $r \to 0$, such an a_0 exists.) Then ϕ_c holomorphically depends on $c \in \Delta$. Define

$$\phi_r(c, z) = \begin{cases} z & z \in S_r \\ \phi_c(z), & z \in T_r. \end{cases}$$

It is a holomorphic motion from $\Delta \times E \to \hat{\mathbb{C}}$. From Theorem 2.1, $\phi(c, z)$ can be extended to a holomorphic motion from $\Delta \times \hat{\mathbb{C}} \to \hat{\mathbb{C}}$, which we still denote by $\phi(c, z)$, such that the quasiconformal dilatation of $\phi(c, \cdot)$ is less than or equal to $(1 + |c|)/(1 - |c|)$. In particular when $c_r = a_0^{-1}a(r)$, $\phi_r(c_r, z)|E = \phi_r$. Let

$$A_{r,j} = \{z \in \mathbb{C} \mid \sigma^{j+1}r \leq |z| \leq \sigma^j r\}.$$

We still use ϕ_r to denote $\phi_r(c_r, \cdot)|A_{r,0}$. For an integer $k > 0$, take $r = r_k = \delta\sigma^k$. Then

$$\Delta_\delta = \cup_{j=-k}^{\infty} A_{r,j} \cup \{0\}.$$

Extend ϕ_r to Δ_δ, which we still denote as ϕ_r, by

$$\phi_r(z) = f^{-j}(\phi_r(\lambda^j z)), \quad z \in A_{r,j}, \quad j = -k, \cdots, -1, 0, 1, \cdots,$$

and $\phi_r(0) = 0$. Since $\phi_r|E$ is a conjugacy from f to λz, ϕ_r is continuous on Δ_δ.

Next we need to estimate the quasiconformal constant of ϕ_r on Δ_δ. We will use the following formula (refer to [1]): If F and G are two quasiconformal maps with the complex dilatations μ_F and μ_G. Then the composition map $G \circ F$ has the complex dilatation

$$\mu_{G\circ F} = \frac{\mu_F + \gamma\mu_G \circ F}{1 + \overline{\mu}_F \gamma \mu_G \circ F}, \quad \text{where} \quad \gamma = \frac{\overline{F_{\overline{z}}}}{F_z}.$$

Thus

$$\|\mu_{G\circ F}\|_\infty \leq (\|\mu_F\|_\infty + \|\mu_G \circ F\|_\infty)(1 - \|\mu_F\|_\infty \|\mu_G \circ F\|_\infty)^{-1}.$$

Suppose, in the beginning of the proof, we pick δ small such that

$$(1 - \omega(\delta)\|\mu_g \circ f\|_\infty)^{-1} \leq \sigma^{-\alpha}$$

for all quasiconformal homeomorphism g. Let $b(r) = (1 + |c_r|)/(1 - |c_r|)$. Then use the above formula for composition, we have that the complex dilatation μ_{ϕ_r} over Δ_δ can be controlled by

$$\|\mu_{\phi_r}\|_\infty \leq \sum_{j=1}^{\infty} \sigma^{-\alpha j}\omega(\sigma^j\delta) + b(r) \leq \tilde{\omega}(\delta) + b(r)$$

and μ_{ϕ_r} over Δ_r can be controlled by

$$\|\mu_{\phi_r}|\Delta_r\|_\infty \leq \sum_{j=1}^{\infty} \sigma^{-\alpha j}\omega(\sigma^j r) + b(r) \leq \tilde{\omega}(r) + b(r).$$

Thus $\{\phi_r\}_{0<r\leq\delta}$ is uniformly K-quasiconformal for a fixed number $K > 1$. Consider $B_r = \Delta_\delta \setminus \Delta_r = \cup_{j=0}^{-k} A_{r,j}$ and $\phi_r(B_r) = \cup_{j=0}^{-k}\phi_r(A_{r,j})$.

Both of the annulli have the same inner circle S_r. Thus the ratio of the modulus of $\phi_r(B_r)$ and the modulus of B_r is controlled by two constants from below and above (independent of r but only depends on K). Therefore, the range of ϕ_r on Δ_δ is a Jordan domain bounded above from ∞ and below from 0 uniformly in $0 < r \leq \delta$. Since, additionally, 0 is fixed by any element in this sequence, $\{\phi_{r_k}\}_{k=1}^{\infty}$ is a compact family (see [1]). Let ϕ be a limit mapping of this family. Then we have

$$f(\phi(z)) = \phi(\lambda z), \quad z \in \Delta_\delta.$$

Since the complex dilatation of ϕ is controlled by $b(r) = \tilde{\omega}(r) + b(r)$, ϕ is asymptotically conformal at 0 and the proof is completed.

Suppose ϕ_1 and ϕ_2 are two asymptotically conformal conjugacies such that

$$\phi_1^{-1} \circ f \circ \phi_1(z) = \lambda z \quad \text{and} \quad \phi_2^{-1} \circ f \circ \phi_2(z) = \lambda z, \quad z \in \Delta_\delta.$$

Then for $\Phi = \phi_2^{-1} \circ \phi_1$, we have $\Phi(\lambda z) = \lambda\Phi(z)$. This implies that the complex dilatation $\mu_\Phi(z) = \mu_\Phi(\lambda z)$, a.e. This in turn implies that $\mu = 0$

a.e. in Δ_δ and thus Φ is conformal. Furthermore, $\Phi(z) = az$ for some $a \neq 0$. This is the uniqueness. □

5. Normal forms for α-asymptotically conformal super-attracting fixed points

The other main result in this article is Theorem 1.2, which says that if $g = f(z^n)$ is a quasiregular map and 0 is an α-asymptotically conformal super-attracting fixed point, then g can be written into the normal form $z :\to z^n$ under some quasiconformal change of coordinate which is asymptotically conformal at 0. The result generalizes the famous Böttcher's Theorem in classical analysis. Again, to present a clear idea about how we get Theorem 1.2, we first use the same idea to give another proof of Böttcher's Theorem. The idea of the new proof follows the viewpoint of holomorphic motions. For the classical proof of Böttcher's Theorem, the reader may refer to Böttcher's original paper [8] or most recent books [9,30]. Actually from the technical point of views, our proof is more complicate and uses a sophistical result. But from the conceptual point of views, our proof gives some inside mechanism of the normal form for a super-attracting fixed point. The idea of the proof is basically the same as that in the previous section, but the actual proof is little bit different. The reason is that in the previous case, f is a homeomorphism so we can iterate both forward and backward, but in Theorem 1.2or Böttcher's Theorem, g is not a homeomorphism.

Theorem 5.1. *[Böttcher's Theorem] Suppose $g(z) = \sum_{j=n}^{\infty} a_j z^j$, $a_n \neq 0$, $n \geq 2$, is analytic on a disk Δ_{δ_0}, $\delta_0 > 0$. Then there exists a conformal map $\phi : \Delta_\delta \to \phi(\Delta_\delta)$ for some $\delta > 0$ such that*

$$\phi^{-1} \circ g \circ \phi(z) = z^n, \quad z \in \Delta_\delta.$$

The conjugacy ϕ^{-1} is unique up to multiplication by $(n-1)^{th}$-roots of the unit.

Proof. [A new proof of Böttcher's Theorem] Conjugating by $z \to bz$, we can assume $a_n = 1$, i.e.,

$$g(z) = z^n + \sum_{j=n+1}^{\infty} a_j z^j.$$

We use $\Delta_r^* = \Delta_r \setminus \{0\}$ to mean a punctured disk of radius $r > 0$. Write

$$g(z) = z^n(1 + \sum_{j=1}^{\infty} a_{j+n} z^j).$$

Assume $0 < \delta_1 < \min\{1/2, \delta_0/2\}$ is small enough such that

$$1 + \sum_{j=1}^{\infty} a_{j+n} z^j \neq 0 \quad \text{and} \quad \frac{1}{\sqrt[n]{|1 + \sum_{j=1}^{\infty} a_{j+n} z^j|}} \geq \frac{1}{2}, \quad z \in \Delta_{2\delta_1}.$$

Then $g : \Delta_{2\delta_1}^* \to f(\Delta_{2\delta_1}^*)$ is a covering map of degree n.

Let $0 < \delta < \delta_1$ be a fixed number such that $g^{-1}(\Delta_\delta) \subset \Delta_{\delta_1}$. Since

$$z \to z^n : \Delta^*_{\sqrt[n]{\delta}} \to \Delta_\delta^* \quad \text{and} \quad g : g^{-1}(\Delta_\delta^*) \to \Delta_\delta^*$$

are both of covering maps of degree n, the identity map of Δ_δ can be lifted to a holomorphic diffeomorphism

$$h : \Delta^*_{\sqrt[n]{\delta}} \to g^{-1}(\Delta_\delta^*),$$

i.e., h is a map such that the diagram

$$\begin{array}{ccc} \Delta^*_{\sqrt[n]{\delta}} & \xrightarrow{h} & g^{-1}(\Delta_\delta^*) \\ \downarrow z \to z^n & & \downarrow g \\ \Delta_\delta^* & \xrightarrow{\text{id}} & \Delta_\delta^* \end{array}$$

commutes. We pick the lift so that

$$h(z) = z\Big(1 + \sum_{j=2}^{\infty} b_j z^{j-1}\Big) = z\psi(z).$$

From

$$g(h(z)) = z^n, \quad z \in \Delta^*_{\sqrt[n]{\delta}},$$

we get

$$|h(z)| = \frac{|z|}{\sqrt[n]{|1 + \sum_{j=1} a_{n+j}(h(z))^j|}} \geq \frac{|z|}{2}.$$

For any

$$0 < r \leq \max\Big\{\Big(\frac{1}{2}\Big)^{\frac{n}{(n-1)}}, \delta^n\Big\},$$

let $S_r = \{z \in \mathbb{C} \mid |z| = r\}$ and $T_r = \{z \in \mathbb{C} \mid |z| = \sqrt[n]{r}\}$. Consider the set $E = S_r \cup T_r$ and the map

$$\phi_r(z) = \begin{cases} z, & z \in S_r \\ z\psi(z), & z \in T_r. \end{cases}$$

Define

$$h_r(c, z) = \begin{cases} z, & z \in S_r \\ z\psi\Big(\frac{cz}{\sqrt[n]{r}}\Big), & z \in T_r \end{cases} : \Delta \times E \to \hat{\mathbb{C}}.$$

Note that

$$z\psi\Big(\frac{cz}{\sqrt[n]{r}}\Big) = \frac{\sqrt[n]{r}}{c} h\Big(\frac{cz}{\sqrt[n]{r}}\Big), \quad z \in T_r,\ c \neq 0.$$

This implies that

$$|h_r(c, z)| = \frac{\sqrt[n]{r}}{|c|} h\Big(\frac{cz}{\sqrt[n]{r}}\Big)\Big| \geq \frac{\sqrt[n]{r}}{|c|} \frac{|cz|}{2\sqrt[n]{r}} \geq \frac{\sqrt[n]{r}}{2} > r, \quad z \in T_r.$$

So images of S_r and T_r under $h_r(c, z)$ do not cross each other.

Now let us check $h_r(c, z)$ is a holomorphic motion. First $h_r(0, z) = z$ for $z \in E$. For fixed $x \in E$, $h_r(c, z)$ is holomorphic on $c \in \Delta$. For fixed $c \in \Delta$, $h_r(c, z)$ restricted to S_r and T_r, respectively, are injective. But the images of S_r and T_r under $h_r(c, z)$ do not cross each other. So $h_r(c, z)$ is injective on E. Thus

$$h_r(c, z) : \Delta \times E \to \hat{\mathbb{C}}$$

is a holomorphic motion. By Theorem 2.1, it can be extended to a holomorphic motion

$$H_r(c, z) : \Delta \times \hat{\mathbb{C}} \to \hat{\mathbb{C}}.$$

And moreover, for each $c \in \Delta$, $H_r(c, \cdot)$ is a quasiconformal map whose quasiconformal dilatation satisfies

$$K(H(c, \cdot)) \leq \frac{1 + |c|}{1 - |c|}.$$

Now consider $H(\sqrt[n]{r}, \cdot)$. It is a quasiconformal map with quasiconformal constant

$$K_r \leq \frac{1 + \sqrt[n]{r}}{1 - \sqrt[n]{r}}.$$

Let

$$A_{r,j} = \{z \in \mathbb{C} \mid \sqrt[n^j]{r} \leq |z| \leq \sqrt[n^{j+1}]{r}\}, \quad j = 0, 1, 2, \cdots.$$

Consider the restriction $\phi_{r,0} = H(\sqrt[n]{r}, \cdot)|A_{r,0}$. It is an extension of ϕ_r, i.e., $\phi_{r,0}|E = \phi_r$.

Let $\tilde{A}_{r,0}$ be the annulus bounded by S_r and $g^{-1}(S_r)$ and define $\tilde{A}_{r,j} = g^{-j}(\tilde{A}_{r,0})$, $j \geq 0$. Since $z \to z^n : A_{r,1} \to A_{r,0}$ and $g : \tilde{A}_{r,1} \to \tilde{A}_{r,0}$ are both covering maps of degree n, so $\phi_{r,0}$ can be lifted to a quasiconformal map $\phi_{r,1} : A_{r,1} \to \tilde{A}_{r,1}$, i.e., the following diagram

$$\begin{array}{ccc} A_{r,1} & \stackrel{\phi_{r,1}}{\longrightarrow} & \tilde{A}_{r,1} \\ \downarrow z \to z^n & & \downarrow g \\ A_{r,0} & \stackrel{\phi_{r,0}}{\longrightarrow} & \tilde{A}_{r,0} \end{array}$$

commutes. We pick the lift $\phi_{r,1}$ such that it agrees with $\phi_{r,0}$ on T_r. The quasiconformal dilatation of $\phi_{r,1}$ is less than or equal to K_r.

For an integer $k > 0$, take $r = r_k = \delta^{n^k}$. Inductively, we can define a sequence of K_r-quasiconformal maps $\{\phi_{r,j}\}_{j=0}^k$ such that

$$\begin{array}{ccc} A_{r,j} & \stackrel{\phi_{r,j}}{\longrightarrow} & \tilde{A}_{r,j} \\ \downarrow z \to z^n & & \downarrow g \\ A_{r,j-1} & \stackrel{\phi_{r,j-1}}{\longrightarrow} & \tilde{A}_{r,j-1} \end{array}$$

commutes and $\phi_{r,j}$ and $\phi_{r,j-1}$ agree on the common boundary of $A_{r,j}$ and $A_{r,j-1}$. Note that

$$\overline{\Delta}_\delta = \Delta_r \cup \cup_{j=0}^k A_{r,j}.$$

Now we can define a quasiconformal map, which we still denote by ϕ_r as follows.

$$\phi_r(z) = \begin{cases} z, & z \in \Delta_r; \\ \phi_{r,j}, & z \in A_{r,j},\ j = 0, 1, \cdots, k. \end{cases}$$

The quasiconformal dilatation of ϕ_r on Δ_δ is less than or equal to K_r and

$$g(\phi_r(z)) = \phi_r(z^n), \quad z \in \cup_{j=1}^k A_{r,j}.$$

Since $g(z) = z^n(1 + O(z))$, $g^k(z) = z^{n^k} \prod_{i=0}^{k-1}(1 + O(z^{n^i}))$. Because $\sqrt[n^k]{r_k} = \delta$, the range of ϕ_{r_k} on Δ_δ is a Jordan domain bounded above from ∞ and below from 0 uniformly in k. In addition, 0 is fixed by ϕ_k and the quasiconformal dilatations of the ϕ_k are uniformly bounded in k. Therefore, the sequence $\{\phi_{r_k}\}_{k=1}^\infty$ is a compact family (see [1]). Let ϕ be a limiting map of this family. Then we have

$$f(\phi(z)) = \phi(z^n), \quad z \in \Delta_\delta.$$

Since the quasiconformal dilatation of ϕ is less than or equal to $(1 + \sqrt[n]{r_k})/(1 - \sqrt[n]{r_k})$ for all $k > 0$, it follows that ϕ is a 1-quasiconformal map, and thus conformal. This is the proof of the existence.

Suppose ϕ_1 and ϕ_2 are two conjugacies such that

$$\phi_1^{-1} \circ g \circ \phi_1(z) = z^n \quad \text{and} \quad \phi_2^{-1} \circ g \circ \phi_2(z) = z^n, \quad z \in \Delta_\delta.$$

For

$$\Phi(z) = \phi_2^{-1} \circ \phi_1(z) = \sum_{j=1}^{\infty} a_j z^j,$$

we have $\Phi(z^n) = (\Phi(z))^n$. This implies $a_1^n = a_1$ and $a_j = 0$ for $j \geq 2$. Since $a_1 \neq 0$, we have $a_1^{n-1} = 1$ and $\phi_2^{-1} = a_1 \phi_1^{-1}$. This is the uniqueness. □

We now prove Theorem 1.2. The proof follows almost the same footsteps of those of Theorem 1.1 and Theorem 5.1.

Proof. [Proof of Theorem 1.2.] Let $g = f \circ q_n$, $n \geq 2$. Conjugating by $z \to bz$, we can assume $f'(0) = \lim_{|z| \to 0} f(z)/z = 1$.

We use $\Delta_r^* = \Delta_r \setminus \{0\}$ to mean a punctured disk of radius $r > 0$. There is a $0 < \delta_1 < 1$ such that $g : \Delta_{2\delta_1}^* \to f(\Delta_{2\delta_1}^*)$ is a covering map of degree n.

Let $0 < \delta < \delta_1$ be a fixed number such that $g^{-1}(\Delta_\delta) \subset \Delta_{\delta_1}$. Since

$$z \to z^n : \Delta_{\sqrt[n]{\delta}}^* \to \Delta_\delta^* \quad \text{and} \quad g : g^{-1}(\Delta_\delta^*) \to \Delta_\delta^*$$

are both of covering maps of degree n, the identity map of Δ_δ can be lifted to a homeomorphism

$$h : \Delta_{\sqrt[n]{\delta}}^* \to g^{-1}(\Delta_\delta^*).$$

Furthermore, h is a quasiconformal map α-asymptotically conformal at 0 such that the diagram

$$\begin{array}{ccc} \Delta_{\sqrt[n]{\delta}}^* & \xrightarrow{h} & g^{-1}(\Delta_\delta^*) \\ \downarrow z \to z^n & & \downarrow g \\ \Delta_\delta^* & \xrightarrow{\text{id}} & \Delta_\delta^* \end{array}$$

commutes. We pick the lift so that

$$h'(z) = \lim_{z \to 0} \frac{h(z)}{z} = 1.$$

These can be seen from the equation

$$g(h(z)) = z^n, \quad z \in \Delta^*_{\sqrt[n]{\delta}}.$$

For any $0 < r \leq \delta$, let $S_r = \{z \in \mathbb{C} \mid |z| = r\}$ and $T_r = \{z \in \mathbb{C} \mid |z| = \sqrt[n]{r}\}$. Consider the set $E = S_r \cup T_r$ and the map

$$\phi_r(z) = \begin{cases} z, & z \in S_r \\ h(z), & z \in T_r. \end{cases}$$

It is clear that

$$g(\phi_r(z)) = \phi_r(z^n)$$

for $z \in S_r$.

Extend h to $\hat{\mathbb{C}}$ by quasiconformal reflection with respect to S_r and $\phi(S_r)$ (see [1]). We still denote this extended map as ϕ. Let $\nu = \phi_{\overline{z}}/\phi_z$ be the complex dilatation of the extended ϕ. Then

$$a(r) = \|\nu\|_\infty = O(\|\mu|\Delta_{\sqrt[n]{r}}\|_\infty) = O(\omega(\sqrt[n]{r})).$$

Assume $h(\sqrt[n]{r}) = \tau_r$. Consider $\nu_c = ca_0a(r)^{-1}\nu$ and the unique solution $\phi_c = \phi^{\nu_c}$ that maps 0, r, and ∞ to 0, τ_r, and ∞, respectively. Here a_0 is a constant independent of r such that $|\phi_c(z)| < r$ for all $|z| \leq \lambda r$ and $|c| < 1$. (Since ϕ_c can be written as a power series in ν_c and $\|\nu_c\| \to 0$ uniformly as $r \to 0$, such an a_0 exists.) Then ϕ_c holomorphically depends on $c \in \Delta$. Define

$$\phi_r(c, z) = \begin{cases} z & z \in S_r \\ \phi_c(z), & z \in T_r. \end{cases}$$

It is a holomorphic motion from $\Delta \times E \to \hat{\mathbb{C}}$. From Theorem 2.1, $\phi(c, z)$ can be extended to a holomorphic motion from $\Delta \times \hat{\mathbb{C}} \to \hat{\mathbb{C}}$, which we still denote by $\phi(c, z)$, such that the quasiconformal dilatation of $\phi(c, \cdot)$ is less than or equal to $(1 + |c|)/(1 - |c|)$. In particular when $c_r = a_0^{-1}a(r)$, $\phi_r(c_r, z)|E = \phi_r$.

Let

$$A_{r,j} = \{z \in \mathbb{C} \mid \sqrt[n^j]{r} \leq |z| \leq \sqrt[n^{j+1}]{r}\}.$$

We still use ϕ_r to denote $\phi_r(c_r, \cdot)|A_{r,0}$. For an integer $k > 0$, take $r = r_k = \sqrt[n^j]{\delta}$. Then

$$\Delta_\delta = \cup_{j=-k}^{\infty} A_{r,j} \cup \{0\}.$$

Extend ϕ_r to Δ_δ by lifting which we still denote as ϕ_r (refer to the proof of Theorem bth). Formally we can use the following formula to define ϕ_r,

$$\phi_r(z) = g^{-j}(\phi_r(z^{n^j})), \quad z \in A_{r,j}, \quad j = -k, \cdots, -1, 0, 1, \cdots,$$

and $\phi_r(0) = 0$. Since $\phi_r|E$ is a conjugacy from g to $q_n(z) = z^n$, ϕ_r is continuous on Δ_δ.

Using the similar argument to that in the proof of Theorem 1.1, we can get some $0 < \sigma < 1$ such that the complex dilatation μ_{ϕ_r} over Δ_δ can be controlled by

$$\|\mu_{\phi_r}\|_\infty \leq \sum_{j=1}^{\infty} \sigma^{-\alpha j} \omega(\sqrt[n^j]{\delta}) + b(r) \leq \tilde{\omega}(\delta) + b(r)$$

and μ_{ϕ_r} over Δ_r can be controlled by

$$\|\mu_{\phi_r}|\Delta_r\|_\infty \leq \sum_{j=1}^{\infty} \sigma^{-\alpha j} \omega(\sqrt[n^j]{\delta}) + b(r) \leq \tilde{\omega}(r) + b(r).$$

Thus $\{\phi_r\}_{0<r\leq\delta}$ is uniformly K-quasiconformal for a fixed number $K > 1$. Consider $B_r = \Delta_\delta \setminus \Delta_r = \cup_{j=0}^{-k} A_{r,j}$ and $\phi_r(B_r) = \cup_{j=0}^{-k} \phi_r(A_{r,j})$. Both of the annulli have the same inner circle S_r. Thus the ratio of the modulus of $\phi_r(B_r)$ and the modulus of B_r is controlled by two constants from below and above (independent of r but only depends on K). Therefore, the range of ϕ_r on Δ_δ is a Jordan domain bounded above from ∞ and below from 0 uniformly in $0 < r \leq \delta$. Since, additionally, 0 is fixed by any element in this sequence the sequence $\{\phi_{r_k}\}_{k=1}^{\infty}$ is a compact family (see [1]). Let ϕ be a limit mapping of this family. Then we have

$$g(\phi(z)) = \phi(z^n), \quad z \in \Delta_\delta.$$

Since the complex dilatation of ϕ is controlled by $b(r_k) = \tilde{\omega}(r_k) + b(r_k)$, ϕ is asymptotically conformal at 0 and the proof is completed.

Suppose ϕ_1 and ϕ_2 are two asymptotically conformal conjugacies such that

$$\phi_1^{-1} \circ g \circ \phi_1(z) = z^n \quad \text{and} \quad \phi_2^{-1} \circ g \circ \phi_2(z) = z^n, \quad z \in \Delta_\delta.$$

Then for $\Phi = \phi_2^{-1} \circ \phi_1$, we have $\Phi(z^n) = (\Phi(z))^n$. This implies that the complex dilatation $\|\mu_\Phi(z)\| = \|\mu_\Phi(z^n)\|$, a.e. This in turn implies that $\mu = 0$ a.e. in Δ_δ and thus Φ is conformal, and therefore, $\Phi(z) = az$ with $a^n = 1$. This is the uniqueness. □

Acknowledgment. The previous version of this paper [18] was first discussed in the complex analysis and dynamical systems seminar at the CUNY Graduate Center in the Spring Semester of 2003. I would like to thank all participants for their patience. During the discussion Professors Linda Keen, Fred Gardiner, and Nikola Lakic provided many useful comments to improve and polish this article. Sudeb Mitra explained to me several points in the development of the measurable Riemann mapping theorem and holomorphic motions. I express my sincere thanks to them. I would like also to thank Professor Weiyuan Qiu to help me to clarify several arguments in this paper.

References

1. L. V. Ahlfors, *Lectures on Quasiconformal Mappings*, Van Nostrand Mathematical studies, **10**, D. Van Nostrand Co. Inc., Toronto-New York-London, 1966.
2. L. V. Ahlfors and L. Bers, Riemann's mapping theorem for variable metrics, *Annals of Math. (2)*, **72** (1960): 385-404.
3. K. Astala, Planar quasiconformal mappings; deformations and interactions, *In Quasiconformal Mappings and Analysis-A collection of papers honoring F. W. Gehring*, 1998, Springer-Verlag, New York, Inc., 33-54.
4. K. Astala and G. J. Martin, Holomorphic motions, *Papers on Analysis*, **27-40**, Rep. Univ. Jyväsklä Dep. Math. Stat., **83**, Univ. Jyväskylä, 2001.
5. L. Bers, On a theorem of Mori and the definition of quasiconformality, *Trans. Amer. Math. Soc.*, **84** (1957): 78-84.
6. L. Bers and H. L. Royden, Holomorphic families of injections, *Acta Math*, **157** (1986): 259-286.
7. V. B. Bojarski, Generalized solutions of a system of differential equations of the first order and elliptic type with discontinuous coefficients, *Math. Sbornik*, **85** (1957): 451-503.
8. L. E. Böttcher, The principal laws of convergence of iterates and their aplication to analysis (Russian), *Izv. Kazan. Fiz.-Mat. Obshch.* **14**) (1904):155-234.
9. L. Carleson and T. Gamelin, *Complex Dynamics*, Universitext: Tracts in Mathematics, Springer-Verlag, New York, 1993
10. S. K. Donaldson and D. P. Sullivan, Quasiconformal -manifolds, *Acta Math.* **163** (1989):181-252.
11. A. Douady, Prolongement de mouvements holomorphes [d'après Slodkowski et autres], *Séminarie N. Bourbaki* (1993/1994), *Astérique* No. **227** (1995) & *Exp.* No. **755** (3): 7-20.
12. P Duren, J. Heinonen, B. Osgood, and B. Palka, *Quasiconformal Mappings and Analysis*, Springer, 1997
13. C. J. Earle, I. Kra, and S. L. Krushkal, Holomorphic motions and Teichmüller spaces, *Trans. Amer. Math. Soc.*, **343** (2) (1994):927-948.
14. C. J. Earle and S. Mitra, Variation of moduli under holomorphic motions, *Comtemp. Math.*, **256** (2000): 39-67.

15. F. Gardiner and L. Keen, Holomorphic motions, *Lipman Bers, Selected Works Part I and II*, Edited by I. Kra and B. Maskit, AMS, 1999.
16. F. Gardiner and N. Lakic, *Quasiconformal Teichmüller Theory*, Mathematicak Survey and Monographs, American Mathematical Society, Rhode Island, **76** (2000).
17. V. Ya. Gulyanskiĭ, O. Martio, V. I. Ryazanov, and M. Vuorinen, On the asymptotic behavior of quasiconformal mappings in space in Quasiconformal Mappings and Analysis (edited by P Duren, J. Heinonen, B. Osgood, and B. Palka), Springer, 1997: 159-180.
18. Y. Jiang, Holomorphic Motions and Normal Forms in Complex Analysis, *Proceesings of ICCM2004*, to appear.
19. G. Königs, Recherches sur les integrals de certains equations fonctionelles, *Ann. Sci. Éc. Norm. Sup.*, (3e ser.) **1** (1884): supplém. 1-41.
20. O. Lehto, Univalent Functions and Teichmüller Spaces, *Springer-Verlag*, New York, Berlin, 1987.
21. C. McMullen and D. Sullivan, Quasiconformal Homeomorphisms and Dynamics III: The Teichmüller space of a holomorphic dynamical systems, *Adv. Math.*, **135** (1998):351-395.
22. Ch. Pommerenke, *Boundary Behaviour of Conformal Maps*, Springer-Verlag, 1992.
23. Yu. G. Reshetnyak, Stability theorems in geometry and analysis, *Nauka, Sibirskoe otdelenie*, Novosibirsk, Russian, 1982.
24. M. A. Lavrent'ev, Sur une classe de représentation continues, *Mat. Sbnornik* **42**(1935):407-423.
25. M. A. Lavrent'ev, A fundamental theorem of the theory of quasi-conformal mapping of plane regions. *Izvestya Akademii Nauk S.S.S.R.* **12** (1948):513-554. (Russian).
26. Z. Li, Quasiconformal mappings and applications in Riemann surfaces, *Science publciation of China*, 1988.
27. G. Lieb, Holomorphic motions and Teichmüller space, *Ph.D. dissertation, Cornell University*, 1990.
28. R. Mañe, P. Sad, and D. Sullivan, On the dynamics of rational maps, *Ann. Éc. Norm. Sup.*, **96** (1983): 193-217.
29. W. de Melo and S. van Strien *One-Dimensional Dynamics*, Springer-Verlag, Berlin, New York, 1993.
30. J. Milnor, *Dynamics in One Complex Variable*, Introductory Lectures, Vieweg, 2nd Edition, 2000.
31. S. Mitra, Techmüller spaces and holomorphic motions, *Journal d'Analyse Mathématique*, **81** (2000): 1-33.
32. C. B. Morrey, On the solutions of quasi-linear elliptic partial differential equations, *Trans. Amer. Math. Math. Soc.*, **43** (1938): 126-166.
33. S. Nag, *The Complex Analytic Theory of Teichmüller Spaces*, John Wiley and Sons, New York, 1988.
34. Z. Slodkowski, Holomorphic motions and polynomial hulls, *Proc. Amer. Math. Soc.*, **111** (1991): 347-355.
35. D. Sullivan and W. Thurston, Extending holomorphic motions, *Acta Math.*, **157** (1986):243-257.

THE HYPER-ORDER OF SOLUTIONS OF CERTAIN HIGH ORDER DIFFERENTIAL EQUATIONS*

CHUNHONG LI

College of Mathematics and Physics
Chongqing University
Chongqing, 400044 P. R. China
and
College of Math. and Information
China West Normal University
Nanchong, 637002 P. R. China
E-mail: chli@cqu.edu.cn

YONGXING GU

College of Mathematics and Physics
Chongqing University
Chongqing, 400044 P. R. China
E-mail: yxgu@cqu.edu.cn

This present paper investigate the growth problem of solution of certain high-order linear differential equations and obtain precise estimates of hyper-order of their solutions with infinite order.

Keywords: Linear differential equations, hyper-order.

2000 Mathematics Subject Classification: 30D35.

In this paper, we will use the standard notations of the Nevanlinna theory(see e.g., [1,2,3]). In addition, we use $\sigma(f), \sigma_2(f)$ to denote respectively the order of growth of $f(z)$.

*Foundation Term: NNSF of China No. 10271122; NNSF of the Education Department of Sichuan No. 2004a104.

J. K. Langley[4], G. Gundersen[5], Z. X. Chen[6] etc. concerned the problem of growth of solutions of the second order differential equation

$$f^{''} + e^{-z}f' + Q(z)f = 0. \tag{1}$$

They proved the following [4,5,6]

Theorem A *Let $Q(z)$ be nonconstant entire function and satisfy one of the following conditions*

(i) $Q(z)$ be nonconstant polynomial;

(ii) $\sigma(Q) < 1$ or$\sigma(Q) > 1$;

(iii) $Q(z) = h(z)e^{-dz}$ where $h(z)(\not\equiv 0)$ be entire function and $\sigma(h) < 1, d(\neq 1)$ be a nonzero complex. Then all solution $f(z)(\not\equiv 0)$ of (0.1) have infinite order.

It is a very important aspect in the complex oscillation theory to more precise estimates for the rate of growth of infinite order solutions of linear differential equations in the complex domain[6]. Z. X. Chen proved the following

Theorem B *Let a be nonconstant polynomial, or $Q(z) = h(z)e^{-dz}$, where $h(z)(\not\equiv 0)$ be nonzero polynomial. Then all solution $f(z)(\not\equiv 0)$ of (0.1) have infinite order and $\sigma_2(f) = 1$ where the hyper-order $\sigma_2(f)$ of $f(z)$ is defined by*[7]

$$\sigma_2(f) = \limsup_{r\to\infty} \frac{loglogT(r,f)}{logr}.$$

A nature problem is what qualities about the growth of solution of the high order differential equation corresponding to (0.1)? C. H. Li[8] and X. J. Huang[8] researched the problem and proved that

Theorem C *Let $A_j(z)(\not\equiv 0)$ be entire functions with $\sigma(A_j) < 1(j = 0, 1, \ldots, k-1; k \geq 2)$, $a_j \in C \setminus \{0\}(j = 0, 1, \ldots, k-1)$ with $a_j = c_j a_0$ $(j = 1, \ldots, k-1)$, and that if $c_{k-1} > c = \max_{1\leq j\leq k-2} c_j$. Then all solutions $f(z)(\not\equiv 0)$of the equation*

$$f^{(k)} + A_{k-1}(z)e^{a_{k-1}z}f^{(k-1)} + \cdots + A_0(z)e^{a_0 z}f = 0 \tag{2}$$

have infinite order.

In this present paper, we consider the hyper-order of infinite order solutions of linear differential equation (0.2) $A_j(z)$ $(j = 0, 1, \ldots, k-1)$ with being nonzero polynomials.

Theorem 1 *Let $A_j(z)(j = 0, 1, \ldots, k-1)$ be nonzero polynomials, $a_j \in C \setminus \{0\}(j = 0, 1, \ldots, k-1), c_j > 1(j = 1, \ldots, k-1)$, and that there exists $m \in \{1, 2, \ldots, k-1\}$, such that $c_m > c = \max_{1 \le j \le k-1, j \ne m} c_j$. Then all solutions $f(z)(\not\equiv 0)$ of the equation (0.2) have infinite order and $\sigma_2(f) = 1$, except at most two exceptional complex a_0.*

Note 1. The condition "$c_j > 1(j = 1, \ldots, k-1)$" in Theorem 1 is necessary.

Example 1. The differential equation

$$f^{(4)} + z^2 e^{2z} f''' + e^{3z} f'' + z e^z f' - e^z f = 0$$

has a solution $f(z) = z$ with nonzero and finite order.

1. Lemmas

Lemma 1[9] *Let $f(z)$ be an entire function of infinite order, with the hyper-order $\sigma_2(f) = \sigma < \infty$, and let $\nu(r)$ be the central index of $f(z)$. Then*

$$\limsup_{r \to \infty} \frac{\log \log \nu(r)}{logr} = \sigma_2(f) = \sigma.$$

Lemma 2[9] *Let $f(z)$ be an entire function with $\sigma(f) = \sigma < +\infty$, then for any $\varepsilon > 0$, there exists a set $E_0 \subset (1, +\infty)$ with finite linear measure $mE_0 < +\infty$, and finite logarithmic measure $lmE_0 < +\infty$, such that for z satisfying $|z| = r \notin [0, 1] \cup E_0$,*

$$|f(z)| \le \exp\{r^{\sigma+\varepsilon}\}.$$

Lemma 3 *Let $B_j(z)(j = 0, 1, \ldots, k-1)$ be entire functions with finite order. If $f(z)$ is any solution of infinite order of the equation*

$$f^{(k)} + B_{k-1}(z) f^{(k-1)} + \cdots + B_0(z) f = 0 \tag{3}$$

then $\sigma_2(f) \le \max_{0 \le j \le k-1} \sigma(B_j)$.

Proof. Set $\max_{0 \le j \le k-1} \sigma(B_j) = \sigma$, by Lemma 2, for any $\varepsilon > 0$, for z satisfying $|z| = r \notin [0, 1] \cup E_0$, we have

$$|B_j(z)| \le e^{r^{\sigma+\varepsilon}} (j = 0, 1, \ldots, k-1). \tag{4}$$

By Wiman-Valiron Theory, there exists a set $E_1 \subset (1, +\infty)$ having finite logarithmic measure $lmE_1 < +\infty$, we can choose $|z| = r \notin [0, 1] \bigcup E_1$ and $|f(z)| = M(r, f)$, we get

$$\frac{f^{(j)}}{f} = (\frac{\nu(r)}{z})^j (1 + o(1))(j = 1, 2, \ldots, k) \tag{5}$$

where $\nu(r)$ is the central index of $f(z)$. From (1.3), we have

$$-\frac{f^{(k)}}{f} = B_{k-1}\frac{f^{(k-1)}}{f} + \cdots + B_1\frac{f'}{f} + B_0. \tag{6}$$

Substituting (1.4)(1.5) into (1.6), for sufficiently large $r(|z| = r \notin [0,1] \bigcup E_0 \bigcup E_1$ and $|f(z)| = M(r,f))$, we have

$$(\frac{\nu(r)}{|z|})^k(1+o(1)) \leq exp\{r^{\sigma+\varepsilon}\}\{(\frac{\nu(r)}{|z|})^{k-1}(1+o(1))+\cdots+(\frac{\nu(r)}{|z|})(1+o(1))+1\}$$

We can further get

$$(\frac{\nu(r)}{|z|})^k \leq 4exp\{r^{\sigma+\varepsilon}\}\{(\frac{\nu(r)}{|z|})^{k-1} + \cdots + (\frac{\nu(r)}{|z|}) + 1\}$$

$$= 4exp\{r^{\sigma+\varepsilon}\}\frac{(\frac{\nu(r)}{|z|})^k - 1}{\frac{\nu(r)}{|z|}) - 1},$$

(notice that $\sigma(f) = \infty$ and let $r \gg 1$), i.e.

$$\nu(r) \leq 5exp\{r^{\sigma+\varepsilon}\}(r \gg 1)$$

i.e.

$$\limsup_{r\notin[0,1]\cup E_0 \bigcup E_1, r\to\infty} \frac{loglog\nu(r)}{logr} \leq \sigma + \varepsilon.$$

Since ε is arbitrary, by Lemma 1.5 in [3] and Lemma 1, we have $\sigma_2(f) \leq \sigma$.

Lemma 4[6] *Suppose that* $p(z) = (\alpha + i\beta)z^n + \cdots (\alpha, \beta \in R, |\alpha| + |\beta| \neq 0)$ *is polynomial with* $degp(z) = n \geq 1$, $A(z)(\not\equiv 0)$ *is an entire function with* $\sigma(A) < n$, $g(z) = A(z)exp\{p(z)\}$, $z = re^{i\theta}$, $\delta(p,\theta) = \alpha\cos n\theta - \beta\sin n\theta$. *Then for any* $\varepsilon > 0$, *there exists* $E_2 \subset (1, +\infty)$ *with* $lmE_2 < +\infty$,*such that for any* $\theta \in [0, 2\pi] \setminus H_1(H_1 = \{\theta \in [0, 2\pi] : \delta(p,\theta) = 0\}$ *is an finite set), and* $|z| = r \notin [0,1] \cup E_2$, *the following are true:*

(i) If $\delta(p,\theta) > 0$, *we have*

$$\exp\{(1-\varepsilon)\delta(p,\theta)r^n\} < |g(re^{i\theta})| < \exp\{(1+\varepsilon)\delta(p,\theta)r^n\};$$

(ii) If $\delta(p,\theta) < 0$, *we have*

$$\exp\{(1+\varepsilon)\delta(p,\theta)r^n\} < |g(re^{i\theta})| < \exp\{(1-\varepsilon)\delta(p,\theta)r^n\}.$$

Lemma 5[6] *Let $f(z)$ be an entire function with $\sigma(f) = +\infty, \sigma_2(f) = \sigma < +\infty$, and a set $E_3 \subset (1, +\infty)$ with $lmE_3 < +\infty$. Then there exists $z_l = r_l e^{i\theta_l} (l = 1, 2, \ldots)$, with $|f(z_l)| = M(r_l, f)$, $\theta_l \in [0, 2\pi), \lim_{l\to+\infty} = \theta_0 \in [0, 2\pi)$, $r_l \notin E_3, r_l \to +\infty$, for $\forall\varepsilon > 0$,and for sufficiently large r_l, we have*

$$\exp\{r_l^{\sigma-\varepsilon}\} < \nu(r_l) < \exp\{r_l^{\sigma+\varepsilon}\}, \limsup_{l\to+\infty} \frac{\log \nu(r_l)}{\log r_l} = +\infty.$$

2. Proof of Theorem 1

By Theorem C, Lemma 3 and combining the conditions of Theorem 1, for all non-trivial solutions $f(z)$ of the equation (0.2), we have $\sigma(f) = \infty, \sigma_2(f) < 1$. We assure that $\sigma_2(f) = 1$, otherwise, suppose that $\sigma_2(f) = \alpha \in [0, 1)$.

By Wiman-Valiron Theory, we have

$$\frac{f^{(j)}(z)}{f(z)} = (\frac{\nu(r)}{|z|})^j (1 + o(1)), j = 1, 2, \ldots, k. \tag{7}$$

where $|z| = r \notin [0, 1] \bigcup E_1$, $lmE_1 < +\infty$, $|f(z)| = M(r, f)$, $\nu(r)$ is the central index of $f(z)$.

By Lemma 5, there exists $z_l = r_l e^{i\theta_l} (l = 1, 2, \ldots)$, with $|f(z_l)| = M(r_l, f)$, $\theta_l \in [0, 2\pi)$, $\lim_{l\to+\infty} = \theta_0 \in [0, 2\pi)$, $r_l \notin [0, 1] \bigcup E_0 \bigcup E_1 \bigcup E_2 \bigcup E_3$, $r_l \to +\infty$, for $\forall\varepsilon > 0$,and for sufficiently large r_l, we have

$$\limsup_{l\to+\infty} \frac{\log \nu(r_l)}{\log r_l} = +\infty, \tag{8}$$

$$\exp\{r_l^{\sigma-\varepsilon}\} < \nu(r_l) < \exp\{r_l^{\sigma+\varepsilon}\}. \tag{9}$$

For above θ_0 , by the assumptions of Theorem 1, we may let $a = r_0 e^{i\phi}$ such that $\delta(a_0 z, \theta_0) = r_0 \cos(\phi + \theta_0) = \delta \neq 0$. Now we divide into the following two cases to prove: case(i)$\delta < 0$, case(ii)$\delta > 0$.

Case (i) $\delta < 0$.

By $\lim_{l\to+\infty} = \theta_0$, for sufficiently large l, we have $\delta(a_0 z, \theta_l) = \delta_l < 0$, notice that $a_j = c_j a_0 (j = 1, \ldots, k-1)$, we can get $\delta(a_j z, \theta_l) = c_j \delta(a_0 z, \theta_l) = c_j \delta_l < 0$, $\delta(-a_0 z, \theta_l) = -\delta_l > 0$,$\delta(a_j z - a_0 z, \theta_l) = \delta(a_0(c_j - 1)z, \theta_l) = (c_j - 1)\delta_l < 0$. From (0.2), we have

$$-e^{-a_0 z} \frac{f^{(k)}}{f} = A_{k-1}(z) e^{a_0(c_{k-1}-1)z} \frac{f^{(k-1)}}{f} + \cdots + A_1(z) e^{a_0(c_1-1)z} \frac{f'}{f} + A_0(z). \tag{10}$$

Substituting (1.7)(1.9) into (1.10), and by Lemma 4, for any given $\varepsilon \in (0, \frac{1-\alpha}{2})$, and for sufficiently large l , we obtain

$$\exp\{(1-\varepsilon)(-\delta_l)r_l\}\exp\{kr_l^{\sigma-\varepsilon}\}r_l^{-k}(1+o(1)) \le |-e^{-a_0 z}\frac{f^{(k)}}{f}|$$

$$\le \exp\{(1-\varepsilon)(c_{k-1}-1)\delta_l r_l\}\exp\{(k-1)r_l^{\sigma+\varepsilon}\}r_l^{-k+1}(1+o(1))$$

$$+\exp\{(1-\varepsilon)(c_{k-2}-1)\delta_l r_l\}\exp\{(k-2)r_l^{\sigma+\varepsilon}\}r_l^{-k+2}(1+o(1))$$

$$+\exp\{(1-\varepsilon)(c_1-1)\delta_l r_l\}\exp\{r_l^{\sigma+\varepsilon}\}r_l^{-1}(1+o(1)) + r^{n+1} \qquad (11)$$

where $n = \max_{0\le j\le k-1} degA_j(z)$. Since $\delta_l < 0$, $0 \le \sigma < 1$, $c_j > 1 (j = 1, \ldots, k-1)$ $\varepsilon \in (0, \frac{1-\alpha)}{2})$, thus (1.11) implies a contradiction.

Case (ii) $\delta > 0$.

By $\lim_{l\to+\infty} = \theta_0$, for sufficiently large l, we have $\delta(a_0 z, \theta_l) = \delta_l > 0$, notice that $a_j = c_j a_0 (j = 1, \ldots, k-1)$, $c_m > c := \max_{1\le j\le k-1, j\ne m} c_j$. From (0.2), we get

$$-A_m(z)e^{a_m z}\frac{f^{(m)}}{f} = \frac{f^{(k)}}{f} + \cdots + A_{m-1}(z)e^{a_{m-1}z}\frac{f^{(m-1)}}{f}$$

$$+ A_{m+1}(z)e^{a_{m+1}z}\frac{f^{(m+1)}}{f} + \cdots + A_0(z)e^{a_0 z}. \qquad (12)$$

Substituting (1.7)(1.9) into (1.12), and by Lemma 4, we get

$$\exp\{(1-\varepsilon)c_m\delta_l r_l\}\exp\{mr_l^{\sigma-\varepsilon}\}r_l^{-m}(1+o(1)) \le |-A_m(z)e^{a_m z}\frac{f^{(m)}}{f}|$$

$$\le \exp\{kr_l^{\sigma+\varepsilon}\}r_l^{-k}(1+o(1)) + \cdots + \exp\{(1+\varepsilon)c_{m-1}\delta_l r_l\}$$

$$\times \exp\{(m-1)r_l^{\sigma+\varepsilon}\}r_l^{-m+1}(1+o(1)) + \exp\{(1+\varepsilon)c_{m+1}\delta_l r_l\}$$

$$\times \exp\{(m+1)r_l^{\sigma+\varepsilon}\}r_l^{-m-1}(1+o(1))\exp\{(1+\varepsilon)\delta_l r_l\}. \qquad (13)$$

Taking sufficiently small such that

$$0 < 2\varepsilon < \min\{1-\sigma, \frac{c_m - c}{c_m + c}\}, \qquad (14)$$

and noticing that $0 \le \sigma < 1$, by (1.13) and (1.14), for sufficiently large l, we have

$$\exp\{\frac{(c_m - c)\delta_l r_l}{2}\} \le 4(k-1)\exp\{kr_l^{\sigma+\varepsilon}\}r_l^m.$$

This is impossible and the Theorem 1 follows.

References

1. W. Hayman, *Meromorphic Function,* Oxford: Clarendon Press, 1964.
2. L. Yang, *Value Distribution Theory,* Springer-Verlag, Berlin, 1993.
3. S.A. Gao, Z.X.Chen and T.W.Chen, *The Complex Oscillation Theory of Linear Differential Equations(in Chinese),* Huazhong University of Science and Techology Press, Wuhan, 1998.
4. J. K. Langley, On Complex Oscillation and a Problem of Ozawa, *Kodai Math.J.,* **9**(1986):430-439.
5. G. Gundersen, On the Question of Whether $f'' + e^{-z}f' + B(z)f = 0$ Can Admit a Solution $f(\not\equiv 0)$ of Finite order, *Proc. R. S. E.,* **102A**(1986):9-17.
6. Z.X.Chen, The Growth of Solutions of Differential Equations $f'' + e^{-z}f' + Q(z)f = 0$ (in Chinese), *Science in China(ser. A),* **31**(9)(2001):775-784.
7. H.X.Yi, C.C.Yang, *The Uniqueness Theory of Meromorphic Functions (in Chinese),* Science Press, Beijing, 1998.
8. C.H. Li, X.J.Huang , The Growth of Solution of a Class of High Order Differential Equations, *Acta Mathematica Scientia,* **5**(2003):613-618.
9. Z.X.Chen, C. C. Yang, Some Further Results on the Zeros and Growths of Entire Solutions of Second Order Linear Differential Equations[J], *Kodai Math. J.* **22**(1999): 273-285.

SHARING VALUES AND NORMAL FAMILIES

JIANG-TAO LI

Department of Mathematics
Chongqing University
Chongqing, 400044 P. R. China
and
Department of Mathematics
Sandong University
Jinan, Sandong 250100 P. R. China
E-mail: ljt@sdu.edu.cn

Let $\mathcal{F}$ be a family of holomorphic functions in a domain D, let k be a positive integer, and let $a(z)$, $b(z)(\neq 0)$ and $c(z)(\neq 0)$ be holomorphic functions in D. If, for each $f \in \mathcal{F}$ and a complex number d, all zeros of $f - d$ have multiplicity at least k, $L[f] = a$ whenever$f = 0$, $f = c$ whenever $L[f] = b$, where $L[f]$ denotes linear differential polynomial in f with coefficients holomorphic in D, then $\mathcal{F}$ is normal in D. This result extends and improves some normality criterions due to Miranda, Chuang and other authors. Some examples are provided to show that our results is sharp.

Keywords: Holomorphic function, normal families, shared values.

2000 Mathematics Subject Classification: 30D35, 30D45

1. Introduction and results

We denote by $\mathbb{C}$ the whole complex plane. Let f and g be two meromorphic functions in a domain $D \subset \mathbb{C}$, and let a and b be complex numbers. If $g(z) = b$ when $f(z) = a$, we write $f(z) = a \Rightarrow g(z) = b$. If $f(z) = a \Rightarrow g(z) = a$ and $g(z) = a \Rightarrow g(z) = a$, we say that f and g share a in D. If f and g share a in $\mathbb{C}$, then we say that f and g share a.

Let h be a meromorphic function in $\mathbb{C}$. h is called a normal function if there exists a positive M such that $h^{\#}(z) \leq M$ for all $z \in \mathbb{C}$, where

$$h^{\#}(z) = \frac{|h'(z)|}{1 + |h(z)|^2}$$

denotes the spherical derivative of h.

Let $\mathcal{F}$ be a family of meromorphic functions in a domain $D \subset \mathbb{C}$. We say that $\mathcal{F}$ is normal in D if every sequence $\{f_n\} \subseteq \mathcal{F}$ contains a subsequence which converges spherically uniformly on compact subsets of D, see [9] or [17].

In 1912, Montel [13] proved

Theorem A (Fundamental Normality Test). *Let $\mathcal{F}$ be a family of holomorphic functions in a domain D. If, for each $f \in \mathcal{F}$, $f \neq 0$, and $f \neq 1$, then $\mathcal{F}$ is normal in D.*

In 1935, Miranda [12] proved

Theorem B. *Let $\mathcal{F}$ be a family of holomorphic functions in a domain D, and let k be a positive integer. If, for each $f \in \mathcal{F}$, $f \neq 0$, and $f^{(k)} \neq 1$, then $\mathcal{F}$ is normal in D.*

Chuang [7] (cf. [17, Theorem 4.4.14]) generalized Theorem B as following

Theorem C. *Let $\mathcal{F}$ be a family of holomorphic functions in a domain D. Set*

$$L[f] = f^{(k)} + \sum_{j=1}^{k} a_{k-j} f^{(k-j)}, \quad f \in \mathcal{F} \tag{1}$$

where k be a positive integer and $a_i(z)$ $(i = 0, 1, \cdots, k-1)$ be holomorphic functions in D. If, for each $f \in \mathcal{F}$, $f \neq 0$, and $L[f] \neq 1$, then $\mathcal{F}$ is normal in D.

In this paper, we prove the following result which is further generalization of the above theorems.

Theorem 1. *Let $\mathcal{F}$ be a family of holomorphic functions in a domain D and $L[f]$ be a differential polynomial of the form (1.1). Let $a(z)$, $b(z)(\neq 0)$ and $c(z)(\neq 0)$ be holomorphic functions in D. If, for each $f \in \mathcal{F}$ and a complex number d, all zeros of $f - d$ have multiplicity at least k, $f = 0 \Rightarrow L[f] = a$, and $L[f] = b \Rightarrow f = c$, then $\mathcal{F}$ is normal in D.*

Remark 1. *When* $\{z \in D : f(z) = 0\} = \emptyset, \{z \in D : L[f](z) - b(z) = 0\} = \emptyset$*, Theorem 1 specializes to Chuang's results, for* $d = 0$*.*

Example 1. Theorem 1 does not hold if the requirement that $\mathcal{F}$ is a family of holomorphic functions is replaced by that $\mathcal{F}$ is a family of meromorphic functions. Indeed, Let

$$\mathcal{F} = \{\frac{n + (nz-1)^2}{n(nz-1)} : n = 1, 2, 3, \cdots\}$$

and $D = \{z : |z| < 1\}$. Then for every $f \in \mathcal{F}$, we have

$$f = \frac{n + (nz-1)^2}{n(nz-1)}, \quad f' = \frac{(nz-1)^2 - n}{(nz-1)^2}$$

Thus $f = 0 \Rightarrow f' = 2$, and $f'(z) \neq 1$ (and hence $f' = 1 \Rightarrow f = c$ for any c). But $\mathcal{F}$ is not normal in D.

Example 2. The following example shows that $b \neq 0$ is necessary in Theorem 1. Let a be a complex number and $D = \{z : |z| < 1\}$. Set

$$\mathcal{F} = \{e^{nz} - \frac{a}{n} : n = 1, 2, 3, \cdots\}$$

Then $\mathcal{F}$ is a family of holomorphic functions on D. For every $f \in \mathcal{F}$, we have $f = 0 \Rightarrow f' = a$, $f'(z) \neq 0$ (and hence $f' = 0 \Rightarrow f = c$ for any c). But $\mathcal{F}$ is not normal in D.

Example 3. The following example shows that $c \neq 0$ is necessary in Theorem 1. Let

$$\mathcal{F} = \{\frac{1}{n^2}(e^{nz} + e^{-nz} - 2) : n = 1, 2, 3, \cdots\}$$

and $D = \{z : |z| < 1\}$. Then $\mathcal{F}$ is a family of holomorphic functions on D. For every $f \in \mathcal{F}$, we have

$$f = \frac{1}{n^2}(e^{nz} + e^{-nz} - 2), \quad f'' = e^{nz} + e^{-nz}$$

Obviously, all zeros of f have multiplicity two, $f = 0 \Rightarrow f'' = 2$, $f'' = 2 \Rightarrow f = 0$. But $\mathcal{F}$ is not normal in D.

Example 4. The restriction in Theorem 1 on the zeros having multiplicity at least k is necessary. In fact, let

$$\mathcal{F} = \{n(e^z - e^{\omega z}) : n = 1, 2, 3, \cdots\}$$

and $D = \{z : |z| < 1\}$, where $\omega \neq 1$, $\omega^k = 1$, $k \geq 2$. Then $\mathcal{F}$ is a family of holomorphic functions on D. Obviously, for each $f \in \mathcal{F}$, $f^{(k)} \equiv f$, so f and $f^{(k)}$ share any complex number in D. But $\mathcal{F}$ is not normal in D.

Xu [19] and Pang [15] proved

Theorem D. *Let $\mathcal{F}$ be a family of holomorphic functions in a domain D, and let a and b be two distinct finite complex numbers. If, for each $f \in \mathcal{F}$, f and f' share a and b in D, then $\mathcal{F}$ is normal in D.*

Chen and Fang [4] extended Theorem D as following.

Theorem E. *Let $\mathcal{F}$ be a family of holomorphic functions in a domain D, let $k \geq 2$ be a positive integer, and let a, $b(\neq a)$ and c be three finite complex numbers. If, for each $f \in \mathcal{F}$, all zeros of $f(z) - c$ have multiplicity at least k, and f and $f^{(k)}$ share a and b in D, then $\mathcal{F}$ is normal in D.*

Recently, Fang and Xu [8] improved Theorem D by proving the the following result.

Theorem F. *Let $\mathcal{F}$ be a family of holomorphic functions in a domain D, and let a and $b(\neq a)$ be two finite complex numbers. If, for each $f \in \mathcal{F}$, f and f' share a in D, and $f = b \Rightarrow f' = b$, then $\mathcal{F}$ is normal in D.*

Theorem G. *Let $\mathcal{F}$ be a family of holomorphic functions in a domain D, and let a and $b(\neq a, 0)$ be two finite complex numbers. If, for each $f \in \mathcal{F}$, f and f' share a in D, and $f' = b \Rightarrow f = b$, then $\mathcal{F}$ is normal in D.*

In this paper, we prove

Theorem 2. *Let $\mathcal{F}$ be a family of holomorphic functions in a domain D and $L[f]$ be a differential polynomial of the form (1.1). Let a be a complex number, and $b(z)(\neq L[a], a)$ be a holomorphic functions in a domain D. If, for each $f \in \mathcal{F}$ and a complex number d, all zeros of $f(z) - d$ have multiplicity at least k, $f = a \Rightarrow L[f] = a$, and $L[f] = b \Rightarrow f = b$, then $\mathcal{F}$ is normal in D.*

As a consequence, we obtain the following result.

Corollary 1. *Let $\mathcal{F}$ be a family of holomorphic functions in a domain D and $L[f]$ be a differential polynomial of the form (1.1) with $a_0 \neq -1$. Let a and b be two complex numbers such that $b \neq a$. If, for each $f \in \mathcal{F}$ and a complex number d, all zeros of $f(z) - d$ have multiplicity at least k, and f and $L[f]$ share a and b in D, then $\mathcal{F}$ is normal in D.*

Proof. Since $a_0 \neq -1$, $a \neq b$, we deduce that $b \neq L[a]$ or $a \neq L[b]$. Noting that $f = a \Leftrightarrow L[f] = a$, and $L[f] = b \Leftrightarrow f = b$. By Theorem 2, $\mathcal{F}$ is normal in D. □

Remark 2. *Obviously, Corollary 1 specializes to Theorem D-E, if we take $a_i \equiv 0$ $(i = 0, 1, \cdots, k-1)$ in $L[f]$.*

Let $k = 1$. From Theorem 2 we immediately obtain the following result which extends and improves theorem F-G.

Corollary 2. *Let $\mathcal{F}$ be a family of holomorphic functions in a domain D, and let a and b be two finite complex numbers such that $b \neq aa_0, a$, where a_0 be a holomorphic function in D. If, for each $f \in \mathcal{F}$, $f = a \Rightarrow f' + a_0(z)f = a$, and $f' + a_0(z)f = b \Rightarrow f = b$, then $\mathcal{F}$ is normal in D.*

It is assumed that the reader is familiar with the basic results and notations of Nevanlinna's value distribution theory, as found in [11]. In particular, we denote by $S(r, f)$ any function satisfying $S(r, f) = o\{T(r, f)\}$ as $r \to \infty$, possibly outside a set of r of finite linear measure.

2. Some lemmas

Lemma 1([16]). *Let $\mathcal{F}$ be a family of holomorphic functions in unit disc Δ with the property that for each $f \in \mathcal{F}$, all zeros of f are of multiplicity at least k. Suppose that there exists a number $A \geq 1$ such that $|f^{(k)}(z)| \leq A$ whenever $f \in \mathcal{F}$ and $f = 0$. If $\mathcal{F}$ is not normal in Δ, then for $0 \leq \alpha \leq k$, there exist*

(1) a number $r \in (0, 1)$;
(2) a sequence of complex numbers z_n, $|z_n| < r$;
(3) a sequence of functions $f_n \in \mathcal{F}$; and
(4) a sequence of positive numbers $\rho_n \to 0$

such that $g_n(\xi) = \rho_n^{-\alpha} f_n(z_n + \rho_n \xi)$ *converges locally uniformly (with respect to the spherical metric) to a nonconstant entire function* $g(\xi)$ *on* $\mathbb{C}$*, and moreover, the zeros of* $g(\xi)$ *are of multiplicity at least* k*,* $g^{\#}(\xi) \le g^{\#}(0) = kA + 1$.

Remark 3. *In Lemma 1, if* $0 \le \alpha < k$*, then the hypothesis on* $f^{(k)}(z)$ *can be dropped, and* $kA + 1$ *can be replaced by an arbitrary positive constant. see [5].*

Lemma 2([6]). *A normal function has order at most two. A normal entire function is of exponential type, and thus has order at most one.*

Lemma 3([3]). *Let* g *be a meromorphic function with finite order. If* g *has only finitely many critical values, then it has only finitely many asymptotic values.*

Lemma 4([1], [2]). *Let* g *be a transcendental meromorphic function such that* $g(0) \ne \infty$ *and the set of finite critical and asymptotic values of* g *is bounded. Then there exists* $R > 0$ *such that*

$$|g'(z)| \ge \frac{|g(z)|}{2\pi |z|} \log \frac{|g(z)|}{R},$$

for all $z \in \mathbb{C} \backslash \{0\}$ *which are not poles of* g.

Lemma 5(Milloux inequality, cf. [11]). *Let* g *be a transcendental meromorphic function and* k *be a positive integer, then for* $b \in \mathbb{C}\backslash\{0\}$*, we have*

$$T(r,g) < \bar{N}(r,g) + N(r,\frac{1}{g}) + N(r,\frac{1}{g^{(k)} - b}) - N(r,\frac{1}{g^{(k+1)}}) + S(r,g).$$

Lemma 6([9]). *Let* $\mathcal{F}$ *be a family of holomorphic functions in a domain* D*. If there exists a positive number* K *such that* $|f(z)| \ge K$ *for all* $f \in \mathcal{F}$ *and* $z \in D$*, then* $\mathcal{F}$ *is normal in* D.

3. Proof of Theorem 1

Proof. It is sufficient to show that $\mathcal{F}$ is normal at each point in D. Let $z' \in D$. Then there exists a disc Δ contained (with its closure) in D such that $z' \in \Delta$ and $b(z)c(z) \ne 0$ for all $z \in \Delta$. Without loss of generality, we assume that $\Delta = \{z : |z| < 1\}$. Suppose that $\mathcal{F}$ is not normal in Δ. We distinguish two cases.

Case 1. Suppose that $d = 0$. Since $a(z)$ is a holomorphic function in $\bar{\Delta}$, there exists a finite number $M > 0$ such that $M = \sup\{|a(z)| : |z| \leq 1\}$. By the assumption we know that $|f^{(k)}(z)| = |L[f](z)| = |a(z)| \leq M$ when $f(z) = 0$. By Lemma 1 (with $\alpha = k$ and $A = M + 1$), there exist $f_n \in \mathcal{F}$, $z_n \to z_0$ $(z_n, z_0 \in \Delta)$, and $\rho_n \to 0^+$ such that

$$g_n(\xi) = \rho_n^{-k} f_n(z_n + \rho_n \xi) \to g(\xi) \tag{2}$$

locally uniformly with respect to the spherical metric on $\mathbb{C}$, where $g(\xi)$ is a nonconstant entire function satisfying that each zero of g has multiplicity at least k, and that

$$g^{\#}(\xi) \leq g^{\#}(0) = k(M + 1) + 1 \tag{3}$$

Thus $g(\xi)$ is of order at most one by Lemma 2.

First we prove that $g = 0 \Rightarrow g^{(k)} = a(z_0)$ on $\mathbb{C}$. Suppose that there exists a point ξ_0 such that $g(\xi_0) = 0$. Then by Hurwitz Theorem, there exist ξ_n, $\xi_n \to \xi_0$ as $n \to \infty$ such that (for n sufficiently large)

$$g_n(\xi_n) = \rho_n^{-k} f_n(z_n + \rho_n \xi_n) = 0 \tag{4}$$

Noting that $\rho_n > 0, f = 0 \Rightarrow L[f] = a$ for all $f \in \mathcal{F}$ and that all zeros of $f(z)$ have multiplicity at least k, from (3.1), (3.3) we have

$$g_n^{(k)}(\xi_n) = f_n^{(k)}(z_n + \rho_n \xi_n) = L[f_n](z_n + \rho_n \xi_n) = a(z_n + \rho_n \xi_n) \tag{5}$$

Thus

$$g^{(k)}(\xi_0) = \lim_{n \to \infty} g_n^{(k)}(\xi_n) = a(z_0)$$

This shows that $g = 0 \Rightarrow g^{(k)} = a(z_0)$.

Next we prove that $g^{(k)}(\xi) \neq b(z_0)$ on $\mathbb{C}$. Suppose that there exists a point ξ_0 such that $g^{(k)}(\xi_0) = b(z_0)$. If $g^{(k)}(\xi) \equiv b(z_0)$, then g is a polynomial of degree k. Since each zero of g has multiplicity at least k, g must has a single zero ξ_1 with multiplicity k, this implies $g(\xi) = \frac{b(z_0)}{k!}(\xi - \xi_1)^k$, which together with the fact that $g = 0 \Rightarrow g^{(k)} = a(z_0)$ gives that $b(z_0) = a(z_0)$. By simple computing we have

$$g^{\#}(0) \leq \begin{cases} k/2, & \text{if } |\xi_1| \geq 1 \\ |a(z_0)|, & \text{if } |\xi_1| < 1 \end{cases}$$

so that $g^{\#}(0) < k(|a(z_0)|+1)+1 \le k(M+1)+1$, which contradicts (3.2). Thus $g^{(k)}(\xi) \not\equiv b(z_0)$. From (1.1) and (3.1) we have

$$\begin{aligned} L[f_n](z_n+\rho_n\xi) &= f_n^{(k)}(z_n+\rho_n\xi) + \sum_{j=1}^{k} a_{k-j}(z_n+\rho_n\xi) f_n^{(k-j)}(z_n+\rho_n\xi) \\ &= g_n^{(k)}(\xi) + \sum_{j=1}^{k} \rho_n^j a_{k-j}(z_n+\rho_n\xi) g_n^{(k-j)}(\xi) \end{aligned} \tag{6}$$

Since

$$\sum_{j=1}^{k} \rho_n^j a_{k-j}(z_n+\rho_n\xi) g_n^{(k-j)}(\xi) \to 0$$

as $n \to \infty$, thus we have

$$\begin{aligned} \lim_{n\to\infty} \{L[f_n](z_n+\rho_n\xi) - b(z_n+\rho_n\xi)\} &= \lim_{n\to\infty} \{g_n^{(k)}(\xi) - b(z_n+\rho_n\xi)\} \\ &= g^{(k)}(\xi) - b(z_0) \end{aligned}$$

Noting that $g^{(k)}(\xi_0) - b(z_0) = 0$, by Hurwitz Theorem, there exist ξ_n, $\xi_n \to \xi_0$ as $n \to \infty$ such that (for n sufficiently large)

$$L[f_n](z_n+\rho_n\xi_n) = b(z_n+\rho_n\xi_n). \tag{7}$$

Since $L[f] = b \Rightarrow f = c$ for all $f \in \mathcal{F}$, from (3.1), (3.6) we have

$$g_n(\xi_n) = \rho_n^{-k} f_n(z_n+\rho_n\xi_n) = \rho_n^{-k} c(z_n+\rho_n\xi_n) \tag{8}$$

But $c(z_n+\rho_n\xi_n) \to c(z_0)$ as $n \to 0$ and $c(z_0) \neq 0$, from (3.7) we get

$$g(\xi_0) = \lim_{n\to\infty} g_n(\xi_n) = \infty$$

which contradicts $g^{(k)}(\xi_0) = b(z_0)$. This shows that $g^{(k)}(\xi) \neq b(z_0)$ on $\mathbb{C}$.

Since g is of order at most one, so is $g^{(k)}$, it follows that

$$g^{(k)}(\xi) = b(z_0) + e^{b_0+b_1\xi} \tag{9}$$

where b_0, b_1 are finite constants. We divide this case into two subcases.

Case 1.1. If $b_1 = 0$, from (3.8) we know that $g(\xi)$ is a polynomial of degree k. Since each zero of g has multiplicity at least k, g must has a single zero ξ_2 with multiplicity k, so that

$$g(\xi) = \frac{b(z_0)+e^{b_0}}{k!}(\xi-\xi_2)^k$$

which together with the fact that $g = 0 \Rightarrow g^{(k)} = a(z_0)$ gives that $b(z_0) + e^{b_0} = a(z_0)$. By simple computing we have

$$g^{\#}(0) \le \begin{cases} k/2, & \text{if } |\xi_2| \ge 1 \\ |a(z_0)|, & \text{if } |\xi_2| < 1 \end{cases}$$

so that $g^{\#}(0) < k(M+1)+1$, which contradicts (3.2).

Case 1.2. If $b_1 \neq 0$, then g is a transcendental entire function with order at most one. Since $g^{(k)} \neq b(z_0)(\neq 0)$, by Lemma 5 we deduce that g has infinitely many zeros $z_1, z_2, \cdots$, and $|z_j| \to \infty$ as $j \to \infty$. Define $h(z) = g^{(k-1)}(z) - b(z_0)z$, then $h'(z) = g^{(k)}(z) - b(z_0) \neq 0$, so h has no critical values. Hence, by Lemma 3, h has only finitely many asymptotic values. Applying Lemma 4 to h, we have

$$\frac{|z_j h'(z_j)|}{|h(z_j)|} \ge \frac{1}{2\pi} \log \frac{|h(z_j)|}{R} = \frac{1}{2\pi} \log \frac{|b(z_0)z_j|}{R}$$

This implies

$$\frac{|z_j h'(z_j)|}{|h(z_j)|} \to \infty \ (\text{as } j \to \infty) \tag{10}$$

Recalling that all zeros of $g(\xi)$ are of multiplicity at least k and that $g = 0 \Rightarrow g^{(k)} = a(z_0)$, we have

$$\frac{|z_j h'(z_j)|}{|h(z_j)|} = |\frac{a(z_0) - b(z_0)}{b(z_0)}| \tag{11}$$

From (3.9) and (3.10) we deduce a contradiction. Thus $\mathcal{F}$ is normal at z'.

Case 2. Suppose that $d \neq 0$. If $k = 1$, then the proof is completed as in Case 1. We now assume that $k \ge 2$. Applying Lemma 1 to $\alpha = 1$, we can find $f_n \in \mathcal{F}$, $z_n \to z_0$ $(z_n, z_0 \in D)$, and $\rho_n \to 0^+$ such that

$$g_n(\xi) = f_n(z_n + \rho_n \xi) - d \to g(\xi) \tag{12}$$

locally uniformly with respect to the spherical metric on $\mathbb{C}$, where $g(\xi)$ is a nonconstant entire function satisfying that each zero of g has multiplicity at least $k \ge 2$, and that g is of order at most one.

We claim that $g(\xi) \neq -d$. For otherwise, there exists ξ_0 such that

$$g(\xi_0) = -d, \quad g(\xi) \not\equiv -d \tag{13}$$

Then by Hurwitz Theorem, there exist ξ_n, $\xi_n \to \ \xi_0$ as $n \to \infty$ such that (for n sufficiently large)

$$-d = g_n(\xi_n) = f_n(z_n + \rho_n \xi_n) - d. \tag{14}$$

Thus we get $f_n(z_n + \rho_n \xi_n) = 0$. Since $f = 0 \Rightarrow L[f] = a$ for all $f \in \mathcal{F}$ and all zeros of $f(z)$ have multiplicity at least k, by (3.11) we have

$$g_n^{(k)}(\xi_n) = \rho_n^k f_n^{(k)}(z_n + \rho_n \xi_n) = \rho_n^k L[f_n](z_n + \rho_n \xi_n) = \rho_n^k a(z_n + \rho_n \xi_n)$$

This implies

$$g^{(k)}(\xi_0) = \lim_{n\to\infty} g_n^{(k)}(\xi_n) = 0. \tag{15}$$

If $g^{(k)}(\xi) \equiv 0$, then g is a polynomial of degree $< k$, and so could not have zeros of multiplicity at least k, a contradiction. Thus $g^{(k)}(\xi) \not\equiv 0$. From (1.1) and (3.11) we have

$$\begin{aligned}
&\lim_{n\to\infty} \rho_n^k \{L[f_n](z_n + \rho_n \xi) - b(z_n + \rho_n \xi)\} \\
&\quad = \lim_{n\to\infty} \rho_n^k L[f_n](z_n + \rho_n \xi) \\
&\quad = \lim_{n\to\infty} \{\rho_n^k f_n^{(k)}(z_n + \rho_n \xi) + \rho_n^k \sum_{j=1}^{k} a_{k-j}(z_n + \rho_n \xi) f_n^{(k-j)}(z_n + \rho_n \xi)\} \\
&\quad = \lim_{n\to\infty} \{g_n^{(k)}(\xi) + \sum_{j=1}^{k-1} \rho_n^j a_{k-j}(z_n + \rho_n \xi) g_n^{(k-j)}(\xi) \\
&\qquad + \rho_n^k a_0(z_n + \rho_n \xi)[g_n(\xi) + d]\} \\
&\quad = \lim_{n\to\infty} g_n^{(k)}(\xi) \\
&\quad = g^{(k)}(\xi)
\end{aligned} \tag{16}$$

By Hurwitz Theorem, from (3.14) and (3.15), we know that there exists $\tilde{\xi}_n$, $\tilde{\xi}_n \to \xi_0$ as $n \to \infty$ such that (for n sufficiently large)

$$\rho_n^k \{L[f_n](z_n + \rho_n \tilde{\xi}_n) - b(z_n + \rho_n \tilde{\xi}_n)\} = 0 \tag{17}$$

which leads to $L[f_n](z_n + \rho_n \tilde{\xi}_n) = b(z_n + \rho_n \tilde{\xi}_n)$. From this and recalling that $L[f] = b \Rightarrow f = c$ for all $f \in \mathcal{F}$, we have

$$g_n(\tilde{\xi}_n) = f_n(z_n + \rho_n \tilde{\xi}_n) - d = c(z_n + \rho_n \tilde{\xi}_n) - d \tag{18}$$

It follows that

$$g(\xi_0) = \lim_{n\to\infty} g_n(\tilde{\xi}_n) = c(z_0) - d \tag{19}$$

Combine (3.12) and (3.18) we get $c(z_0) = 0$, a contradiction. This shows that $g(\xi) \neq -d$. Since g is of order at most one, we have $g(\xi) = -d + e^{d_0 + d_1 z}$, where d_0 and $d_1 (\neq 0)$ are two constants. Noting that $d \neq 0$, thus g has zeros and only simple zeros. This contradicts the fact that all zeros of g have multiplicity at least $k \geq 2$. Thus $\mathcal{F}$ is normal at $z' \in D$.

The proof of Theorem 1 is completed. □

4. Proof of Theorem 2

Proof. Set $\tilde{\mathcal{F}} = \{\tilde{f} : \tilde{f} = f - a; f \in \mathcal{F}\}$. Then for each $\tilde{f} \in \tilde{\mathcal{F}}$, all zeros of $\tilde{f} - (d - a)$ have multiplicity $\geq k$, $\tilde{f} = 0 \Rightarrow L[\tilde{f}] = a - L[a]$, and $L[\tilde{f}] = b - L(a) \Rightarrow \tilde{f} = b - a$. Since $b - L[a] \neq 0$, $b - a \neq 0$, applying Theorem 1 to $\tilde{\mathcal{F}}$, we know that $\tilde{\mathcal{F}}$ is normal in D. It follows that $\mathcal{F}$ is normal in D.

The proof of Theorem 2 is completed. □

References

1. W. Bergweiler, On the zeros of certain homogeneous differential polynomials, *Arch. Math.* **64** (1995):199–202.
2. W. Bergweiler, Normality and exceptional values of derivatives, *Proc. Amer. Math. Soc.* **129** (2001):121–129.
3. W. Bergweiler and A. Eremenko, On the singularities of the inverse to a meromorphic function of finite order, *Rev. Mat. Iberoamericana* **11** (1995):355–373.
4. H. H. Chen and M. L. Fang, Shared values and normal families of meromorphic functions, *J. Math. Anal. Appl.* **260** (2001):124–132.
5. H. H. Chen and Y. X. Gu, An improvement of Marty's criterion and its applications, *Sci. China Ser. A* *36*(1993):674–681.
6. J. Clunie and W. K. Hayman, The spherical derivatives of integeral and meromorphic functions, *Comment. Math. Hev.* *40*(1996):117–148.
7. C. T. Chuang, Sur les fonctions holomorphes dans le cercle unite, *Bull. Soc. Math. France.* *68*(1940):11–41.
8. M. L. Fang and Y. Xu, Normal families of holomorphic functions and shared values, *Israel J. Math.* **129** (2002):125–141.
9. Y. X. Gu, *Normal Families of Meromorphic Functions (in Chinese),* Sichuan Education Press, Sichuan, China, 1991.
10. G.G. Gundersen, Meromorphic functions that shaare finite values and their derivatives, *J. Math. Anal. Appl.* *75*(1980): 441–446.
11. W. K. Hayman, *Meromorphic Functions,* Clarendon Press Oxford 1964.
12. C. Miranda, Sur un nouveau critere de normalite pour les familles des fonctions holomorphes, *Bull. Sci. Math Fr.* **63** (1935):185–196.
13. P. Montel, *Lecons sur les familles normales de fonctions analytiques et leurs applications,* Gauthier-Villars, Paris, 1927.
14. E. Mues, and N. Steinmetz, *Meromorphe Funktionen, die mit ihrer Ableitung Werte teilen,* Manuscripta Math. *29* (1979):195–206.
15. X. C. Pang, Shared values and normal families,*Analysis,* *22*(2002):175–182.

16. X. C. Pang and L. Zalcman, Normal families and shared values, *Bull. London Math. Soc. 32*(2000):325–331.
17. J. Schiff, *Normal Families,* Springer-verlag, Berlin, 1993.
18. W. Schwick, Sharing values and normality, *Arch. Math. 59*(1992):50–54.
19. Y. Xu, Normality criteria concerning sharing values, *Indian J. Pure Appl. Math. 30* (1999):287–293.
20. L. Zalcman, Normal families: new perspectives, *Bull. Amer. Math. Soc. 35*(1998):215-230.

THE BLOCH TYPE SPACES VIA THE CESÀRO MEANS*

JING LIN

Department of Mathematics, Shantou University
Shantou, Guangdong, 515063 P. R. China.
E-mail: g_jlin@stu.edu.cn

Using the Cesàro means of Taylor extensions of holomorphic functions f on the unit disc, an equivalent condition for f to belong to the Bloch type spaces is given.

2000 Mathematics Subject Classification: 30H05 and 46E15

1. Introduction

Throughout this paper we denote $H(\mathbb{D})$ the set of holomorphic functions on the unit disc $\mathbb{D}$. For $f(z) = \sum_{n=0}^{\infty} a_n z^n \in H(\mathbb{D})$ the Cesàro means of the power series of f is defined by

$$\sigma_n(f)(z) = \sum_{k=0}^{n} \frac{s_k(f)(z)}{n+1} = \sum_{k=0}^{n}(1 - \frac{k}{n+1})a_k z^k, \qquad n = 0, 1, 2, ...,$$

where

$$s_n(f)(z) = \sum_{k=0}^{n} a_k z^k, \qquad n = 0, 1, 2, ...,$$

is the partial sums of f.

Holland and Walsh [1] gave criteria for a function f in $H(\mathbb{D})$ to belong to the Bloch space $\mathcal{B}$ in terms of the Cesàro means of the Taylor series of f. In this paper we consider the Bloch type spaces by studying the Cesàro means. For $0 < \alpha < \infty$, a function f in $H(\mathbb{D})$ is said to belong to the α-Bloch space, denoted by $f \in \mathcal{B}^\alpha$, if

$$\|f\|_{\mathcal{B}^\alpha} = \sup_{z\in\mathbb{D}}(1 - |z|^2)^\alpha |f'(z)| < \infty.$$

*This work is supported by NNSF of China (No. 10371069) and the NSF of Guangdong (No. 04011000).

Note that 1-Bloch space $\mathcal{B}^1$ is just the classical Bloch space. Denote by H^∞ the space of all functions in $H(\mathbb{D})$ with

$$\|f\|_\infty = \sup_{z\in\mathbb{D}} |f(z)| < \infty.$$

Our main result in this note is the following theorem.

Theorem Let $f \in H(\mathbb{D})$ and $0 < \alpha < \frac{3}{2}$. The following statements are equivalent:

(i) $f \in \mathcal{B}^\alpha$;

(ii) $$\|(n+m+1)\sigma_{n+m}(f) - (n+1)\sigma_n(f) - (m+1)\sigma_m(f)\|_\infty = O(\sqrt{nm}(n^{\frac{\alpha}{2}-\frac{1}{2}}m^{\frac{\alpha}{2}-\frac{1}{2}} + n^{\alpha-1} + m^{\alpha-1})) \qquad (n,m\to\infty);$$

(iii) $\|\sigma_{2n}(f) - \sigma_n(f)\|_\infty = O(n^{\alpha-1})$.

If α=1, the conclusion is Theorem 2 in [1].

2. Preliminaries

To prove our main theorem we consider the Hadamard product of two functions $f(z) = \sum_{n=0}^{\infty} a_n z^n$ and $g(z) = \sum_{n=0}^{\infty} b_n z^n$ in $H(\mathbb{D})$ as the following

$$f * g(z) = \sum_{n=0}^{\infty} a_n b_n z^n.$$

Using the Hadamard product we know that the Cesàro means of $f \in H(\mathbb{D})$ can be written as

$$\sigma_n(f) = \sigma_n(k) * f,$$

where $k(z) = 1/(1-z) = \sum_{n=0}^{\infty} z^n, z \in \mathbb{D}$. Also, we have

$$\sigma_n(f) = \frac{1}{2\pi}\int_0^{2\pi} K_n(\theta) f(ze^{-i\theta})d\theta,$$

where

$$K_n(\theta) = \sum_{k=-n}^{n} (1 - \frac{|k|}{n+1})e^{ik\theta} = \frac{1-\cos(n+1)\theta}{(n+1)(1-\cos\theta)}$$

is Fejér's kernel. Note that $K_n(\theta) \geq 0$ and $\sigma_n(1) = 1$ for n= 0,1, ...

Let A_α, $0 \le \alpha < 2$, denote the space of functions f in $H(\mathbb{D})$ for which

$$|||g|||_{A_\alpha} = |g(0)| + \|g\|_{A_\alpha} < \infty,$$

where

$$\|g\|_{A_\alpha} = \frac{1}{\pi}\int_0^1 \int_0^{2\pi} |re^{i\theta}g'(re^{i\theta}) + g(re^{i\theta}) - g(0)|(1-r^2)^{1-\alpha} dr d\theta. \quad (2.1)$$

Note that A_α contains constant functions only for $\alpha \ge 2$. The following result will be used in the proof of our main theorem and it is also an extension of Theorem 2.3 of [2].

Lemma 1 Suppose $0 < \alpha < 2$. Let $f = \sum_{n=0}^{\infty} a_n z^n \in \mathcal{B}^\alpha$ and $g(z) = \sum_{n=0}^{\infty} b_n z^n \in A_\alpha$. Then

$$|f * g(z)| \le C\|f\|_{\mathcal{B}^\alpha}\|g\|_{A_\alpha}. \quad (2.2)$$

Proof. For $z = re^{i\theta} \in \mathbb{D}$ and $\xi \in \mathbb{D}$, it is easy to see that

$$\sum_{n=1}^{\infty} a_n b_n \xi^{n-1} = \frac{1}{\pi}\int_0^1 \int_0^{2\pi} (1-r^2) f'(\xi\overline{z}) \frac{d}{dz}[(zg(z) - g(0))]e^{-i\theta} d\theta dr.$$

Since $f \in \mathcal{B}^\alpha$, we have

$$\left|\sum_{n=1}^{\infty} a_n b_n \xi^n\right|$$
$$\le \|f\|_{\mathcal{B}^\alpha} \frac{1}{\pi}\int_0^1 \int_0^{2\pi} |re^{i\theta}g'(re^{i\theta}) + g(re^{i\theta}) - g(0)|(1-r^2)^{1-\alpha} d\theta dr,$$

and so

$$\left|\sum_{n=0}^{\infty} a_n b_n \xi^n\right| \le C\|f\|_{\mathcal{B}^\alpha}\|g\|_{A_\alpha},$$

which completes the proof. □

The following lemma will also be used in the proof of our main result (see [5]).

Lemma 2 (Bernstein Theorem): Suppose T(x) is a real polynomial of order n, and $M = \max|T(x)|$. Then

$$|T'(x)| \le 2nM.$$

3. The Proof of the Main Theorem

From now on, we begin to prove the main theorem.

Proof. (i)⇒(ii). Our proof will be given by Lemma 2. Noting that $\sigma_n(f) = \sigma_n(k) * f$, we write that

$$(n+m+1)\sigma_{n+m}(f) - (n+1)\sigma_n(f) - (m+1)\sigma_m(f) \equiv g_{n,m} * f,$$

where

$$g_{n,m} = g_{m,n} = (n+m+1)\sigma_{n+m}(k) - (n+1)\sigma_n(k) - (m+1)\sigma_m(k).$$

Now for $n = 0, 1, 2, \ldots$,

$$\begin{aligned}\sigma_n(k)(z) &= \sum_{i=0}^{n}(1 - \frac{i}{n+1})z^i, \qquad n = 0, 1, 2, \ldots, \\ &= -\frac{z(1-z^{n+1})}{(n+1)(1-z)^2} + \frac{1}{1-z}.\end{aligned}$$

Hence

$$g_{n,m}(z) = \frac{z(1-z^{n+1})(1-z^{m+1})}{(1-z)^2}.$$

To establish the assertion, it suffices, in view of Lemma 1, to show that

$$\begin{aligned}\|g_{n,m}\|_{A_\alpha} &= \frac{1}{\pi}\int_0^1\int_0^{2\pi} |re^{i\theta}g'(re^{i\theta}) + g(re^{i\theta}) - g(0)|(1-r^2)^{1-\alpha}drd\theta \\ &= O(\sqrt{nm}(n^{\frac{\alpha}{2}-\frac{1}{2}}m^{\frac{\alpha}{2}-\frac{1}{2}} + n^{\alpha-1} + m^{\alpha-1}))\end{aligned}$$

as $n, m \to \infty$. With this in mind, we observe that, if $n, m = 1, 2, \ldots$, then

$$\begin{aligned}&zg'_{n-1,m-1}(z) + g_{n-1,m-1}(z) \\ &= \frac{d}{dz}zg_{n-1,m-1}(z) \\ &= \frac{2z(1-z^n)(1-z^m)}{(1-z)^3} - \frac{nz^{n+1}(1-z^m)}{(1-z)^2} - \frac{mz^{m+1}(1-z^n)}{(1-z)^2}.\end{aligned}$$

According to the definition of the norm of A_α, we have

$$\begin{aligned}\|g_{n,m}\|_{A_\alpha} &\approx \frac{1}{\pi}\int_0^1\int_0^{2\pi}\left|\frac{2z(1-z^n)(1-z^m)}{(1-z)^3} - \frac{nz^{n+1}(1-z^m)}{(1-z)^2}\right.\\ &\quad \left. - \frac{mz^{m+1}(1-z^n)}{(1-z)^2}\right|(1-r^2)^{1-\alpha}drd\theta \\ &\leq \frac{2}{\pi}\int_0^1\int_0^{2\pi}\frac{|z(1-z^n)(1-z^m)|}{|1-z|^3}(1-r^2)^{1-\alpha}drd\theta + \\ &\quad +\frac{n}{\pi}\int_0^1\int_0^{2\pi}\frac{|z^{n+1}(1-z^m)|}{|(1-z)^2|}(1-r^2)^{1-\alpha}drd\theta + \\ &\quad +\frac{m}{\pi}\int_0^1\int_0^{2\pi}\frac{|z^{m+1}(1-z^n)|}{|(1-z)^2|}(1-r^2)^{1-\alpha}drd\theta \\ &= I_1(n,m) + I_2(n,m) + I_3(n,m).\end{aligned}$$

Now we estimate $I_1(n,m)$, $I_2(n,m)$ and $I_3(n,m)$, respectively.

To deal with $I_1(n,m)$, the Hölder inequality shows that

$$\begin{aligned}I_1(n,m) &= \frac{2}{\pi}\int_0^1\int_0^{1\pi}\frac{|z(1-z^n)(1-z^m)|}{|1-z|^3}(1-r^2)^{1-\alpha}drd\theta \\ &\leq \frac{2}{\pi}\int_0^1\int_0^{2\pi}\frac{|(1-z^n)(1-z^m)|}{|1-z|^3}(1-r^2)^{1-\alpha}rdrd\theta \\ &\leq (I_1(n))^{\frac{1}{2}}(I_1(m))^{\frac{1}{2}},\end{aligned}$$

where

$$\begin{aligned}I_1(n) &= \int_0^1\int_0^{2\pi}\frac{|1-z^n|^2}{|1-z|^3}(1-r^2)^{1-\alpha}rdrd\theta, \\ I_1(m) &= \int_0^1\int_0^{2\pi}\frac{|1-z^m|^2}{|1-z|^3}(1-r^2)^{1-\alpha}rdrd\theta.\end{aligned}$$

Now we estimate $I_1(n)$. If λ is not zero or a negative, then (also see [7])

$$\frac{1}{(1-z\overline{w})^\lambda} = \sum_{n=0}^{+\infty}\frac{\Gamma(n+\lambda)}{n!\Gamma(\lambda)}z^n\overline{w}^n.$$

Letting $\lambda = \frac{3}{2}$, we obtain

$$\frac{1}{(1-z)^{3/2}} = \sum_{i=0}^{\infty}c_i z^i = \sum_{i=0}^{\infty}\gamma(i)z^i.$$

Define $c_s = 0$ if $s < 0$, and we have

$$\frac{(1-z^n)}{(1-z)^{3/2}} = \sum_{i=0}^{\infty}(c_i - c_{i-n})z^i.$$

Hence

$$\begin{aligned} I_1(n) &= 4\int_0^1 \sum_{i=0}^{\infty} |c_i - c_{i-n}|^2 r^{2i+1}(1-r^2)^{1-\alpha}dr \\ &= 4\sum_{i=0}^{\infty} |c_i - c_{i-n}|^2 \int_0^1 r^{2i+1}(1-r^2)^{1-\alpha}dr \\ &= 2\sum_{i=0}^{\infty} |c_i - c_{i-n}|^2/(i+1)^{2-\alpha} \\ &= 2\sum_{i=0}^{n-1} |c_i|^2/(i+1)^{2-\alpha} + 2\sum_{i=n}^{\infty} |c_{i-n} - c_i|^2/(i+1)^{2-\alpha} \\ &= 2(J_1 + J_2). \end{aligned}$$

We know that

$$c_i = \gamma(i), \qquad i = 0, 1, 2, \ldots.$$

For $t \geq 0$, let

$$\gamma(t) = \frac{\Gamma(t+3/2)}{\Gamma(3/2)\Gamma(t+1)}.$$

By the properties of the Γ function (also see [1]), it is easy to show that

$$\begin{aligned} \gamma'(t) &= \gamma(t)\left\{\frac{\Gamma'(t+3/2)}{\Gamma(t+3/2)} - \frac{\Gamma'(t+1)}{\Gamma(t+1)}\right\} \\ &= \gamma(t)\sum_{t=1}^{\infty} \frac{1/2}{(n+t)(n+t+1/2)} \\ &\leq \gamma(t)\sum_{t=1}^{\infty} \frac{1/2}{(n+t-1/2)(n+t+1/2)} \\ &= \gamma(t)/(1+2t). \end{aligned}$$

By the above inequality, we can show that $\gamma(t)/\sqrt{2t+1}$ is decreasing on $[0, \infty)$. Therefore, for $t \geq 0$,

$$\gamma(t) \leq \sqrt{2t+1}$$

and

$$\gamma'(t) \leq 1/\sqrt{2t+1}.$$

Particularly, $\gamma(i) = c_i \leq \sqrt{2(1+i)}$. Thus

$$J_1 = \sum_{i=0}^{n-1} \frac{|c_i|^2}{(i+1)^{2-\alpha}} = O(n^\alpha).$$

Now we begin to handle J_2. Note that if $i \geq n$, then

$$\begin{aligned} 0 \leq c_i - c_{i-n} &= \int_{i-n}^{i} \gamma'(t)dt \\ &\leq \int_{i-n}^{i} \frac{dt}{\sqrt{(1+2t)}} \\ &= \sqrt{(1+2i)} - \sqrt{(1+2i-2n)} \\ &\leq \frac{2n}{\sqrt{1+2i}}. \end{aligned}$$

Therefore

$$\begin{aligned} J_2 &= \sum_{i=n}^{\infty} |c_{i-n} - c_i|^2/(i+1)^{2-\alpha} \\ &\leq \sum_{i=n}^{\infty} \frac{4n^2}{(1+2i)(i+1)^{2-\alpha}} \\ &\leq 4n^2 \sum_{i=n}^{\infty} \frac{1}{(i+1)^{3-\alpha}} \\ &= 4n^2 \int_{n}^{\infty} (x+1)^{\alpha-3} dx \\ &= O(n^\alpha). \end{aligned}$$

Hence

$$I_1(n) = 2(J_1 + J_2) = O(n^\alpha),$$

and then by the Cauchy-Schwarz inequality, we obtain that

$$I_1(n,m) = O(n^{\frac{\alpha}{2}} m^{\frac{\alpha}{2}}).$$

Next we begin to estimate $I_2(n,m)$. Indeed, we can write $I_2(n,m)$ as

$$\begin{aligned}
I_2(n,m) &= \frac{n}{\pi}\int_0^1\int_0^{2\pi}\frac{|1-z^m|}{|1-z|^2}(1-r^2)^{1-\alpha}r^{n+1}drd\theta\\
&= \frac{n}{\pi}\int_0^1(1-r^2)^{1-\alpha}r^{n+1}\int_0^{2\pi}\frac{|1-r^me^{im\theta}|}{|1-re^{i\theta}|^2}d\theta dr\\
&= \frac{n}{\pi}\int_0^1(1-r^2)^{1-\alpha}r^{n+1}\left\{\int_0^{2\pi}\frac{|1-r^me^{im\theta}|^2}{|1-re^{i\theta}|^2}d\theta\int_0^{2\pi}\frac{1}{|1-re^{i\theta}|^2}d\theta\right\}^{1/2}dr\\
&\leq Cn\sqrt{m}\int_0^1 r^{n+1}(1-r)^{\frac{1}{2}-\alpha}dr\\
&= Cn\sqrt{m}B(n+2,\frac{3}{2}-\alpha)\\
&= O(n^{\alpha-\frac{1}{2}}m^{\frac{1}{2}}).
\end{aligned}$$

Similarly we can obtain that

$$I_3(n,m) = O(m^{\alpha-\frac{1}{2}}n^{\frac{1}{2}}).$$

We can summarize that

$$\|g_{n,m}\|_{A_\alpha} = O(\sqrt{nm}(n^{\frac{\alpha}{2}-\frac{1}{2}}m^{\frac{\alpha}{2}-\frac{1}{2}}+n^{\alpha-1}+m^{\alpha-1})).$$

(ii)⇒(iii). Assume that (ii) holds and take $m=n$ in (ii). The triangle inequality gives

$$(2n+1)\|\sigma_{2n}(f)-\sigma_n(f)\|_\infty \leq \|\sigma_n(f)\|_\infty + O(n^\alpha).$$

To estimate $\|\sigma_n(f)\|_\infty$, we obtain by letting $m=1$ in (ii),

$$\begin{aligned}
(n+2)\|\sigma_{n+1}(f)\|_\infty &\leq (n+1)\|\sigma_n(f)\|_\infty + \|\sigma_1(f)\|_\infty + O(n^{\frac{\alpha}{2}}+n^{\alpha-\frac{1}{2}})\\
&\leq (n+1)\|\sigma_n(f)\|_\infty + O(n^{\frac{\alpha}{2}}+n^{\alpha-\frac{1}{2}}).
\end{aligned}$$

That is to say

$$\begin{aligned}
\|\sigma_{n+1}(f)\|_\infty &\leq \frac{(n+1)}{(n+2)}\|\sigma_n(f)\|_\infty + O(n^{\frac{\alpha}{2}-1}+n^{\alpha-\frac{3}{2}})\\
&\leq \|\sigma_n(f)\|_\infty + O(n^{\frac{\alpha}{2}-1}+n^{\alpha-\frac{3}{2}})\\
&\leq \|\sigma_{n-1}(f)\|_\infty + 2O(n^{\frac{\alpha}{2}-1}+n^{\alpha-\frac{3}{2}})\\
&\leq \cdots\\
&\leq O((n^{\frac{\alpha}{2}-1}+n^{\alpha-\frac{3}{2}})\cdot n)\\
&= O(n^{\frac{\alpha}{2}}+n^{\alpha-\frac{1}{2}}).
\end{aligned}$$

Thus,

$$(2n+1)\|\sigma_{2n}(f)-\sigma_n(f)\|_\infty \le \|\sigma_n(f)\|_\infty + O(n^\alpha) = O(n^\alpha).$$

Hence

$$\|\sigma_{2n}(f)-\sigma_n(f)\|_\infty = O(n^{\alpha-1}).$$

(iii)⇒(i). Suppose (iii) holds. That is to say ,

$$\|\sigma_{2n}(f)-\sigma_n(f)\|_\infty = O(n^{\alpha-1}).$$

Since $\sigma_{2n}(f)-\sigma_n(f)$ is a polynomial of degree at most $2n$, Lemma 2 shows that

$$\|\sigma'_{2n}(f)-\sigma'_n(f)\|_\infty = O(n^{\alpha-1}\cdot n) = O(n^\alpha) \qquad (n\to\infty).$$

Taking $n=2^j$ and summing on j from $j=0$ to $j=m$, we obtain

$$\begin{aligned}\|\sigma'_{2^m}(f)\|_\infty &\le \|\sigma'_{2^m}(f)-\sigma'_{2^{m-1}}(f)\|_\infty + \|\sigma'_{2^{m-1}}(f)-\sigma'_{2^{m-2}}(f)\|_\infty \\ &\quad + \cdots + \|\sigma'_{2^1}(f)-\sigma'_{2^0}(f)\|_\infty + \|\sigma'_{2^0}(f)\|_\infty \\ &= O((2^{m-1})^\alpha) + O((2^{m-2})^\alpha) + \ldots + O((2^0)^\alpha) + \frac{1}{2} \\ &= O((2^m)^\alpha).\end{aligned}$$

Let

$$V_m = (2+2^{-m})\sigma_{2^{m+1}}(k) - (1+2^{-m})\sigma_{2^m}(k).$$

Then

$$V_m * f = (2+2^{-m})\sigma_{2^{m+1}}(f) - (1+2^{-m})\sigma_{2^m}(f).$$

Hence

$$\begin{aligned}\|(V_m * f)'\|_\infty &= \|(2+2^{-m})\sigma'_{2^{m+1}}(f) - (1+2^{-m})\sigma'_{2^m}(f)\|_\infty \\ &= \|\sigma'_{2^{m+1}}(f) + (1+2^{-m})(\sigma'_{2^{m+1}}(f)-\sigma'_{2^m}(f)\|_\infty \\ &\le \|\sigma'_{2^{m+1}}(f)\|_\infty + (1+2^{-m})\|(\sigma'_{2^{m+1}}(f)-\sigma'_{2^m}(f)\|_\infty \\ &= O((2^m)^\alpha) + (1+2^{-m})O((2^m)^\alpha) \\ &= O((2^m)^\alpha).\end{aligned}$$

Note that

$$\sigma_n(\mu_0 f') = \mu_0\sigma'_n(f)$$

holds for the identity function μ_0 on $\mathbb{D}$ and then $\|\sigma_n(h)\|_\infty \le \|h\|_\infty$ whenever $h \in H^\infty$. Furthermore, the first 2^m Taylor coefficients of f and $V_m * f$ are equal.

We put these three facts together to prove that $\|\sigma_n'(f)\|_\infty = O(n^\alpha)$ as $n \to \infty$. So , let $n \geq 1$ and choose m so that $2^{m-1} \leq n < 2^m$. Then

$$\begin{aligned}\|\mu_0\sigma_n'(f)\|_\infty &= \|\sigma_n(\mu_0 f')\|_\infty = \|\sigma_n(\mu_0(V_m * f)')\|_\infty \\ &\leq \|(V_m * f)'\|_\infty = O((2^m)^\alpha) = O(n^\alpha).\end{aligned}$$

For $f(z) = \sum\limits_{n=0}^{\infty} a_n z^n$ and $z = r\xi$, it is easy to obtain

$$zf'(z) = (1-r)^2 \sum_{n=0}^{\infty}(n+1)\xi\sigma_n'(f)(\xi)r^n.$$

Furthermore

$$\begin{aligned}r|f'(z)| &\leq (1-r)^2 \sum_{n=0}^{\infty}(n+1)|\sigma_n'(f)(\xi)|r^n \\ &\leq C(1-r)^2 \sum_{n=0}^{\infty}(n+1)n^\alpha r^n \\ &\approx \frac{C}{(1-r)^\alpha}.\end{aligned}$$

That is to say $f \in \mathcal{B}^\alpha$. We have completed the proof of (iii)⇒(i). □

References

1. F. Holland and D. Walsh, Criteria for membership of Bloch spaces and its subspaces BMOA, Math. Ann, **273** (1986), 317-335.
2. J. M. Anderson, Clunie and Ch.Pommerenke, On Bloch functions and normal functions, J.Reine Angew. Math, **270**, (1974), 12-37.
3. H. Wulan and J. Zhou, Q_K type spaces of analytic functions. Journal of Function Spaces and Applications, 4(2006), 73-84.
4. E. Titchmarsh, Theory of functions, 2^{nd}ed, Oxford: University Press 1939.
5. A. Zygmund, Trigonometric Series, I, II. Cambridge: University Press 1968.
6. T. Körner, Fourier Analysis, Cambridge University Press, London , 1988.
7. K. Zhu, Operator theory in function spaces, New York, Marcel Dekker, 1990.

UNIQUENESS OF MEROMORPHIC FUNCTIONS CONCERNING WEAKLY WEIGHTED-SHARING

SHANHUA LIN

Department of Mathematics
Fujian Normal University
Fuzhou, 350007 P. R. China
E-mail: jian076@163.com

WEICHUAN LIN

Department of Mathematics
Fujian Normal University
Fuzhou 350007, P. R. China
and
Department of Mathematics
Shantou University
Shantou 515063, P. R. China
E-mail: sxlwc936@fjnu.edu.cn

In this paper, we introduce the definition of weakly weighted-sharing which is between "CM" and "IM". Using the notion of weakly weighted-sharing, we study the uniqueness problems on meromorphic function and its kth order derivative $f^{(k)}$. As consequences, we are able to answer questions posed by Kit-wing Yu, which were also studied by L. P. Liu and Y. X. Gu. Our results sharpen the above results.

Keywords: Weakly weighted-sharing, uniqueness, meromorphic function.

2000 Mathematics Subject Classification: 30D35, 30D45

1. Introduction and Main Results

In this paper, we assume that the reader is familiar with the basic concepts of Nevanlinna value distribution theory and the notations, see e.g. [1, 2]. We denote $S(f)$ the set of all small functions of f.

For any two nonconstant meromorphic functions f and g, and $a \in S(f) \cap$

$S(g)$. Let $N_E(r, a)$ be the counting function of all common zeros of $f - a$ and $g - a$ with the same multiplicities, and $N_0(r, a)$ be the counting functions of all common zeros of $f - a$ and $g - a$ ignoring multiplicities. Denote by $\overline{N}_E(r, a)$ and $\overline{N}_0(r, a)$ the reduced counting functions of f and g corresponding to the counting functions $N_E(r, a)$ and $N_0(r, a)$, respectively. If $\overline{N}(r, \frac{1}{f-a}) + \overline{N}(r, \frac{1}{g-a}) - 2\overline{N}_E(r, a) = S(r, f) + S(r, g)(\overline{N}(r, \frac{1}{f-a}) + \overline{N}(r, \frac{1}{g-a}) - 2\overline{N}_0(r, a) = S(r, f) + S(r, g))$, then we say that f and g share a "CM"("IM"). When $\overline{N}(r, \frac{1}{f-a}) + \overline{N}(r, \frac{1}{g-a}) = 2\overline{N}_E(r, a)(\overline{N}(r, \frac{1}{f-a}) + \overline{N}(r, \frac{1}{g-a}) = 2\overline{N}_0(r, a))$, then we say that f and g share a CM(IM).

In 2003, Kit-wing Yu[3] proved the following results.

Theorem A. *Let $k \geq 1$. Let f be a non-constant entire function, $a \in S(f)$ and $a \not\equiv 0, \infty$. If f, $f^{(k)}$ share a CM and $\delta(0, f) > \frac{3}{4}$, then $f \equiv f^{(k)}$.*

Theorem B. *Let $k \geq 1$. Let f be a non-constant non-entire meromorphic function, $a \in S(f)$ and $a \not\equiv 0, \infty$, f and a do not have any common pole. If f, $f^{(k)}$ share a CM and $4\delta(0, f) + 2(8+k)\Theta(\infty, f) > 19 + 2k$, then $f \equiv f^{(k)}$.*

In the same paper, Kit-wing Yu posed the following open questions:

Question 1. Can a CM shared value be replaced by an IM shared value in Theorem A?

Question 2. Is the condition $\delta(0, f) > \frac{3}{4}$ sharp in Theorem A?

Question 3. Is the condition $4\delta(0, f)+2(8+k)\Theta(\infty, f) > 19+2k$ sharp in Theorem B?

Question 4. Can the condition "f and a do not have any common pole" be deleted in Theorem B?

In 2004, L. P. Liu and Y. X. Gu[4] obtained the following results which answered questions 2 and 3.

Theorem C. *Let f be a non-constant meromorphic function, $a \in S(f)$ and $a \not\equiv 0, \infty$. If f, $f^{(k)}$ share a CM, $f^{(k)}$ and a do not have any common pole of same multiplicity and $2\delta(0, f) + 4\Theta(\infty, f) > 5$, then $f \equiv f^{(k)}$.*

Theorem *D*. *Let f be a non-constant entire function, $a \in S(f)$ and $a \not\equiv 0, \infty$. If f, $f^{(k)}$ share a CM and $\delta(0, f) > \frac{1}{2}$, then $f \equiv f^{(k)}$.*

In this paper, we introduce the definition of weakly weighted-sharing. By the new definition, we obtain uniqueness theorems which answer the questions posed by Kit-wing Yu. Moreover, our results improve Theorem A, B, C, D mentioned above.

Next, we introduce some notations for our definition.

Definition 1. *Let f and g be two nonconstant meromorphic functions sharing "IM", for $a \in S(f) \cap S(g)$, and a positive integer k or ∞.*

(i) $\overline{N}^{E}_{k)}(r, a)$ denotes the counting function of those common a-points of f and g with the same multiplicities, both of their multiplicities are not greater than k, where each a-point is counted only once.

(ii) $\overline{N}^{o}_{(k}(r, a)$ denotes the counting function of those common a-points of f and g, both of their multiplicities are not less than k, where each a-point is counted only once.

Definition 2. *For $a \in S(f) \cap S(g)$, if k is a positive integer or ∞, and*

$$\overline{N}_{k)}(r, \frac{1}{f-a}) - \overline{N}^{E}_{k)}(r, a) = S(r, f), \quad \overline{N}_{k)}(r, \frac{1}{g-a}) - \overline{N}^{E}_{k)}(r, a) = S(r, g);$$

$$\overline{N}_{(k+1}(r, \frac{1}{f-a}) - \overline{N}^{o}_{(k+1}(r, a) = S(r, f),$$

$$\overline{N}_{(k+1}(r, \frac{1}{g-a}) - \overline{N}^{o}_{(k+1}(r, a) = S(r, g).$$

Or if $k = 0$ and

$$\overline{N}(r, \frac{1}{f-a}) - \overline{N}_0(r, a) = S(r, f), \quad \overline{N}(r, \frac{1}{g-a}) - \overline{N}_0(r, a) = S(r, g),$$

then we say f and g weakly share a with weight k. Here, we write f, g share "(a, k)" to mean that f, g weakly share a with weight k.

Obviously, if f and g share "(a, k)", then f and g share "(a, p)" for any $p(0 \le p \le k)$. Also, we note that f and g share a "IM" or "CM" if and only if f and g share "$(a, 0)$" or "(a, ∞)", respectively.

Now, we state the main results of this paper.

Theorem 1. *Let $k \ge 1$ and $2 \le m \le \infty$. Let f be a non-constant meromorphic function, $a \in S(f)$ and $a \not\equiv 0, \infty$. If f, $f^{(k)}$ share "(a, m)" and $2\delta(0, f) + 4\Theta(\infty, f) > 5$, then $f \equiv f^{(k)}$.*

Theorem 2. *Let $k \geq 1$, and let f be a non-constant meromorphic function, $a \in S(f)$ and $a \not\equiv 0, \infty$. If f, $f^{(k)}$ share a "IM" and $5\delta(0,\ f) + (2k+7)\Theta(\infty,\ f) > 2k+11$, then $f \equiv f^{(k)}$.*

If f is a nonconstant entire function, then $\Theta(\infty,\ f) = 1$. So we have the following results.

Corollary 1. *Let $k \geq 1$ and $2 \leq m \leq \infty$. Let f be a non-constant entire function, $a \in S(f)$ and $a \not\equiv 0, \infty$. If f, $f^{(k)}$ share "$(a,\ m)$" and $\delta(0,\ f) > \frac{1}{2}$, then $f \equiv f^{(k)}$.*

Corollary 2. *Let $k \geq 1$, and let f be a non-constant entire function, $a \in S(f)$ and $a \not\equiv 0, \infty$. If f, $f^{(k)}$ share a "IM" and $\delta(0,\ f) > \frac{4}{5}$, then $f \equiv f^{(k)}$.*

Remark 1: *Theorem 1 and Corollary 1 improve Theorem A–D. Theorem 2 and Corollary 2 answer question 1. Meanwhile, we give an affirmative answer to the fourth question.*

2. Some Lemmas

Lemma 1[5]. *Let f be a nonconstant meromorphic function and let k be a positive integer. Then*

(i) $N(r,\ \frac{1}{f^{(k)}}) \leq N(r,\ \frac{1}{f}) + k\overline{N}(r,\ f) + S(r,\ f)$.

(ii) $N(r,\ \frac{1}{f^{(k)}}) \leq T(r,\ f^{(k)}) - T(r,\ f) + N(r,\ \frac{1}{f}) + S(r,\ f)$.

Next, we introduce some notations for the following lemma.

When f and g share 1 "IM", $\overline{N}^L(r,\ \dfrac{1}{f-1})$ denotes the counting function of the 1-points of f whose multiplicities are greater than 1-points of g, where each zero is counted only once. Similarly, we have $\overline{N}^L(r,\ \dfrac{1}{g-1})$. $N_{11}(r,\ \dfrac{1}{f-1})$ denotes the counting function of common simple 1-points of f and g.

Lemma 2. *Let m be a nonnegative integer or ∞. Let F and G be two nonconstant meromorphic functions, and F, G share "$(1,\ m)$". Let*

$$H = (\frac{F''}{F'} - 2\frac{F'}{F-1}) - (\frac{G''}{G'} - 2\frac{G'}{G-1}).$$

If $H \not\equiv 0$, then

(i) If $2 \le m \le \infty$, then

$$T(r,\ F) \le N_2(r,\ F) + N_2(r,\ \frac{1}{F}) + N_2(r,\ G) + N_2(r,\ \frac{1}{G}) + S(r,\ F) + S(r,\ G).$$

(ii) If $m = 1$, then

$$T(r,F) \le N_2(r,\ F) + N_2(r,\ \frac{1}{F}) + N_2(r,\ G) + N_2(r,\ \frac{1}{G}) + \overline{N}^L(r,\ \frac{1}{F-1})$$
$$+S(r,\ F) + S(r,\ G).$$

(iii) If $m = 0$, then

$$T(r,\ F) \le N_2(r,\ F) + N_2(r,\ \frac{1}{F}) + N_2(r,\ G) + N_2(r,\ \frac{1}{G}) + 2\overline{N}^L(r,\ \frac{1}{F-1})$$
$$+\overline{N}^L(r,\ \frac{1}{G-1}) + S(r,\ F) + S(r,\ G).$$

The same inequalities holds for $T(r,\ G)$.

Proof. (i) If $2 \le m \le \infty$, then by the Second Fundamental Theorem, we have

$$T(r,\ F) \le \overline{N}(r,\ F) + \overline{N}(r,\ \frac{1}{F}) + \overline{N}(r,\ \frac{1}{F-1}) - N_0(r,\ \frac{1}{F'}) + S(r,\ F), \quad (2.1)$$

where $N_0(r,\ \frac{1}{F'})$ is the counting function of those zeros of F' which are not the zeros of $F(F-1)$. $N_0(r,\ \frac{1}{G'})$ can be defined similarly.

By a simple calculation, any pole of F is not a pole of $\frac{F''}{F'} - 2\frac{F'}{F-1}$, any pole of G is not a pole of $\frac{G''}{G'} - 2\frac{G'}{G-1}$. Furthermore, let z_1 be a common zero of $F-1$ and $G-1$ with multiplicity t, where $1 \le t \le 2$. We know that H is analytic at z_1. Therefore, by $H \not\equiv 0$, we have

$$N_{1)}(r,\ \frac{1}{F-1}) \le \overline{N}(r,\ \frac{1}{H}) + S(r,\ F) + S(r,\ G) \le T(r,\ H) + S(r,\ F) + S(r,\ G). \quad (2.2)$$

Note that $m(r,\ H) = S(r,\ F) + S(r,\ G)$ and

$$N(r,\ H) \le \overline{N}_{(2}(r,\ F) + \overline{N}_{(2}(r,\ G) + \overline{N}_{(2}(r,\ \frac{1}{F}) + \overline{N}_{(2}(r,\ \frac{1}{G}) +$$

$$N_0(r,\ \frac{1}{F'}) + N_0(r,\ \frac{1}{G'}) + \overline{N}^L(r,\ \frac{1}{F-1}) + \overline{N}^L(r,\ \frac{1}{G-1}) + S(r,\ F) + S(r,\ G).$$

By (2.2), we have

$$N_{1)}(r,\ \frac{1}{F-1}) \le \overline{N}_{(2}(r,\ F) + \overline{N}_{(2}(r,\ G) + \overline{N}_{(2}(r,\ \frac{1}{F}) + \overline{N}_{(2}(r,\ \frac{1}{G}) + N_0(r,\ \frac{1}{F'})$$

$$+N_0(r, \frac{1}{G'}) + \overline{N}^L(r, \frac{1}{F-1}) + \overline{N}^L(r, \frac{1}{G-1}) + S(r, F) + S(r, G). \quad (2.3)$$

Since F and G share "(1, m)", we have
$\overline{N}_{(2}(r, \frac{1}{G-1}) + \overline{N}^L(r, \frac{1}{F-1}) + \overline{N}^L(r, \frac{1}{F-1}) + N_0(r, \frac{1}{G'}) + N(r, \frac{1}{G}) - \overline{N}(r, \frac{1}{G}) \leq N(r, \frac{1}{G'})$.

It follows from Lemma 1 that

$$\overline{N}_{(2}(r, \frac{1}{G-1}) + \overline{N}^L(r, \frac{1}{F-1}) + \overline{N}^L(r, \frac{1}{G-1}) + N_0(r, \frac{1}{G'})$$

$$\leq \overline{N}(r, \frac{1}{G}) + \overline{N}(r, G) + S(r, G). \quad (2.4)$$

In addition, we have

$$\overline{N}(r, \frac{1}{F-1}) = N_{1)}(r, \frac{1}{F-1}) + \overline{N}_{(2}(r, \frac{1}{F-1})$$

$$= N_{1)}(r, \frac{1}{F-1}) + \overline{N}_{(2}(r, \frac{1}{G-1}). \quad (2.5)$$

Combining (2.1), (2.3), (2.4) and (2.5), we obtain

$$T(r, F) \leq N_2(r, F) + N_2(r, \frac{1}{F}) + N_2(r, G) + N_2(r, \frac{1}{G}) + S(r, F) + S(r, G).$$

(ii) If $m = 1$, then (2.4) is replaced by

$$\overline{N}_{(2}(r, \frac{1}{G-1}) + \overline{N}^L(r, \frac{1}{G-1}) + N_0(r, \frac{1}{G'}) \leq \overline{N}(r, \frac{1}{G}) + \overline{N}(r, G) + S(r, G).$$

Similar to the arguments in (i), we see that (ii) holds.

(iii) If $m = 0$, then by a simple calculation, any common simple zero of $F - 1$ and $G - 1$ is a zero of H. Therefore, by $H \not\equiv 0$, we have

$$N_{11}(r, \frac{1}{F-1}) \leq \overline{N}(r, \frac{1}{H}) + S(r, F) + S(r, G) \leq T(r, H) + S(r, F) + S(r, G). \quad (2.6)$$

Thus

$$N_{11}(r, \frac{1}{F-1}) \leq \overline{N}_{(2}(r, F) + \overline{N}_{(2}(r, G) + \overline{N}_{(2}(r, \frac{1}{F}) + \overline{N}_{(2}(r, \frac{1}{G}) + N_0(r, \frac{1}{F'})$$

$$+N_0(r, \frac{1}{G'}) + \overline{N}^L(r, \frac{1}{F-1}) + \overline{N}^L(r, \frac{1}{G-1}) + S(r, F) + S(r, G). \quad (2.7)$$

By the Second Fundamental Theorem, we have

$$T(r,\ F)+T(r,\ G)\leq \overline{N}(r,\ F)+\overline{N}(r,\ \frac{1}{F})+\overline{N}(r,\ \frac{1}{F-1})-N_0(r,\ \frac{1}{F'})$$
$$+\overline{N}(r,\ G)+\overline{N}(r,\ \frac{1}{G})+\overline{N}(r,\ \frac{1}{G-1})-N_0(r,\ \frac{1}{G'})+S(r,\ F)+S(r,\ G).\quad (2.8)$$

In addition, we have

$$\overline{N}(r,\ \frac{1}{F-1})+\overline{N}(r,\ \frac{1}{G-1})$$
$$=2\overline{N}(r,\ \frac{1}{G-1})+S(r,\ F)+S(r,\ G)$$
$$\leq N_{11}(r,\ \frac{1}{F-1})+\overline{N}^L(r,\ \frac{1}{F-1})+N(r,\ \frac{1}{G-1})+S(r,\ F)+S(r,\ G)$$
$$\leq N_{11}(r,\ \frac{1}{F-1})+\overline{N}^L(r,\ \frac{1}{F-1})+T(r,\ G)+S(r,\ F)+S(r,\ G).$$

Combining (2.6), (2.7) and (2.8), we obtain

$$T(r,\ F)\leq N_2(r,\ F)+N_2(r,\ \frac{1}{F})+N_2(r,\ G)+N_2(r,\ \frac{1}{G})+2\overline{N}^L(r,\ \frac{1}{F-1})$$
$$+\overline{N}^L(r,\ \frac{1}{G-1})+S(r,\ F)+S(r,\ G).$$

Lemma 3[6]. *Let f be a transcendental meromorphic function and $\alpha(\not\equiv 0,\infty)$ be a meromorphic function such that $T(r,\ \alpha)=S(r,\ f)$. Let b and c are any two finite nonzero distinct complex numbers. If $\psi=\alpha(f)^n(f^{(k)})^p$, where $n(\geq 0)$, $p(\geq 1)$ and $k(\geq 1)$ are integers, then*

$$(p+n)T(r,\ f)\leq (p+n)N(r,\ \frac{1}{f})+N(r,\ \frac{1}{\psi-b})+N(r,\ \frac{1}{\psi-c})$$
$$-N(r,\ f)-N(r,\ \frac{1}{\psi'})+S(r,\ f).$$

3. Proofs of Main Theorems

Proof of Theorem 1

Let

$$F=\frac{f}{a},\qquad G=\frac{f^{(k)}}{a}.\tag{3.1}$$

Then it is easy to verify F and G share "(1, m)".

Let H be defined as in Lemma 2. Suppose that $H\not\equiv 0$. It follows from Lemma 2 that

$$T(r,\ G)\leq N_2(r,\ F)+N_2(r,\ \frac{1}{F})+N_2(r,\ G)+N_2(r,\ \frac{1}{G})+S(r,\ F)+S(r,\ G).$$

Using Lemma 1, we have

$$T(r,\ f^{(k)}) \le N_2(r,\ f) + N_2(r,\ \frac{1}{f}) + N_2(r,\ f^{(k)}) + N_2(r,\ \frac{1}{f^{(k)}}) + S(r,\ f)$$
$$\le N(r,\ \frac{1}{f}) + T(r,\ f^{(k)}) - T(r,\ f) + N(r,\ \frac{1}{f}) + 4\overline{N}(r,\ f) + S(r,\ f)$$

i.e.

$$T(r,\ f) \le 2N(r,\ \frac{1}{f}) + 4\overline{N}(r,\ f) + S(r,\ f).$$

It follows that $2\delta(0,\ f) + 4\Theta(\infty,\ f) \le 5$, which contradicts $2\delta(0,\ f) + 4\Theta(\infty,\ f) > 5$. Therefore $H \equiv 0$. That is

$$\frac{F''}{F'} - 2\frac{F'}{F-1} \equiv \frac{G''}{G'} - 2\frac{G'}{G-1}.$$

It follows that

$$\frac{1}{F-1} = \frac{A}{G-1} + B,$$

where $A(\neq 0)$ and B are constants. Therefore,

$$F = \frac{(B+1)G + (A-B-1)}{BG + (A-B)}. \tag{3.2}$$

and

$$T(r,\ F) = T(r,\ G) + S(r,\ f).$$

Now we distinguish the following two cases.

Case 1. Suppose that $B \neq -1, 0$.

If $A - B - 1 \neq 0$, then from (3.2), we have $\overline{N}(r,\ \frac{1}{G + \frac{A-B-1}{B+1}}) = \overline{N}(r,\ \frac{1}{F})$.

By the Second Fundamental Theorem, we have $T(r,\ G) < \overline{N}(r,\ G) + \overline{N}(r,\ \frac{1}{G}) + \overline{N}(r,\ \frac{1}{G + \frac{A-B-1}{B+1}}) + S(r,\ G)$, i.e.

$$T(r,\ f^{(k)}) < \overline{N}(r,\ f) + \overline{N}(r,\ \frac{1}{f^{(k)}}) + \overline{N}(r,\ \frac{1}{f}) + S(r,\ f)$$
$$\le \overline{N}(r,\ f) + T(r,\ f^{(k)}) - T(r,\ f) + N(r,\ \frac{1}{f}) + \overline{N}(r,\ \frac{1}{f}) + S(r,\ f),$$

and so

$$T(r,\ f) < 2N(r,\ \frac{1}{f}) + \overline{N}(r,\ f) + S(r,\ f).$$

It follows that $2\delta(0, f) + \Theta(\infty, f) \le 2$, which contradicts $2\delta(0, f) + 4\Theta(\infty, f) > 5$. Therefore, $A - B - 1 = 0$. From (3.2), we obtain $\overline{N}(r, \frac{1}{G + \frac{1}{B}}) = \overline{N}(r, F)$.

Similar to the arguments in the above, we also have a contradiction.

Case 2. Suppose that $B = -1$.

If $A + 1 \ne 0$. Then from (3.2), we have $\overline{N}(r, \frac{1}{G - (A+1)}) = \overline{N}(r, F)$.

Similar to the arguments in Case 1, we can get a contradiction. Therefore $A + 1 = 0$, then from (3.2), we have $FG \equiv 1$. From (3.1), we have

$$ff^{(k)} \equiv a^2. \tag{3.3}$$

In the following, we distinguish two subcases.

a) If f is a rational function, then a becomes a nonzero constant. So from (3.1), we see that f has no zero and pole. Since f is nonconstant, this is a contradiction.

b) If f is transcendental then by Lemma 3, we get in view of (3.1)

$$\begin{aligned} 2T(r, f) &\le 2N(r, \frac{1}{f}) + 2T(r, ff^{(k)}) + S(r, f) \\ &\le 2N(r, \frac{1}{f}) + S(r, f) \\ &\le 2N(r, \frac{1}{a^2}) + S(r, f) = S(r, f), \end{aligned}$$

This is a contradiction.

Case 3. Suppose that $B = 0$.

If $A - 1 \ne 0$, then from (3.2), we have $\overline{N}(r, \frac{1}{G + (A-1)}) = \overline{N}(r, \frac{1}{F})$.

Similar to the arguments in case 1, we also have a contradiction. Therefore $A - 1 = 0$. From (3.2), we have $F \equiv G$, this implies $f \equiv f^{(k)}$.

This completes the proof of the Theorem 1.

Proof of Theorem 2. Using Lemma 2(iii), note that

$$\begin{aligned} \overline{N}^{L}(r, \frac{1}{G-1}) &\le N(r, \frac{G}{G'}) \le N(r, \frac{G'}{G}) + S(r, f) \\ &\le \overline{N}(r, G) + \overline{N}(r, \frac{1}{G}) + S(r, f) \\ &\le \overline{N}(r, f) + \overline{N}(r, \frac{1}{f^{(k)}}) + S(r, f). \end{aligned}$$

and

$$\overline{N}^L(r,\ \frac{1}{F-1}) \le N(r,\ \frac{F}{F'}) \le N(r,\ \frac{F'}{F}) + S(r,\ f)$$
$$\le \overline{N}(r,\ F) + \overline{N}(r,\ \frac{1}{F}) + S(r,\ f)$$
$$\le \overline{N}(r,\ f) + \overline{N}(r,\ \frac{1}{f}) + S(r,\ f).$$

Similar to the arguments in Theorem 1, we see that Theorem 2 holds.

References

1. W. K. Hayman, *Meromorphic Functions*, Clarendon Press, Oxford, 1964.
2. L. Yang, *Value Distribution Theory*, Springer-Verlag, Berlin, 1993.
3. Kit-wing Yu, On entire and meromorphic functions that share small functions with their derivatives, *J. Inequal. Pure and Appl. Math.*, **4**(1)(2003):1-7.
4. L. P. Liu and Y. X. Gu, Uniqueness of meromorphic functions that share one small function with their derivatives, *Kodai. Math. J.* , **27**(2004):272-279.
5. H. X. Yi and C. C. Yang, *Uniqueness Theory of Meromorphic Functions*, Science Press, Beijing, 1995(In Chinese).
6. I. Lahiri and S. Dewan, Inequealities arising out of the value distribution of a differential monomial, *J. Inequal. Pure Appl. Math.*, **4**(2)(2003): Art.27.

UNIQUENESS THEORY OF MEROMORPHIC FUNCTIONS IN AN ANGULAR DOMAIN

WEICHUAN LIN

Department of Mathematics
Shantou University
Shantou, 515063 P. R. China
and
Department of Mathematics
Fujian Normal University
Fuzhou, 350007 P. R. China
E-mail: linwei936@263.net

SEIKI MORI

Department of Mathematical Sciences, Faculty of Science
Yamagata University
Yamagata, 990-8560, Japan
E-mail: mori@sci.kj.yamagata-u.ac.jp

In this paper, we deal with a problem of uniqueness for meromorphic functions defined in the whole plane **C** under certain value/set-sharing conditions in a sector instead of the plane.

Keywords: Shared-set, uniqueness, meromorphic function, angular domain.

2000 Mathematics Subject Classification: 30D35

1. Introduction and Main Results

In this paper, for a discrete subset S of $C\bigcup\{\infty\}$ we put $E(S,f) := \bigcup_{a\in S} E(a,f)$ with $E(a,f) := \{z \in C | f_a(z) = 0\}$, where $f_a(z) = f(z) - a$ when $a \in C$ and $f_\infty(z) = 1/f(z)$, and all the roots are counted according to its multiplicity (CM).

Given a domain $X \subset C$, a set $S \subset C\bigcup\{\infty\}$ and a meromorphic function

f in C, we define a 'set' $E_X(S,f) \subset C$ similarly to the above when $X = C$:

$$E_X(S,f) = \bigcup_{a \in S} \{z \in \overline{X} | f_a(z) = 0, CM\},$$

where $\overline{X}$ is the closure of the domain X in C. Let f and g be two non-constant meromorphic functions. If $E_X(S,f) = E_X(S,g)$, we say f and g share the set S CM (counting multiplicity) in X. When $S = \{a\}$, we also say f and g share a CM in X.

We assume that the reader is familiar with the basic results and notations of Nevanlinna's value distribution theory (see [1, 2]), such as $T(r,f)$, $N(r,f)$ and $m(r,f)$. Meanwhile, the order ρ and hyper-order σ_2 of f are defined by

$$\rho := \rho(f) = \limsup_{r \to \infty} \frac{\log T(r,\ f)}{\log r},$$

$$\sigma_2 := \sigma_2(f) = \limsup_{r \to \infty} \frac{\log \log T(r,f)}{\log r}.$$

Here, we denote $M(\sigma_2)(E(\sigma_2))$ by the set of transcendental meromorphic (entire) functions of finite hyper-order.

Since R. Nevanlinna proved his celebrated 'four-CM' and 'five-IM' theorems, there have been given a lot of results on the uniqueness of meromorphic functions in the plane C under a variety of value-sharing conditions. Although examples such as e^z and e^{-z} with values $0, 1$ and ∞ show that 'four' and 'five' in those theorems cannot be replaced by any smaller number, the Weierstrass Product Theorem says that 'three-CM' is in general a sufficient condition to deduce the uniqueness of meromorphic functions in C with no exceptional values.

In connection to this fact, F. Gross (see [3, 4]) proposed an idea of set-sharing conditions and asked about the minimum number or the smallest cardinality of the sets which imply the uniqueness of meromorphic functions sharing those sets.

Let $S_j (j = 1,2,3)$ as follows: $S_1 = \{0\}$, $S_2 = \{\infty\}$ and $S_3 = \{w | w^n(w + a) - b = 0\}$, where $n \in N$, and the algebraic equation $w^n(w+a) - b = 0$ has no multiple roots.

In 1998, H. X. Yi (see [5]) answer the question posed by Gross (see [4]) and proved the following:

Theorem A. Let $n \in N \setminus \{1\}$. If f and g are two entire functions satisfying $E_C(S_j, f) = E_C(S_j, g)$ for $j = 1, 3$, then $f \equiv g$.

We shall deal with Theorem A under certain value/set-sharing conditions in a sector instead of the plane C.

Theorem 1. Let $f \in M(\sigma_2)$, $\rho(f) = \infty$, and $\delta(\infty, f) > 0$. Then there exists a direction $\arg z = \alpha$ $(0 \le \alpha < 2\pi)$ such that for any ε $(0 < \varepsilon < \frac{\pi}{2})$, if a meromorphic function $g \in M(\sigma_2)$ satisfies the condition $E_C(S_1, f) = E_C(S_1, g)$ and $E_X(S_j, f) = E_X(S_j, g)$ for $j = 2, 3$ where $n \ge 3$ and $X = X(\alpha - \varepsilon,\ \alpha + \varepsilon)$, then $f \equiv g$.

Theorem 2. Let $f \in E(\sigma_2)$ and $\rho(f) = \infty$. Then there exists a direction $\arg z = \alpha$ $(0 \le \alpha < 2\pi)$ such that for any ε $(0 < \varepsilon < \frac{\pi}{2})$, if an entire function $g \in E(\sigma_2)$ satisfies the condition $E_C(S_1, f) = E_C(S_1, g)$ and $E_X(S_3, f) = E_X(S_3, g)$ where $n \in N \setminus \{1\}$ and $X = X(\alpha - \varepsilon,\ \alpha + \varepsilon)$, then $f \equiv g$.

2. Some Lemmas

We shall prove the Theorems by using the Nevanlinna theory in an angular domain. First of all, we recall some notations and definitions. Let $f(z)$ be a meromorphic function on the angular domain $\overline{X}(\alpha,\ \beta) = \{z | \alpha \le \arg z \le \beta\}$, where $0 < \beta - \alpha \le 2\pi$. Nevanlinna defined the following notations (see [6, 7]).

$$A_{\alpha,\beta}(r, f) := \frac{\omega}{\pi} \int_1^r \left(\frac{1}{t^\omega} - \frac{t^\omega}{r^{2\omega}} \right) \{\log^+ |f(te^{i\alpha})| + \log^+ |f(te^{i\beta})|\} \frac{dt}{t}$$

$$B_{\alpha,\beta}(r, f) := \frac{2\omega}{\pi r^\omega} \int_\alpha^\beta \log^+ |f(re^{i\theta})| \sin \omega(\theta - \alpha) d\theta$$

$$C_{\alpha,\beta}(r, f) := 2 \sum_{1<|b_m|<r} \left(\frac{1}{|b_m|^\omega} - \frac{|b_m|^\omega}{r^{2\omega}} \right) \sin \omega(\theta_m - \alpha),$$

where $\omega = \pi/(\beta - \alpha)$, $1 \le r < \infty$ and $b_m = |b_m| e^{i\theta_m}$ are the poles of $f(z)$ on $X(\alpha, \beta)$ appeared according to their multiplicities. $C_{\alpha,\beta}(r, f)$ is called the angular counting function of the poles of f on $\overline{X}$ and the Nevanlinna angular characteristic is defined as follows:

$$S_{\alpha,\beta}(r, f) := A_{\alpha,\beta}(r, f) + B_{\alpha,\beta}(r, f) + C_{\alpha,\beta}(r, f).$$

Similarly, for $a \ne \infty$, we can define $A_{\alpha,\beta}(r, 1/(f - a))$, $B_{\alpha,\beta}(r, 1/(f - a))$, $C_{\alpha,\beta}(r, 1/(f - a))$, $S_{\alpha,\beta}(r, 1/(f - a))$ and so on. For the sake of simplicity, we omit the subscript of all the notations and use the notations $A(r, a)$, $B(r, a)$, $C(r, a)$ and $S(r, a)$ instead of $A_{\alpha,\beta}(r, 1/(f-a))$, $B_{\alpha,\beta}(r, 1/(f-a))$,

$C_{\alpha,\beta}(r, 1/(f-a))$ and $S_{\alpha,\beta}(r, 1/(f-a))$ if $a \neq \infty$. We shall give some properties of $S(r, f)$ as follows.

Lemma 1 [7]. *Let $f(z)$ be a meromorphic function on $\overline{X}(\alpha, \beta)$. Then for arbitrary finite complex number a, we have*

$$S(r, a) = S(r, f) + \varepsilon(r, a),$$

where $\varepsilon(r, a) = O(1)$ as $r \to \infty$.

Lemma 2 [7]. *Let $P(z)$ be a polynomial of degree $d \geq 1$, and $f(z)$ be a meromorphic function on $\overline{X}(\alpha, \beta)$. Then*

$$S(r, P(f)) = dS(r, f) + O(1).$$

Lemma 3 [7]. *Let $f(z)$ be a meromorphic function in $\mathbf{C}$, $\overline{X}(\alpha, \beta) \subseteq \mathbf{C}$.*

Then

$$A\left(r, \frac{f'}{f}\right) + B\left(r, \frac{f'}{f}\right) = Q(r, f),$$

where $Q(r, f)$ satisfies

i). $Q(r, f) = O(1)$ as $r \to \infty$ if $\rho(f) < \infty$;
ii). $Q(r, f) = O(\log rT(r, f))$ as $r \to \infty$ and $r \notin E$ if $\rho(f) = \infty$, where E is the set of finite linear measure.

Lemma 4. *Let $f(z)$ be a meromorphic function in $\mathbf{C}$, $\overline{X}(\alpha, \beta) \subseteq \mathbf{C}$. Then*

$$S(r, f) \leq \overline{C}(r, f) + \overline{C}\left(r, \frac{1}{f}\right) + \overline{C}\left(r, \frac{1}{f-1}\right) - C_0(r, f) + Q(r, f),$$

where $Q(r, f) = O(\log rT(r, f))$, $r \notin E$ and E is a set of finite linear measure. $\overline{C}(r, f)$ is the reduced counting function of poles of f, the number of each distinct pole of $f(z)$ in $X \cap \{z||z| < r\}$ be counted only once; $C_0(r, 1/f')$ is the counting function of the zeros of f' but not the zeros of f and $f-1$ in $X \cap \{z||z| < r\}$.

Next, we introduce some notations for the following main lemma. Let f be a meromorphic function in an angular domain $X(\alpha, \beta)$. We denote by $C_2(r, f)$ the counting function of the poles of f in $\{z \in X(\alpha, \beta)||z| < r\}$, where a simple pole is counted once and a multiple pole is counted twice. In the same way, we can define $C_2(r, 1/f)$.

Lemma 5. *Let F and G be two nonconstant meromorphic functions such that F and G share $1, \infty$ CM in $X(\alpha, \beta)$. Then one of the following three cases holds:*

i). $S(r) \leq C_2(r, 1/F) + C_2(r, 1/G) + 2\overline{C}(r, F) + Q(r, F) + Q(r, G)$;
ii). $F \equiv G$;
iii). $FG \equiv 1$, *where* $S(r) = \max\{S(r, F), S(r, G)\}$, $Q(r, F)$ *and* $Q(r, G)$ *as defined in Lemma 3.*

Let f and g be two nonconstant meromorphic functions in $\mathbf{C}$, and $S_3 = \{w | w^n(w + a) - b = 0\}$, where $n(\geq 2)$ is an integer, a and b are two nonzero constants such that the algebraic equation $w^n(w+a)-b = 0$ has no multiple roots. We denote

$$F = \frac{f^n(f + a)}{b}, \quad G = \frac{g^n(g + a)}{b}. \tag{1}$$

Obviously, if $E_X(S_3, f) = E_X(S_3, g)$ then F and G share 1 CM in X. In the following, we shall give some lemmas relating to F and G.

Lemma 6. *Suppose that $\overline{E}_C(\{0\}, f) = \overline{E}_C(\{0\}, g)$ and $\delta(\infty, f) > 0$. If $F \equiv G$, where F and G are defined as (1), then $f \equiv g$.*

Lemma 7. *Let S_j $(j = 1, 2, 3)$ be defined as in Theorem 1, and let F and G be defined as (1) such that $\overline{E}_X(S_1, f) = \overline{E}_X(S_1, g)$, $E_X(S_2, f) = E_X(S_2, g)$ and $E_X(S_3, f) = E_X(S_3, g)$. If $F \not\equiv G$, then*

$$\overline{C}\Big(r, \frac{1}{f}\Big) = \overline{C}\Big(r, \frac{1}{g}\Big) = Q(r, f) + Q(r, g).$$

Lemma 8. *Under the condition of Lemma 7, we have*

$$\overline{C}(r, f) = \overline{C}(r, g) \leq \frac{1}{n}(S(r, f) + S(r, g)) + Q(r, f) + Q(r, g). \tag{2}$$

Finally, we need the following important lemma in [8, Theorem VII. 3]. We first introduce some notations.

Let $f(z)$ be a meromorphic function in an angular domain $\Delta_0 : |\arg z| \leq \alpha_0$ and $\Delta : |\arg z| \leq \alpha(< \alpha_0)$ be an angular domain contained in Δ_0. Let $\Delta_0(r)$, $\Delta(r)$ be the part of Δ_0, Δ, which is contained in $|z| \leq r$, respectively. We put

$$S^*(r, \Delta) = \frac{1}{\pi} \int\int_{\Delta(r)} \Big(\frac{|f'(z)|}{1 + |f(z)|^2}\Big)^2 r dr d\theta, \quad z = re^{i\theta},$$

$$T^*(r, \Delta) = \int_0^r \frac{S^*(r, \Delta)}{r} dr,$$

which is called as Ahlfors-Shimizu characteristics. We write the above characteristic functions of $f(z)$ in the whole complex plain as $S^*(r, f)$, $T^*(r, f)$.

Let $\overline{n}(r, \Delta_0, a)$ be the number of zeros of $f(z) - a$ contained in Δ_0, where each multiple zero is counted only once. We put

$$\overline{N}(r, \Delta_0, a) := \overline{N}(r, \Delta_0, f = a) = \int_1^r \frac{\overline{n}(r, \Delta_0, a)}{r} dr.$$

Then we shall give the following analogue of the second fundamental theorem.

Lemma 9. *Let $f(z)$ be a meromorphic function in the complex plane. Then for any three distinct points a_1, a_2, a_3 on $\hat{\mathbf{C}}$, we have*

$$S^*(r, \Delta) \leq 3 \sum_{i=1}^{3} \overline{n}(2r, \Delta_0, a_i) + O(\log r),$$

and

$$T^*(r, \Delta) \leq 3 \sum_{i=1}^{3} \overline{N}(2r, \Delta_0, a_i) + O((\log r)^2).$$

Finally, from [1, Theorem 1.4], we have

Lemma 10. *Let $f(z)$ be a meromorphic function in the complex plane. Then*

$$|T(r, f) - T^*(r, f) - \log^+ |f(0)|| \leq \frac{1}{2} \log 2.$$

Lemma 11 [9, Lemma 1.1.1]. *Let $g : (0, +\infty) \to \mathbf{R}$, $h : (0, +\infty) \to \mathbf{R}$ be monotone increasing functions such that $g(r) \leq h(r)$ outside of an exceptional set E of finite linear measure. Then, for any $\alpha > 1$, there exists $r_0 > 0$ such that $g(r) \leq h(\alpha r)$ for all $r > r_0$.*

3. Proof of Theorems

We only prove Theorem 1 since we can obtain Theorem 2 by the similar proof method.

Suppose that Theorem 1 does not hold. Then for any $\alpha \in [0, 2\pi)$, we have a constant $\varepsilon_\alpha \in (0, \pi/2)$ and a meromorphic function $g = g_{[\alpha]}$ in $\mathbf{C}$

such that $E_C(S_1, f) = E_C(S_1, g_{[\alpha]})$ and $E_{X(\alpha)}(S_j, f) = E_{X(\alpha)}(S_j, g_\alpha)$ for $j = 2, 3$, but $f \not\equiv g_{[\alpha]}$, where $X(\alpha) := \{z \,|\, |\arg z - \alpha| < \varepsilon_\alpha\}$. First we define F and G as (1), then F and G share 1 and ∞ CM in X.

By Lemma 6, we deduce that $F \not\equiv G$. Thus, Lemma 7 implies that

$$\overline{C}(r, 1/f) = \overline{C}(r, 1/g_{[\alpha]}) = Q(r, f) + Q(r, g_{[\alpha]}). \tag{3}$$

Therefore, by the definition of F and G and (3) we have

$$\begin{aligned} C_2\Big(r, \frac{1}{F}\Big) + C_2\Big(r, \frac{1}{G}\Big) + 2\overline{C}(r, F) \\ \le C\Big(r, \frac{1}{f+a}\Big) + C\Big(r, \frac{1}{g_{[\alpha]}+a}\Big) + 2\overline{C}(r, f) \\ + Q(r, f) + Q(r, g_{[\alpha]}). \end{aligned} \tag{4}$$

Set $S_1(r) := \max\{S(r, f), S(r, g_{[\alpha]})\}$. Then, from (1) and Lemma 2, we have

$$S(r) = (n+1)S_1(r) + O(1), \tag{5}$$

where $S(r) = \max\{S(r, F), S(r, G)\}$. By the estimate (2) obtained in Lemma 8 and (4) we deduce that

$$C_2\Big(r, \frac{1}{F}\Big) + C_2\Big(r, \frac{1}{G}\Big) + 2\overline{C}(r, F) \le \Big(2 + \frac{4}{n}\Big) S_1(r) + Q(r, f) + Q(r, g_{[\alpha]}). \tag{6}$$

Suppose that $FG \equiv 1$. From (1) we obtain

$$f^n(f+a)g_{[\alpha]}^n(g_{[\alpha]} + a) \equiv b^2,$$

which implies that $0, -a$ and ∞ are all Picard exceptional values of f. This is a contradiction. Therefore, $FG \not\equiv 1$, and hence, by Lemma 5 and noting that $n \ge 3$, we have from (5) and (6), $S_1(r) \le Q(r, f) + Q(r, g_{[\alpha]})$. Therefore, by Lemma 3, we have

$$S(r, f) = O\Big(\log rT(r, f)T(r, g_{[\alpha]})\Big), \quad r \notin E_\alpha, \tag{7}$$

where E_α is a set of r of finite linear measure possibly depending on α. Therefore, by Lemma 1 and (7), for any complex number $a \in \hat{\mathbf{C}}$, we have

$$C_{X(\alpha)}(r,\ a) := C_{\alpha-\varepsilon_\alpha,\ \alpha+\varepsilon_\alpha}(r,\ a) = O\big(\log rT(r, f)T(r, g_{[\alpha]})\big), \quad r \notin E_\alpha. \tag{8}$$

On the other hand, we define $X(\alpha)_1 = \{z| \,|\arg z - \alpha| < \varepsilon_\alpha/2\}$. Noting that if an a-point $b_m = |b_m|e^{i\theta_m}$ of f is in $X(\alpha)_1$,

$$\sin\omega\big(\theta_m - (\alpha - \varepsilon_\alpha)\big) \ge \sin(\omega\frac{\varepsilon_\alpha}{2}) = \sqrt{2}/2$$

since $\omega = \omega_\alpha := \pi \,/\, (2\varepsilon_\alpha)\,(> 1)$, we have

$$\begin{aligned}
C_{X(\alpha)}(2r,\ a) &= 2 \sum_{\substack{1<|b_m|<2r \\ |\theta_m-\alpha|<\varepsilon_\alpha}} \left\{ \frac{1}{|b_m|^\omega} - \frac{|b_m|^\omega}{(2r)^{2\omega}} \right\} \sin\omega\Big(\theta_m - (\alpha - \varepsilon_\alpha)\Big) \\
&\geq \sqrt{2} \sum_{\substack{1<|b_m|<2r \\ |\theta_m-\alpha|<\varepsilon_\alpha/2}} \left\{ \frac{1}{|b_m|^\omega} - \frac{|b_m|^\omega}{(2r)^{2\omega}} \right\} \\
&= \sqrt{2} \left\{ \frac{n\big(2r, X(\alpha)_1, a\big)}{(2r)^\omega} + \omega \int_1^{2r} \frac{n\big(t, X(\alpha)_1, a\big)}{t^{\omega+1}} dt \right. \\
&\quad \left. - \frac{1}{(2r)^\omega} n\big(2r, X(\alpha)_1, a\big) + \frac{\omega}{(2r)^{2\omega}} \int_1^{2r} n\big(t, X(\alpha)_1, a\big) t^{\omega-1} dt \right\} \\
&\geq \sqrt{2} \left\{ \frac{n\big(r, X(\alpha)_1, a\big)}{(2r)^{\omega+1}} (2r - r) + n\big(r, X(\alpha)_1, a\big) r^{\omega-1}(2r - r) \right\} \\
&\geq \sqrt{2} n\big(r, X(\alpha)_1, a\big) \left\{ \frac{1}{2^{\omega+1}} r^{-\omega} + r^\omega \right\}.
\end{aligned}$$

It follows from (8) that

$$n\big(r, X(\alpha)_1, a\big) = O(r^\omega \log r T(2r, f) T(2r, g_{[\alpha]})), \quad r \notin E_\alpha. \tag{9}$$

If we identify the interval $[0, 2\pi)$ with the unit circle and $(\alpha - \varepsilon_\alpha/4,\ \alpha + \varepsilon_\alpha/4)$ with the corresponding open arc on the unit circle. Then since the unit circle is compact and

$$[0,\ 2\pi) \subseteq \bigcup_{\alpha\in[0,2\pi)} \left(\alpha - \frac{\varepsilon_\alpha}{4},\ \alpha + \frac{\varepsilon_\alpha}{4}\right),$$

we can choose finite many coverings $(\alpha_1 - \varepsilon_{\alpha_1}/4,\ \alpha_1 + \varepsilon_{\alpha_1}/4)$, $(\alpha_2 - \varepsilon_{\alpha_2}/4,\ \alpha_2 + \varepsilon_{\alpha_2}/4)$, $\ldots$, $(\alpha_k - \varepsilon_{\alpha_k}/4,\ \alpha_k + \varepsilon_{\alpha_k}/4)$ of the interval $[0, 2\pi)$. Therefore, using Lemma 9 and (9), for any three distinct complex numbers $a_j,\ j = 1, 2, 3$, we have

$$\begin{aligned}
S^*(r, f) &\leq \sum_{i=1}^{k} S^*(r, \Delta_i, f) \\
&\leq \sum_{i=1}^{k} \Big\{ 3 \sum_{j=1}^{3} \overline{n}(2r, \Delta_i, f = a_j) \Big\} + O(\log r) \\
&= O\big(r^\Omega \log r T(4r, f) T(4r, g_{[\alpha]})\big), \quad r \notin E_0,
\end{aligned}$$

where
$\Delta_i = \Delta_{(\alpha_i - \varepsilon_{\alpha_i}/4,\ \alpha_i + \varepsilon_{\alpha_i}/4)}$ $(i = 1, \ldots, k)$, $\Omega := \max\{\omega(\alpha_1), \ldots, \omega(\alpha_k)\}$ and $E_0 := \cup_1^k E_{\alpha_i}$.

Thus,

$$T^*(r, f) = O\big(r^{\Omega+1} \log r T(4r, f) T(4r, g_{[\alpha]})\big), \quad r \notin E_0. \tag{10}$$

By Lemma 10, it follows from (10) that

$$T(r, f) = O\big(r^{\Omega+1} \log r T(4r, f) T(4r, g_{[\alpha]})\big), \quad r \notin E_0.$$

It follows from Lemma 11 that there exists $r_0 > 0$ such that for all $r > r_0$, we have

$$T(r, f) = O\big(r^{\Omega+1} \log r T(5r, f) T(5r, g_{[\alpha]})\big),$$

which contradicts $\rho(f) = \infty$ since

$$\limsup_{r\to\infty} \frac{\log\log T(r, f)}{\log r} < \infty \quad \text{and} \quad \limsup_{r\to\infty} \frac{\log\log T(r, g_{[\alpha]})}{\log r} < \infty.$$

This completes the proof of Theorem 1.

References

1. W. K. Hayman, *Meromorphic Functions,* Clarendon, Oxford, 1964.
2. H. X. Yi and C. C. Yang, Uniqueness Theory of Meromorphic Functions, *Pure and Applied Math. Monographs No.32,* Science Press, Beijing, 1995.
3. F. Gross, On the distribution of values of meromorphic functions, *Trans. Amer. Math. Soc.,* **131** (1968):199-214.
4. F. Gross, Factorization of meromorphic functions and some problems in Complex Analysis (Proc. Conf. Univ. Kentucky, Lexington, KY, 1976), *Lecture Notes in Math.,* **599** (Springer-Verlag, Berlin, Heidelberg, New York, 1977), pp. 51-69.
5. H. X. Yi, On a question of Gross concerning uniqueness of entire functions, *Bull. Austral. Math. Soc.,* **57** (1998):343-349.
6. R. Nevalinna, Uber die Eigenschaften meromorpher Funktionen in einem Winkelraum, *Acta Soc. Sci. Fenn.,* **50** (12)(1925):1-45.
7. A. A. Gol'dberg and I. V. Ostrovskii, *The Distribution of Values of Meromorphic Functions (in Russian),* Izdat. Nauk. Moscow 1970.
8. M. Tsuji, *Potential theory in mordern function theory,* Maruzen Co. Ltd, Tokyo, 1959.
9. I. Laine, *Nevanlinna Theory and Complex Differential Equations,* De Gruyter, New York, 1993.

THE CONVERGENCE OF LAPLACE-STIELTJES TRANSFORMS*

LUO XI and SUN DAOCHUN

Department of Mathematics
South China Normal University
Guangzhou, 510631 P. R. China
E-mail: luoxi8100@yahoo.com.cn

In this paper,we study the convergent properties of Laplace-Stieltjes transforms and obtain some completely different results.

Keywords: Laplace-Stieltjes transform; abscissa of convergence; abscissa of uniform convergence; abscissa of absolute convergence.

2000 Mathematics Subject Classification: 30D35

1. Introduction

Yu Jiarong [1] combined Valion [3]and Knopp [4]'s ideas,and extended the Cauchy-Hadamard formulas of powder series to the Laplace-Stieltjes transforms. Yu Jiarong [5] had the corespondent results about the double Dirichlet series and double Laplace-Stieltjes transforms. The results about Laplace-Stieltjes transforms were relied on a sequence of $\{\lambda_n\}$, which was useful to apply some results about Dirichlet series.

In the paper,we manage to obtain some new results which are not relied on $\{\lambda_n\}$, but just relied on a constant.

*The Project supported by the National Natural Science Foundation of China.

2. Main Results

We define Laplace-Stieltjes transform as

$$F(s)=\int_0^{+\infty} e^{-sy}d\alpha(y),$$

Where $s=\sigma+it$ is a complex variable,and $\alpha(y)$ is a real or complex function and of bounded variation on any interval $[0,X], X\in(0,+\infty)$.

We define the abscicca of convergence and abscicca of uniform convergence and abscicca of absolute convergence respectively as

$\sigma_c^F=\inf\{\sigma_0;$F(s) is convergent for $\sigma>\sigma_0\}$;

$\sigma_u^F=\inf\{\sigma_0;$ F(s) is uniformly convergent for $\sigma\geq\sigma_0\}$;

$\sigma_a^F=\inf\{\sigma_0;$F(s) is absolutely convergent for $\sigma>\sigma_0\}$;

Before giving the formulas of convergence, we introduce a lemma.

Lemma 1 *If F(s)is convergent at the point $s_0=\sigma_0+it_0$, then*

1) for all $r\in[0,\frac{\pi}{2}), F(s)$ is uniformly convergent on $E_r=\{s|\,|\arg(s-s_0)|\leq r\}$;

2) $F(s)$ is convergent for $\sigma>\sigma_0$.

Proof From the condition, we know, for all $\varepsilon>0$, there exists $M>0$, such that

$$|\int_u^{+\infty} e^{-s_0y}d\alpha(y)|<\frac{\varepsilon}{2}$$

for $u>M$.

Then, for all $u_1,u_2>M$,

$$|\int_{u_1}^{u_2} e^{-s_0y}d\alpha(y)|\leq|\int_{u_1}^{+\infty} e^{-s_0y}d\alpha(y)|+|\int_{u_2}^{+\infty} e^{-s_0y}d\alpha(y)|<\frac{\varepsilon}{2}+\frac{\varepsilon}{2}=\varepsilon.$$

Denote $I_u(x;s_0)=\int_u^x e^{-s_0y}d\alpha(y)$.

For all $r\in[0,\frac{\pi}{2})$, taking $s\in E_r$, since

$$\int_u^x e^{-sy}d\alpha(y)=\int_u^x e^{-s_0y}e^{-(s-s_0)y}d\alpha(y)$$

$$=\int_u^x e^{-(s-s_0)y}dI_u(y;s_0)=I_u(x;s_0)e^{-(s-s_0)x}-\int_u^x I_u(y;s_0)de^{-(s-s_0)y},$$

taking $u>M, x>u$, then we have

$$|\int_u^x e^{-sy}d\alpha(y)|\leq|I_u(x;s_0)|e^{-(\sigma-\sigma_0)x}+\int_u^x |I_u(y;s_0)|e^{-(\sigma-\sigma_0)y}|s-s_0|dy$$

$$\leq \varepsilon e^{-(\sigma-\sigma_0)x} + \varepsilon|\frac{s-s_0}{\sigma-\sigma_0}||[e^{-(\sigma-\sigma_0)x} + e^{-(\sigma-\sigma_0)u}] \leq \varepsilon + 2\varepsilon\cos r.$$

Therefore, $F(s)$ is uniformly convergent on E_r.

According to 1), we can obtain 2) easily.

Theorem 1

$$\sigma_u^F = \overline{\lim_{x\to+\infty}}\frac{\ln p(x)}{x},$$

where $p(x) = \sup_{x<z\leq x+K} |\int_x^z d\alpha(y)|$, *$K$ is a positive constant.*

Proof We write

$$A = \overline{\lim_{x\to+\infty}}\frac{\ln p(x)}{x}, q_x(y) = \int_x^z d\alpha(y).$$

At first, we consider that A is finite.

On one hand, we take

$$\sigma_0 > A, \varepsilon = \frac{\sigma_0 - A}{2} > 0,$$

then there exists$X > 0$, such that $\frac{\ln p(x)}{x} < A + \frac{\sigma_0 - A}{2} = \frac{\sigma_0 + A}{2}$for $x > X$.

That is $p(x) < e^{x\frac{\sigma_0+A}{2}}$for $x < X$.

Taking $x < X, x < z \leq x + K$, we have

$$|\int_x^z d\alpha(y)| \leq p(x) < e^{x\frac{\sigma_0+A}{2}}.$$

Since

$$|\int_x^{x+K} e^{-y\sigma_0}d\alpha(y)| = |\int_x^{x+K} e^{-y\sigma_0}d_y q_x(y)|$$

$$= |e^{-(x+K)\sigma_0}q_x(x+K) + \sigma_0\int_x^{x+K} q_x(y)e^{-y\sigma_0}dy|$$

$$\leq e^{-(x+K)\sigma_0}e^{x\frac{\sigma_0+A}{2}} + e^{x\frac{\sigma_0+A}{2}}|e^{-x\sigma_0} - e^{-(x+K)\sigma_0}|$$

$$\leq K(\sigma_0)e^{-x\frac{\sigma_0-A}{2}},$$

$K(\sigma_0) = 2e^{-K\sigma_0}$ when $\sigma_0 < 0$; $K(\sigma_0) = 1$, when $\sigma_0 \geq 0$.

Hence

$$|\int_X^{+\infty} e^{-y\sigma_0}d\alpha(y)| \leq \sum_{n=1}^{+\infty}|\int_{X+(n-1)K}^{X+nK} e^{-y\sigma_0}d\alpha(y)|$$

$$\leq K(\sigma_0)\sum_{n=1}^{+\infty} e^{-[X+(n-1)K]\frac{\sigma_0-A}{2}}$$

$$= K(\sigma_0)e^{-X\frac{\sigma_0-A}{2}}\sum_{n=1}^{+\infty} e^{-(n-1)K\frac{\sigma_0-A}{2}}$$

$$= K(\sigma_0)e^{-X\frac{\sigma_0-A}{2}}\frac{1}{1-e^{-K\frac{\sigma_0-A}{2}}} < +\infty.$$

Now, we see $F(s)$ is convergent at $s = \sigma_0$.

By Lemma 1, $F(s)$ is convergent for $\sigma > A$. So

$$\sigma_u^F \leq A.$$

On the other hand, we need prove

$$\sigma_u^F \geq A.$$

Suppose $F(s)$ is convergent at $s = \sigma_0$, then there exists a positive real number M, such that for all $x_1, x_2 \in (0, +\infty)$,

$$|\int_{x_1}^{x_2} e^{-\sigma_0 y} d\alpha(y)| < M.$$

Denoting

$$p_x(z;\sigma_0) = \int_x^z e^{-\sigma_0 y} d\alpha(y),$$

we have

$$|p_x(y;\sigma_0)| < M.$$

Taking $x < z \leq x + K$,

$$|\int_x^z d\alpha(y)| = |\int_x^z e^{\sigma_0 y} e^{-\sigma_0 y} d\alpha(y)|$$

$$= |\int_x^z e^{\sigma_0 y} dp_x(y;\sigma_0)| = |p_x(z;\sigma_0)e^{\sigma_0 z} - \sigma_0 \int_x^z e^{\sigma_0 y} p_x(y;\sigma_0)dy|$$

$$\leq Me^{\sigma_0 z} + M|e^{\sigma_0 z} - e^{\sigma_0 x}| \leq K(\sigma_0)e^{\sigma_0 x}.$$

$K(\sigma_0) = 2Me^{K\sigma_0}$, when $\sigma_0 > 0 . K(\sigma_0) = M$, when $\sigma_0 \leq 0$.

So

$$\sup_{x<z\leq x+k} |\int_x^z d\alpha(y)| \leq K(\sigma_0)e^{\sigma_0 x}.$$

That is

$$p(x) \leq K(\sigma_0)e^{\sigma_0 x}.$$

Now we obtain

$$\overline{\lim_{x\to+\infty}} \frac{\ln p(x)}{x} \leq \sigma_0.$$

Since σ_0 is arbitrary,so

$$\sigma_c^F \leq A.$$

When $A = +\infty$, by the above proof, we know $\sigma_c^F \geq +\infty$. That is,$\sigma_c^F = +\infty$.

When $A = -\infty$, for all $\sigma < 0$, there exists$X > 0$, such that

$$|\int_x^z d\alpha(y)| < e^{2\sigma x}$$

for $x > X, z > x$.

For all $x > X$, we have

$$|\int_x^{x+K} e^{-y\sigma} d\alpha(y)| \leq e^{2\sigma x}e^{-\sigma(x+K)} + e^{2\sigma x}[e^{-\sigma(x+K)} - e^{-\sigma x}]$$

$$\leq 2e^{2\sigma x}e^{-\sigma(x+K)} = 2e^{2\sigma x - K\sigma}.$$

What is more,

$$\int_X^{+\infty} e^{-y\sigma} d\alpha(y)| \leq \sum_{n=1}^{+\infty} |\int_{X+(n-1)K}^{X+nK} e^{-y\sigma} d\alpha(y)|$$

$$2e^{-K\sigma} \sum_{n=1}^{+\infty} e^{\sigma[X+(n-1)K]} = 2e^{-K\sigma+X\sigma} \frac{1}{1-e^{K\sigma}} < +\infty.$$

It shows that $\sigma_c^F \leq \sigma$.

Since σ is arbitrary, $\sigma_c^F = -\infty$.

Similarly, we obtain

Lemma 2 *If $F(s)$ is uniformly convergent on the line $\sigma = \sigma_0$, then $F(s)$ is uniformly convergent on the right half plane $\sigma \geq \sigma_0$.*

Proof From the condition, we see, for all $\varepsilon > 0$, there exists $M > 0$, such that

$$|\int_u^{+\infty} e^{-(\sigma_0+it)y} d\alpha(y)| < \frac{\varepsilon}{2}$$

for $u > M$ and $t \in R$

Then, for all $u_1, u_2 > M$,

$$|\int_{u_1}^{u_2} e^{-(\sigma_0+it)y} d\alpha(y)| \le |\int_{u_1}^{+\infty} e^{-(\sigma_0+it)y} d\alpha(y)|$$

$$+|\int_{u_2}^{+\infty} e^{-(\sigma_0+it)y} d\alpha(y)| < \tfrac{\varepsilon}{2} + \tfrac{\varepsilon}{2} = \varepsilon.$$

Denote $I_u(x;\sigma_0+it) = \int_u^x e^{-(\sigma_0+it)y} d\alpha(y)$. Take $s = \sigma + it$, since

$$\begin{aligned} \int_u^x e^{-sy} d\alpha(y) &= \int_u^x e^{-(\sigma_0+it)y} e^{-(\sigma-\sigma_0)y} d\alpha(y) \\ &= \int_u^x e^{-(\sigma-\sigma_0)y} dI_u(y;\sigma_0+it) \\ &= I_u(x;\sigma_0+it)e^{-(\sigma-\sigma_0)x} - \int_u^x I_u(y;\sigma_0+it) de^{-(\sigma-\sigma_0)y}, \end{aligned}$$

taking $u > M, x > u, \sigma \ge \sigma_0$, then we have

$$\begin{aligned} |\int_u^x e^{-sy} d\alpha(y)| &\le |I_u(x;\sigma_0+it)| e^{-(\sigma-\sigma_0)x} \\ &\quad + \int_u^x |I_u(y;\sigma_0+it)| e^{-(\sigma-\sigma_0)y} |\sigma-\sigma_0| dy \\ &\le \varepsilon e^{-(\sigma-\sigma_0)x} + \varepsilon[e^{-(\sigma-\sigma_0)x} + e^{-(\sigma-\sigma_0)u}] \\ &\le \varepsilon + 2\varepsilon = 3\varepsilon. \end{aligned}$$

Therefore, $F(s)$ is uniformly convergent for $\sigma \ge \sigma_0$.

Theorem 2

$$\sigma_u^F = \overline{\lim_{x\to+\infty}} \frac{\ln n(x)}{x},$$

where $n(x) = \sup_{t\in(-\infty,+\infty), x<z\le x+K} |\int_x^z e^{-ity} d\alpha(y)|$, *$K$ is a positive constant.*

Lemma 3 *If $F(s)$ is absolutely convergent at the point $s = \sigma_0$, then $F(s)$ is absolutely convergent for $\sigma \ge \sigma_0$.*

Theorem 3

$$\sigma_a^F = \overline{\lim_{x\to+\infty}} \frac{\ln m(x)}{x},$$

where $m(x) = \int_x^{x+K} |d\alpha(y)|$, *$K$ is a positive constant.*

References

1. J. R. Yu, The Borel Line Of Entire Function Defined By The Laplace-Stieltjes Transform, *Maths Journal,* 1963:471-484
2. J. R. Yu, *Dirichlet Series And Random Dirichlet Series,* Science Press,1997.
3. G. Valiron, Entire functions and Borel's directions, *Proc. Nat. Acad. Sc., U.S.A.,* **20**(1934):211-215.
4. K. Knopp, Uber die Knovergenzabscisse des Laplace-Integrals, *Math Zetts,* **54**(1951):291-296.
5. J. R. Yu, The convergent property of double Dirichlet series and double Laplace-Stieltjes transforms, *Jounal of Wuhan university(science),* 1962:1-17.

A NEW DECOMPOSITION FOR THE HARDY SPACE ON DOMAINS*

ZENGJIAN LOU

Department of Mathematics
Shantou University
Shantou, Guangdong, 515063, P. R. China
Email: zjlou@stu.edu.cn

We give a new atomic decomposition for the Hardy space on domains.

Keywords: Hardy space; atomic decomposition; domains.

2000 Mathematics Subject Classification: 42B30

1. Introduction

Let φ be an infinitely differentiable function with compact support in $\mathbb{R}^n$ such that $\int_{\mathbb{R}^N} \varphi(x)\, dx = 1$. For $t > 0$, define $\varphi_t(x) = t^{-N}\varphi(x/t)$. A locally integrable function f on $\mathbb{R}^N$ is said to be in the *Hardy space* $\mathcal{H}^1(\mathbb{R}^N)$ if the maximal function

$$M(f)(x) = \sup_{t>0} |\varphi_t * f(x)|$$

belongs to $L^1(\mathbb{R}^N)$ ([9]). If this is the case, define

$$\|f\|_{\mathcal{H}^1(\mathbb{R}^N)} = \|M(f)\|_{L^1(\mathbb{R}^N)}.$$

The theory of Hardy spaces on domains $\Omega \subset \mathbb{R}^N$ was developed, among other places, in [7] (arbitrary domains) and in [3] (bounded Lipschitz domains). Let Ω be an open domain in $\mathbb{R}^n$. A function f on Ω belongs to

*This work was supported by Natural Science Foundation of Guangdong Province (Grant No. 032038), National Natural Science Foundation of China (Grant No.10371069) and SRF for ROCS, State Education Ministry.

$\mathcal{H}_z^1(\Omega)$ if the function F defined by

$$F(x) = \begin{cases} f(x), & \text{if } x \in \Omega; \\ 0, & \text{if } x \in \Omega^c. \end{cases}$$

belongs to $\mathcal{H}^1(\mathbb{R}^n)$ with $\|f\|_{\mathcal{H}_z^1(\Omega)} = \|F\|_{\mathcal{H}^1(\mathbb{R}^N)}$.

A Lebesgue measurable function a is said to be an $\mathcal{H}_z^1(\Omega)$-*atom* if there exists a cube Q with $8Q \subset \Omega$ and supp $a \subset Q$ such that a satisfies the moment condition

$$\int_Q a(x)\, dx = 0$$

and the size condition

$$\|a\|_{L^2(\mathbb{R}^N)} \leq |Q|^{-1/2}.$$

Those atoms we defined are different from the ordinary ones, since supports of them are away from the boundary of the domain Ω. This is convenient to use when we deal with problems related to the boundary $\partial\Omega$. In this note using an idea in [6], we prove that any function in $\mathcal{H}_z^1(\Omega)$ can be decomposed into $\mathcal{H}_z^1(\Omega)$-atoms.

2. Theorem and its Proof

Theorem 1. *Suppose Ω is a Lipschitz domain in $\mathbb{R}^n$. A function f on Ω belongs to $\mathcal{H}_z^1(\Omega)$ if and only if it has an atomic decomposition*

$$f = \sum_{k=0}^{\infty} \lambda_k a_k,$$

where the a_k's are $\mathcal{H}_z^1(\Omega)$-atoms and $\sum_{k=0}^{\infty} |\lambda_k| < \infty$.

For the proof of Theorem 1, we need the following lemma of Nečas in [8, Lemma 7.1 in Chapter 3].

Lemma 1. *Let Ω be a bounded Lipschitz domain in $\mathbb{R}^n$. Then the divergence operator is a continuous map from $H_0^1(\Omega, \mathbb{R}^N)$ onto $L_0^2(\Omega) = \{f \in L^2(\Omega) : \int_\Omega f\, dx = 0\}$, that is, there exists a constant C depending only on the domain Ω and the dimension N such that for any $f \in L_0^2(\Omega)$, there exists φ in the Sobolev space $W_0^{1,2}(\Omega, \mathbb{R}^N)$ such that $f = div\, \varphi$ and*

$$\|\nabla \varphi\|_{L^2(\Omega,\mathbb{R}^N)} \leq C\|f\|_{L^2(\Omega)}.$$

Proof of Theorem 1. We only prove the necessity. From Theorem 3.2 in [3] (see also, [2], [5], and [1]), any $f \in \mathcal{H}^1_z(\Omega)$ can be written as

$$f = \sum_{k=0}^{\infty} \epsilon_k \alpha_k,$$

where supp $\alpha_k \subset Q_k$, cubes in Ω such that $\int_{Q_k} \alpha_k \, dx = 0$, $\|\alpha_k\|_{L^2(Q_k)} \leq |Q_k|^{-1/2}$, and

$$\sum_{k=0}^{\infty} |\epsilon_k| \leq C \|f\|_{\mathcal{H}^1_z(\Omega)}.$$

For convenience we drop the subscript k temporarily. Applying Nečas' lemma to the function $\alpha = \alpha_k$, one finds that there exists $\varphi \in W_0^{1,2}(Q, \mathbb{R}^N)$ and a constant C independent of φ and Q such that $\alpha = \text{div}\ \varphi$ and

$$\| \nabla \varphi \|_{L^2(Q,\mathbb{R}^N)} \leq C \|\alpha\|_{L^2(Q)}.$$

Applying a Whitney decomposition with respect to the boundary ∂Q, Q can be decomposed into a family of subcubes: $Q = \bigcup_{i=0}^{\infty} Q_i$ such that $8Q_i \subset Q$ and $|Q| = \sum_{i=0}^{\infty} |Q_i|$. Using this decomposition we construct a smooth partition of unity $\sum_{i=0}^{\infty} \eta_i(x) \equiv 1$ in Q, where $\eta_i(x) = 1$ in Q_i; $\eta_i(x) = 0$ outside $2Q_i$ and $| \nabla \eta_i| \leq C\ l(Q_i)^{-1}$. We have

$$\begin{aligned} \alpha =&\ \text{div} \left(\textstyle\sum_{i=0}^{\infty} \eta_i \varphi\right) \\ =&\ \textstyle\sum_{i=0}^{\infty} \text{div}\ (\eta_i \varphi) =: \sum_{i=0}^{\infty} \tau_i \alpha_i, \end{aligned}$$

where $\alpha_i = \frac{\text{div}\ (\eta_i\varphi)}{|2Q_i|^{1/2} \|\text{div}\ (\eta_i \varphi)\|_{L^2(2Q_i)}}$ and $\tau_i = |2Q_i|^{1/2} \|\text{div}\ (\eta_i\varphi)\|_{L^2(2Q_i)}$. It is obvious that α_i is an $\mathcal{H}^1_z(\Omega)$-atom supported in $2Q_i$ with $8Q_i \subset \Omega$. So Theorem 1 is proved if we can show that there exists a constant C independent of α and Q such that $\sum_{i=0}^{\infty} \tau_i \leq C < \infty$. Write

$$\begin{aligned} \textstyle\sum_{i=0}^{\infty} \tau_i &= \textstyle\sum_{i=0}^{\infty} |2Q_i|^{1/2} \|\text{div}\ (\eta_i \varphi)\|_{L^2(2Q_i)} \\ &\leq 2^{1/2} \textstyle\sum_{i=0}^{\infty} |2Q_i|^{1/2} \Big(\|\eta_i \text{div}\ \varphi\|_{L^2(2Q_i)} + \| \nabla \eta_i \cdot \varphi\|_{L^2(2Q_i)} \Big) \\ &=: I + II. \end{aligned}$$

Note that $0 \leq \eta_i \leq 1$. By the Cauchy-Schwartz inequality, we have

$$\begin{aligned} I &\leq 2^{1/2} \textstyle\sum_{i=0}^{\infty} |2Q_i|^{1/2} \|\text{div}\ \varphi\|_{L^2(2Q_i)} \\ &\leq 2^{1/2} \left(\textstyle\sum_{i=0}^{\infty} |2Q_i|\right)^{1/2} \left(\textstyle\sum_{i=0}^{\infty} \int_{2Q_i} |\alpha|^2 \, dx \right)^{1/2} \\ &\leq C |Q|^{1/2} \|\alpha\|_{L^2(Q)} \leq C. \end{aligned}$$

We now deal with the second part II. Note that $| \bigtriangledown \eta_i| \leq C\ l(Q_i)^{-1} \leq Cd(x,\partial Q)^{-1}(x \in 2Q_i)$ and Hardy's inequality (see, for example, [4, Chapter 1, Section 5]) give

$$\begin{aligned}
II &\leq C|Q|^{1/2}\left(\sum_{i=0}^{\infty}\int_{2Q_i}\left|\bigtriangledown \eta_i\cdot\varphi\right|^2 dx\right)^{1/2}\\
&\leq C|Q|^{1/2}\left(\sum_{i=0}^{\infty}\int_{2Q_i}\left|\frac{\varphi(x)}{d(x,\partial Q)}\right|^2 dx\right)^{1/2}\\
&\leq C|Q|^{1/2}\left(\int_{Q}\left|\frac{\varphi(x)}{d(x,\partial Q)}\right|^2 dx\right)^{1/2}\\
&\leq C|Q|^{1/2}\|\bigtriangledown\varphi\|_{L^2(Q,\mathbb{R}^N)}\\
&\leq C|Q|^{1/2}\|\alpha\|_{L^2(Q)}\leq C,
\end{aligned}$$

where C does not depend on Q and α, Thus each atom α_k can be decomposed into $\mathcal{H}^1_z(\Omega)$-atoms α^i_k and the supports of these atoms can be away from the boundary $\partial\Omega$.

References

1. P. Auscher, E. Russ, Hardy spaces and divergence operators on strongly Lipschitz domains of $\mathbb{R}^N$, *J. Funct. Anal.* **201**(2003):148-184.
2. D. C. Chang, The dual of Hardy spaces on a domain in $\mathbb{R}^n$, *Forum Math.* **6**(1994):65-81.
3. D. C. Chang, S. G. Krantz, E. M. Stein, $\mathcal{H}^p$ theory on a smooth domain in $\mathbb{R}^N$ and elliptic boundary value problems, *J. Funct. Anal.* **114**(1993):286-347.
4. B. Davies, *Heat Kernels and Spectral Theory,* Cambridge University Press, 1989.
5. A. Jonsson, P. Sjögren, H. Wallin, Hardy and Lipschitz spaces on subsets of $\mathbb{R}^n$, *Studia Math.* **80**(1984):141-166.
6. Z. J. Lou, A. McIntosh, Hardy spaces of exact forms on Lipschitz domains in $\mathbb{R}^n$, *Indiana Univ. Math. J.* **53**(2004):583-611.
7. A. Miyachi, $\mathcal{H}^p$ spaces over open subsets of $\mathbb{R}^n$, *Studia Math.* **95**(1990):205-228.
8. J. Nečas, *Les méthodes directes en théorie des équations elliptiques,* Masson et Cie, Eds., Paries; Academia, Editeurs, Prague 1967.
9. E. M. Stein, *Harmonic Analysis, real-variable methods, orthogonality, and oscillatory integrals,* Princeton Univ. Press, New Jersey, 1993.

TRANSVERSALITY ON COORDINATED MANIFOLDS

TSOY-WO MA
School of Mathematics and Statistics
University of Western Australia
Nedlands, W.A., 6907, Australia
E-mail: twma@maths.uwa.edu.au

We develop elementary properties of locally compact morphisms from coordinated manifolds transversal to submanifolds of coordinated manifolds as extension of regular values and submersions. Transversal submanifolds are treated as a special case. Examples are given at the end.

2000 Mathematics Subject Classification: 57N75, 58B15

1. Introduction

The difficulty of many definitions of infinite dimensional differential calculus beyond Banach spaces is summarized in [8]. Global analysis [5] based on convenient spaces and also products of generalized functions [2] in terms of bornological vector spaces make use of differentiable maps which need not be continuous. Quantum probability is used to study infinite dimensional analysis [11]. However, we prefer to work in infinite dimensional complex analysis on traditional locally convex spaces. In contrast to various definitions of holomorphic maps given in [3], all our holomorphic maps must be locally bounded. Our first application is to provide a simple framework for products of distributions [10]. With holomorphic locally compact perturbations of continuous linear maps, we initiate a primitive theory of infinite dimensional holomorphic manifolds [8].

A complex quasi-complete separated locally convex space E is called a *coordinate space* if every compact set is contained in a closed vector subspace on which there is a continuous *norm*. This better definition is equivalent to [7] in terms of a coordination $\{E_j : j \in \Delta\}$ if we let E_j be the smallest closed vector subspace containing j for each j in the family Δ of all compact subsets of E. Coordinate spaces are important because the inverse mapping theorem is available to enrich the primitive theory on coordinated manifolds which are holomorphic manifolds modelled on coordinate spaces. Immersions, embeddings, submersions, regular values, regular points, level sets, locally linear maps and subimmersions have been established in [9]. *This paper extends these results to the context of transversality* similar to the treatment in [4] for finite dimensional manifolds and in [1,6] for Banach manifolds but the generalization of Whitney topology [4] will be considered later. We shall use the notation of [8] and [9] without any further explanation.

2. Transversal Maps

Let M, N be holomorphic manifolds modelled on coordinate spaces E, F respectively, $f : M \to N$ a *locally compact* morphism and Y a submanifold of N. Then f is said to be *transversal* to Y at $m \in f^{-1}(Y)$ if the following conditions hold.

(a) $df(m)(T_mM) + T_nY = T_nN$ where $n = f(m)$,
(b) $df(m)^{-1}(T_nY)$ splits in T_mM.

The first condition is also called the *transversal equation.* The map f is *transversal* to Y if it is transversal to Y at every point in $f^{-1}(Y)$. Clearly if $f(M)$ is disjoint from Y, then f is transversal to Y. We shall assume that Y is modelled on a split subspace G of F and $H = F \ominus G$ is any topological complement.

Lemma 1 *If every point of Y is regular value of f, then f is transversal to Y.*

Proof *Take any $m \in f^{-1}(Y)$. Since $df(m)(T_mM) = T_nN$, the transversal equation holds. Let (W, ψ) be an adapted chart of N for Y at $n = f(m)$, that is $\psi(W \cap Y) = \psi(W) \cap G$. By Local Submersion Theorem [9], choose a chart (V, φ) of M at m such that $\varphi(V) \subset W$ and $\psi f \varphi^{-1}(x) = J(x)$ for all $x \in \varphi(V)$ where $J : E \to F$ is a submersive linear map. Let $E_2 = ker(J)$ and $E_1 = E \ominus E_2$. Then $J_1 = J|E_1$ is a topological isomorphism onto F. From*

$$\begin{aligned} T_mM \simeq E &= E_1 \oplus E_2 \\ &\simeq [J_1^{-1}(G) \oplus J_1^{-1}(H)] \oplus E_2 \\ &\simeq J_1^{-1}(H) \oplus [J_1^{-1}(G) \oplus E_2] \\ &= J_1^{-1}(H) \oplus J^{-1}(G) \end{aligned}$$

$$\simeq [d\varphi(m)]^{-1}J_1^{-1}(H) \oplus [d\varphi(m)]^{-1}J^{-1}d\psi(n)(T_nY)$$
$$\simeq [d\varphi(m)]^{-1}J_1^{-1}(H) \oplus d(\psi^{-1}J\varphi)(m)^{-1}(T_nY)$$
$$\simeq [d\varphi(m)]^{-1}J_1^{-1}(H) \oplus df(m)^{-1}(T_nY)\,,$$

$df(m)^{-1}(T_nY)$ splits in T_mM. Consequently, f is transversal to Y.

Submanifold Lemma *Every submanifold is locally the inverse of a regular value. More precisely, for every adapted chart (W,ψ) at any point $n \in Y$, there is a submersion g on W into H such that $g(n) = 0$, $W \cap Y = g^{-1}(0)$ and $dg(n)(T_nN) = H$.*

Proof Let $\pi : F = G \oplus H \to H$ be the projection. Then $g : W \to H$ given by $g(w) = \pi[\psi(w) - \psi(n)]$ for every $w \in W$ is a locally compact morphism with $g(n) = 0$ and $dg(n)(T_nN) = d\pi(0)d\psi(n)(T_nN) = \pi F = H$. Since $ker(\pi) = G$ splits, g is a submersion on W. In particular, 0 is a regular value of g. Finally, take any $y \in W$. Then $y \in W \cap Y$ iff $\psi(y) \in \psi(W \cap Y) = \psi(W) \cap G$ iff $\psi(y) \in G$ iff $\psi(y) - \psi(n) \in G$ iff $g(y) = 0$ iff $y \in g^{-1}(0)$. This completes the proof.

The following elementary observation is useful. Let $A : E \to F$ be a continuous linear map and $B : F \to H$ be a continuous linear surjection. Then BA is surjective iff $AE + ker(B) = F$. We also have $ker(BA) = A^{-1}[ker(B)]$ and $E/ker(BA) \simeq BA(E)$.

Lemma 2 *Let W be an open neighborhood of $n \in Y$ and g a locally compact morphism on W into a coordinate space with $g(n) = 0$ and $W \cap Y = g^{-1}(0)$. Suppose that g is submersive on $W \cap Y$. Then f is transversal to Y at $m \in f^{-1}(n)$ iff the composite gf is submersive at m.*

Proof Since g is submersive on $g^{-1}(0)$, n is a regular point of g. Hence $dg(n) : T_nN \to H$ is surjective and we have $ker[dg(n)] = T_n(W \cap Y) = T_nY$. Clearly, $d(gf)(m) = dg(n)df(m)$ is surjective iff

$$df(m)(T_mM) + ker[dg(n)] = T_nN$$

iff the transversal equation holds. We also have

$$ker[d(gf)(m)] = df(m)^{-1}\{ker[dg(n)]\} = df(m)^{-1}(T_nY).$$

Therefore $ker[d(gf)(m)]$ splits iff $df(m)^{-1}(T_nY)$ splits. This completes the proof.

Transversal Mapping Theorem *Let $f : M \to N$ be a locally compact morphsim transversal to a submanifold Y of N. Then*

(a) $X = f^{-1}(Y)$ is a submanifold of M.

(b) $T_mX = df(m)^{-1}(T_nY)$ for all $m \in X$.

(c) $T_mM \ominus T_mX$ and $T_nN \ominus T_nY$ are topologically isomorphic.

(d) If Y has finite codimension in N, then codim X = codim Y.

Proof Take any $m \in X$. Then $n = f(m) \in Y$. Choose W and g by Submanifold Lemma 2. Let V be an open neighborhood of m with $f(V) \subset W$ and let $h = f|V$. Since f is transversal to Y, gh is submersive at every point of $V \cap f^{-1}(n)$, or m is a regular point of gh. By Level Set Theorem [9], $X \cap V = (gh)^{-1}(0)$ is a submanifold of V and we also have $T_m(X \cap V) = ker[d(gh)(m)]$. Hence X is a submanifold of M. Clearly, we have

$$\begin{aligned} T_mX &= T_m(X \cap V) \\ &= ker[d(gh)(m)] \\ &= ker[d(gf)(m)] \\ &= df(m)^{-1}(T_nY). \end{aligned}$$

Part (c) follows from the calculation:

$$\begin{aligned} T_mM \ominus T_mX &\simeq T_mM/ker[d(gf)(m)] \\ &\simeq dg(n)df(m)(T_mM) \\ &= dg(n)[df(m)(T_mM) + ker\, dg(n)] \\ &= dg(n)T_nN = H\,, \qquad \text{by Submanifold Lemma}\,. \end{aligned}$$

Part (d) follows by taking dimensions of (c). This completes the proof.

Theorem 1 *Let P, M, N be holomorphic manifolds and $h : P \to M$, $f : M \to N$ locally compact morphisms. Suppose that f is transversal to a submanifold Y of N. Then h is transversal to the submanifold $X = f^{-1}(Y)$ of M iff fh is transversal to Y.*

Proof Take any $z \in (fh)^{-1}(Y)$. Let $m = h(z)$ and $n = f(m)$. Construct (W, ψ) and g by Submanifold Lemma Choose an open neighborhood V of m with $f(V) \subset W$. Then $(gf)(m) = g(n) = 0$. Pick any $v \in V$. Then $v \in (gf)^{-1}(0)$ iff $(gf)(v) = 0$ iff $f(v) \in g^{-1}(0) = W \cap Y$ iff $v \in f^{-1}(Y) = X$ iff $v \in V \cap X$. Thus $V \cap X = (gf)^{-1}(0)$. Since f is transversal to Y, gf is submersive at every point of $V \cap X$. Therefore h is transversal to X iff

$(gf)h = g(fh)$ is submersive at every point $z \in X$ iff fh is transversal to Y.

3. Transversal Submanifolds

Let X, Y be submanifolds of a holomorphic manifold M modelled on a coordinate space E. Clearly the inclusion map $f : X \to M$ is a locally compact morphism and its derivative $df(m) : T_m X \to T_m M$ is an inclusion map for every $m \in X$. Now X is said to be *transversal* to Y if f is transversal to Y, that is for all $m \in X \cap Y$,

(a) $T_m X + T_m Y = T_m M$,

(b) $T_m X \cap T_m Y$ splits in $T_m X$, or equivalently in $T_m M$.

Theorem 2 *Let X, Y be transversal submanifolds of M. Then*

(a) $X \cap Y$ is also a submanifold of M.

(b) For every $m \in X \cap Y$, we have $T_m(X \cap Y) = T_m X \cap T_m Y$.

(c) $codim[T_m(X \cap Y)] = codim(T_m X) + codim(T_m Y)$ if the right hand side is finite.

Proof Only part (c) requires a proof. For convenience, write $F = T_m M$, $A = T_m X$ and $B = T_m Y$. Since the restriction α to B of the quotient map $F \to F/A$ is continuous linear surjective, we have

$$F/A \simeq B/ker(\alpha) = B/(A \cap B).$$

By $F = A + B$, we obtain $F/A \cap B = A/A \cap B + B/A \cap B$. Clearly this is an algebraic direct sum. Therefore we get

$$\begin{aligned} codim[T_m(X \cap Y)] &= dim\ F/A \cap B \\ &= dim\ A/A \cap B + dim\ B/A \cap B \\ &= dim\ F/B + dim\ F/A \\ &= codim(T_m Y) + codim(T_m X). \end{aligned}$$

4. Examples

Let E_n be the set of sequences $x = (x_1, x_2, \cdots)$ with $x_j = 0$ for all $j > n$. As a finite dimensional vector space, each E_n is equipped with a unique locally convex topology. The strict inductive limit $E = \bigcup_{n=1}^{\infty} E_n$ consisting of all finite sequences is a coordinate space. It is well known that E is not metrizable and hence not a Banach space.

If A_j is an open subset of then the set

$$\prod_{j=1}^{\infty} A_j = \{x \in E : x_j \in A_j \,,\ \forall j\}$$

is open in E. In fact, take any $a \in A = \prod_{j=1}^{\infty} A_j$. Choose $\delta_j > 0$ such that $a_j + \mathbb{B}(\delta_j) \subset A_j$ where $\mathbb{B}(\delta_j) = \{t \in \mathbb{C} : |t| < \delta_j\}$. Then the convex balanced absorbing set $V = \prod_{j=1}^{\infty} \mathbb{B}(\delta_j)$ is a 0-neighborhood of E such that $a + V \subset A$. Therefore A is open in E. In particular, the set $M = \{x \in E : |x_j| < 1/2^j\}$ is open in E. Therefore it is a coordinated manifold. For morphisms f between open sets of coordinate spaces, we identify the differential $df(m)$ and the total derivative $Df(m)$.

The function f on M given by

$$f(x) = \sum_{j=1}^{\infty} \frac{x_j}{2 - x_j}$$

is holomorphic. In fact, each composite function $x \to x_j \to x_j/(2 - x_j)$ is holomorphic on M. Since $|x_j/(2 - x_j)| \leq |x_j| < 1/2^j$, the uniform limit f is also holomorphic. For each $h = (h_1, h_2, \cdots) \in E$, we have

$$Df(x)h = \frac{1}{dt}\, f(x + th)\bigg|_{t=0} = \sum_{j=1}^{\infty} \frac{h_j}{(2 - x_j)^2}\ .$$

Since $Df(x) \neq 0$, it is surjective onto $\mathbb{C}$. Because $Df(x)$ is a continuous linear form on E, its kernel splits. Therefore every point in M is regular. The level set $X = f^{-1}(0)$ is a submanifold of M.

Consider the locally compact map K on M into E_2 given by

$$K(x) = (x_1 e^{x_3} - x_2, x_1 e^{x_3} + x_2, 0, 0, \cdots)$$

Then we have

$$DK(x)h = (h_1 e^{x_3} - h_2 + h_3 x_1 e^{x_3}, h_1 e^{x_3} + h_2 + h_3 x_1 e^{x_3}, 0, 0, \cdots)\,.$$

Because $DK(x)$ carries E_2 onto E_2, the linear map $DK(x) : E \to E_2$ is surjective. Since the kernel of $DK(x)$ is a closed vector subspace with finite codimension, it splits. Therefore level set $Y = K^{-1}(0)$ is also submanifold of M where 0 is the zero vector of E.

Take any $m = (m_1, m_2, m_3, \cdots) \in X \cap Y$. From $K(m) = 0$, we have $m_1 = m_2 = 0$. For each $h \in E$, we need to find $a \in T_m X = ker[Df(m)]$ and $b \in T_m Y = ker[DK(m)]$ such that $h = a + b$. To satisfy $DK(m)b = 0$, let $b_1 = b_2 = 0$ and $b_j = h_j$ for all $j > 3$. Set $a_1 = h_1$, $a_2 = h_2$, $a_j = 0$ for all $j > 3$. Solve for a_3 from

$$Df(m)a = \frac{a_1}{(2-0)^2} + \frac{a_2}{(2-0)^2} + \frac{a_3}{(2-m_3)^2} = 0\,.$$

Define $b_3 = h_2 - a_3$. Hence $T_mX + T_mY = T_mM$. Next, $T_mX \cap T_mY$ is a closed vector subspace with finite codimension, it splits in T_mM. Consequently X, Y are transversal submanifolds of M.

The left shift L defined by $L(x) = (x_2, x_3, \cdots)$ is a continuous linear map on E. Hence $g = L + K$ given by

$$g(x) = (x_1 e^{x_3}, x_1 e^{x_3} + x_2 + x_3, x_4, x_5, \cdots)$$

is a locally compact morphism from M into E. Furthermore, we have

$$\begin{aligned} Dg(x)h &= L(h) + DK(x)h \\ &= (h_1 e^{x_3} + h_3 x_1 e^{x_3}, h_1 e^{x_3} + h_2 + h_3 x_1 e^{x_3} + h_3, h_4, h_5, \cdots). \end{aligned}$$

For each $k \in E$, define $h \in E$ by $h_1 = k_1 e^{-x_3}$, $h_2 = k_2 - k_1$, $h_3 = 0$ and $h_j = k_{j-1}$ for all $j > 3$. Then $Df(x)h = k$. Hence $Dg(x) : E \to E$ is surjective. Since $ker[Dg(x)]$ is finite dimensional, it splits. Therefore every point in M is regular. The level set $Z = g^{-1}(0)$ is a submanifold of M.

Take any $m \in X \cap Z$. From $g(m) = 0$, we have $m_2 + m_3 = 0$ and $m_1 = m_j = 0$ for all $j > 3$. Coupling with $f(m) = 0$ gives $m_2 = m_3 = 0$. Hence $X \cap Z = \{0\}$. Let $m = 0$. For each $h \in E$, we have $a \in T_mX$, $b \in T_mZ$ and $h = a + b$ where $a_2 = h_2$, $b_2 = 0$, $a_j = 0$ and $b_j = h_j$ for all $j \neq 2$. Thus $T_mX + T_mZ = T_mM$. Because $T_mX \cap T_mZ$ is finite dimensional, it splits. Therefore X, Z are also transversal submanifolds.

References

1. A. Abraham, J.E. Marsden and T. Ratiu, *Manifolds, tensor analysis, and applications. Second edition,* Springer-Verlag, 1988.
2. J. F. Columbeau, Differential calculus and holomorphy, *North Holland Math. Studies* **64**(1982).
3. S. Dineen, Complex analysis in locally convex spaces, *North Holland Math. Studies* **57**(1981).
4. M. Golubitsky and V. Guillemin, *Stable mappings and their singularities,* Springer-Verlag, 1973.
5. A. Kriegl and P.W. Michor, The convenient setting of global analysis, *Math. Surveys and Monographs, Amer. Math. Soc.,* 53(1997).
6. S. Lang, *Fundamentals of differential geometry,* Springer-Verlag, 1999.
7. T. W. Ma, Inverse mapping theorem, Bull. London Math. Soc. **33** (2001): 473-482.
8. T. W. Ma, Holomorphic manifolds on locally convex spaces, *accepted by Analysis in Theory and Application,* 2004.
9. T. W. Ma, Locally linear maps on coordinated manifolds, *accepted by Proceedings, 12-th Finite or infinite dimensional complex analysis,* Tokyo, 2004.

10. T. W. Ma, Infinite dimensional complex analysis as a framework for products of distributions, *accepted by Bull. Institute of Math., Academia Sinica,* 2005.
11. M. Schurmann and U. Franz, *Quantum probability and infinite dimensional analysis from foundations to applications, World Scientific,* 2005.

ON LOGARITHMIC (α, β)-BLOCH SPACES*

XIAOGE MENG

Department of Mathematics
Shantou University
Shantou 515063, Guangdong, P. R. China
E-mail address: mengxiaoge@126.com

Define the logarithmic (α, β)-Bloch space and characterize the logarithmic (α, β)-Bloch functions by their Taytor coefficients. The relationship between logarithmic (α, β)-Bloch space and Cesàro means is given.

1. Introduction

Let $D = \{z \in \mathbb{C} : |z| < 1\}$ be the open unit disk in the complex plane $\mathbb{C}$, $T = \{z \in \mathbb{C} : |z| = 1\}$ its boundary, and denote $H(D)$ the class of analytic functions on D.

Let L^1 be the space of Lebesgue integrable functions on D and L^1_a be the closed subspace of L^1 consisting of analytic functions. For $f \in L^1$, the Toeplitz operator $T_f : L^1_a \to L^1_a$ and Hankel operator $H_f : L^1_a \to L^1$ are defined as follows:

$$T_f(g) = P(fg), \qquad H_f(g) = (I - P)(fg),$$

where P is the usual Bergman projection and I is the identity operator, respectively. Attele [1] proved the following theorem.

Theorem A (i) If f is in L^1_a, then $H_{\overline{f}}$ is bounded on L^1_a if and only if

$$\sup_{z \in D}(1 - |z|^2)|f'(z)| \log \frac{1}{1 - |z|^2} < \infty.$$

(ii) If f in L^1 is real-valued and harmornic, then T_f is bounded on L^1_a if

*This research is supported in part by the National Natural Science Foundation of China (No. 10371051, No. 10371069) and the NSF of Guangdong Province of China (No. 04011000).

and only if f is bounded and

$$\sup_{z\in D}(1-|z|^2)|\nabla f(z)|\log\frac{1}{1-|z|^2}<\infty.$$

For $0<\alpha<\infty$, $-\infty<\beta<\infty$, we define the logarithmic (α,β)-Bloch space, consisting of all functions $f\in H(D)$, for which

$$\|f\|_{\log B_{(\alpha,\beta)}}\sup_{z\in D}(1-|z|^2)^{\alpha}|f'(z)|\left(\log\frac{1}{1-|z|^2}\right)^{\beta}<\infty,$$

denoted by $\log B_{(\alpha,\beta)}$. By choosing different α and β, it will be B, B^{α}, $\log B$, $\log B^{\alpha}$ respectively. A function $f\in\log B_{(\alpha,\beta)}$ is said to belong to $\log B_{0(\alpha,\beta)}$ if

$$\lim_{|z|\to 1}(1-|z|^2)^{\alpha}|f'(z)|\left(\log\frac{1}{1-|z|^2}\right)^{\beta}=0.$$

Note that $\|.\|_{\log B_{(\alpha,\beta)}}$ defines a seminorm while the natural norm is given by $\|f\|_{\log}=|f(0)|+\|f\|_{\log B_{(\alpha,\beta)}}$, which makes $\log B_{(\alpha,\beta)}$ into a Banach space.

2. The coefficients of $\log B_{(\alpha,\beta)}$ functions

Theorem 2.1. *Let $0<\alpha<\infty$, $-\infty<\beta<\infty$ and $f(z)=\sum_{n=1}^{\infty}a_nz^n$ in $H(D)$.*

i) If $f\in\log B_{(\alpha,\beta)}$, then $n^{1-\alpha}|a_n|(\log n)^{\beta}=O(1)\quad(n\to\infty)$;

ii) If $f\in\log B_{0(\alpha,\beta)}$, then $n^{1-\alpha}|a_n|(\log n)^{\beta}=o(1)\quad(n\to\infty)$.

Proof. i) Assume $f\in\log B_{(\alpha,\beta)}$. Let $z=r\xi$, $0<r<1$, $\xi\in T$. Using Cauchy's integral formula, we obtain

$$\begin{aligned}|a_n|&=\frac{1}{2n\pi}\left|\int_T f'(r\xi)(r\xi)^{1-n}|d\xi|\right|\\&\leq\frac{1}{2n\pi}\int_T|f'(r\xi)|r^{1-n}|d\xi|\\&=\frac{1}{2n\pi}\int_T(1-r^2)^{\alpha}|f'(r\xi)|\left(\log\frac{1}{1-r^2}\right)^{\beta}\frac{r^{1-n}}{(1-r^2)^{\alpha}}\left(\log\frac{1}{1-r^2}\right)^{-\beta}|d\xi|\\&\leq\frac{\|f\|_{\log B_{(\alpha,\beta)}}r^{1-n}}{n(1-r^2)^{\alpha}}\left(\log\frac{1}{1-r^2}\right)^{-\beta}.\end{aligned}$$

Choosing $r = (1-\frac{1}{n})^{\frac{1}{2}}$, $n \geq 2$, we have

$$|a_n| \leq \|f\|_{\log B_{(\alpha,\beta)}} n^{\alpha-1}\left(1-\frac{1}{n}\right)^{\frac{1-n}{2}} (\log n)^{-\beta}.$$

Since

$$\lim_{n\to\infty}\left(1-\frac{1}{n}\right)^{\frac{1-n}{2}} = e^{\frac{1}{2}},$$

we get

$$n^{1-\alpha}|a_n|(\log n)^{\beta} = O(1), \quad n\to\infty.$$

ii) Assume $f \in \log B_{0(\alpha,\beta)}$. Then for any $\epsilon > 0$, there exists $l \in (0,1)$, and for any $r \in (l,1)$, we have

$$(1-r^2)^{\alpha}|f^{'}(r\xi)|\left(\log\frac{1}{1-r^2}\right)^{\beta} < \frac{\epsilon}{e+1}.$$

For all $n \in \mathbb{N}$, we have

$$\begin{aligned}|a_n| &= \frac{1}{2n\pi}\left|\int_T f^{'}(r\xi)(r\xi)^{1-n}|d\xi|\right| \\ &\leq \frac{1}{2n\pi}\int_T (1-r^2)^{\alpha}|f^{'}(r\xi)|\left(\log\frac{1}{1-r^2}\right)^{\beta}\frac{r^{1-n}}{(1-r^2)^{\alpha}}\left(\log\frac{1}{1-r^2}\right)^{-\beta}|d\xi|.\end{aligned}$$

Consider $r = (1-\frac{1}{n})^{\frac{1}{2}}$, $n \geq 2$. Since

$$\lim_{n\to\infty}\left(1-\frac{1}{n}\right)^{\frac{1-n}{2}} = e^{\frac{1}{2}},$$

for the above ϵ, there exists $N \in \mathbb{N}$, with $N \geq \frac{1}{1-l} > 0$, such that for all $n > N$

$$(1-\frac{1}{n})^{\frac{1-n}{2}} < e^{\frac{1}{2}} + 1 < e+1.$$

Notice that $r = (1-\frac{1}{n})^{\frac{1}{2}} \in (l,1)$. For all $n > N$

$$|a_n| < \frac{\epsilon}{e+1} n^{\alpha-1}\left(1-\frac{1}{n}\right)^{\frac{1-n}{2}} (\log n)^{-\beta} < n^{\alpha-1}(\log n)^{-\beta}\epsilon.$$

This means that

$$n^{1-\alpha}|a_n|(\log n)^{\beta} = o(1), \quad n\to\infty.$$

3. $\log B_{(\alpha,\beta)}$ space and Cesàro means

For given a function $f(z) = \sum_{n=0}^{\infty} a_n z^n \in H(D)$, we write

$$s_n(f)(z) = \sum_{k=0}^{n} a_k z^k, \qquad n = 0, 1, 2, ...,$$

and

$$\sigma_n(f)(z) = \sum_{k=0}^{n} \frac{s_k(f)(z)}{n+1} = \sum_{k=0}^{n} \left(1 - \frac{k}{n+1}\right) a_k z^k, \qquad n = 0, 1, 2, ...,$$

for the partial sums and the Cesàro means of the power series for f. If $n \geq 0$ and $z \in D$, then (see [2])

$$\sigma_n(f)(z) = \frac{1}{2\pi} \int_T K_n(\xi) f(z\xi) |d\xi|,$$

where

$$K_n(\xi) = \sum_{k=-n}^{n} \left(1 - \frac{|k|}{n+1}\right) \bar{\xi}^k, \quad \xi \in T$$

is Fejèr kernel with the properties as follows (see [3]):

1) $K_n(\xi) \geq 0$;
2) $\frac{1}{2\pi} \int_T K_n(\xi) |d\xi| = 1$.

Theorem 3.1. *Let $0 < \alpha < \infty$, $-\infty < \beta < \infty$, and $f(z) = \sum_{n=0}^{\infty} a_n z^n$ in $H(D)$. Then $f \in \log B_{(\alpha,\beta)}$ if and only if*

$$\sup\{\|\sigma_n(f)\|_{\log B_{(\alpha,\beta)}} : n \in \mathbb{N} \cup \{0\}\} < \infty.$$

Moreover,

$$\sup\{\|\sigma_n(f)\|_{\log B_{(\alpha,\beta)}} : n \in \mathbb{N} \cup \{0\}\} = \|f\|_{\log B_{(\alpha,\beta)}}.$$

Proof. Since

$$\sigma_n(f)(z) = \frac{1}{2\pi} \int_T K_n(\xi) f(z\xi) |d\xi|,$$

we get

$$(\sigma_n(f))'(z) = \frac{1}{2\pi} \int_T K_n(\xi) f'(z\xi) \xi |d\xi|.$$

Assume that $f \in \log B_{(\alpha,\beta)}$. Then

$$\begin{aligned}
&\sup_{z\in D}(1-|z|^2)^{\alpha}|(\sigma_n(f))^{'}(z)|\left(\log\frac{1}{1-|z|^2}\right)^{\beta}\\
&\leq \frac{1}{2\pi}\int_T \sup_{z\in D}(1-|z|^2)^{\alpha}|f^{'}(z\xi)|\left(\log\frac{1}{1-|z|^2}\right)^{\beta} K_n(\xi)|d\xi|\\
&= \|f\|_{\log B_{(\alpha,\beta)}}\frac{1}{2\pi}\int_T K_n(\xi)|d\xi|\\
&= \|f\|_{\log B_{(\alpha,\beta)}}.
\end{aligned}$$

This implies that $\sigma_n(f) \in \log B_{(\alpha,\beta)}$ and

$$\|\sigma_n(f)\|_{\log B_{(\alpha,\beta)}} \leq \|f\|_{\log B_{(\alpha,\beta)}}.$$

Conversely, we suppose

$$\sup\{\|\sigma_n(f)\|_{\log B_{(\alpha,\beta)}} : n \in \mathbb{N}\cup\{0\}\} < \infty.$$

And then

$$\begin{aligned}
f_r^{'}(z) &= f^{'}(rz)\\
&= z^{-1}(1-r)^2\left(\sum_{n=0}^{\infty}(n+1)r^n\right)\left(\sum_{n=0}^{\infty}na_nr^nz^n\right)\\
&= z^{-1}(1-r)^2\sum_{n=0}^{\infty}(n+1)\left(\sum_{k=0}^{n}ka_kz^k\frac{n+1-k}{n+1}\right)r^n\\
&= z^{-1}(1-r)^2\sum_{n=0}^{\infty}(n+1)z\sigma_n^{'}(f)(z)r^n\\
&= (1-r)^2\sum_{n=0}^{\infty}(n+1)\sigma_n^{'}(f)(z)r^n.
\end{aligned}$$

Hence,

$$|f_r^{'}(z)| \leq (1-r)^2\sum_{n=0}^{\infty}(n+1)|\sigma_n^{'}(f)(z)|r^n.$$

Consequently,

$$\begin{aligned}
&(1-|z|^2)^\alpha |f_r'(z)| \left(\log\frac{1}{1-|z|^2}\right)^\beta \\
&\le (1-|z|^2)^\alpha (1-r)^2 \left(\sum_{n=0}^{\infty}(n+1)|\sigma_n'(f)(z)|r^n\right)\left(\log\frac{1}{1-|z|^2}\right)^\beta \\
&= (1-r)^2 \sum_{n=0}^{\infty}(n+1)r^n (1-|z|^2)^\alpha |\sigma_n'(f)(z)| \left(\log\frac{1}{1-|z|^2}\right)^\beta \\
&\le (1-r)^2 \sum_{n=0}^{\infty}(n+1)r^n \|\sigma_n(f)\|_{\log B_{(\alpha,\beta)}} \\
&\le \sup_{n\in N\cup\{0\}} \|\sigma_n(f)\|_{\log B_{(\alpha,\beta)}}.
\end{aligned}$$

As a result, letting $r \to 1$,

$$\|f\|_{\log B_{(\alpha,\beta)}} \le \sup_{n\in N\cup\{0\}} \|\sigma_n(f)\|_{\log B_{(\alpha,\beta)}} < \infty,$$

and then $f \in \log B_{(\alpha,\beta)}$.

Summarizing,

$$\sup\{\|\sigma_n(f)\|_{\log B_{(\alpha,\beta)}} : n \in \mathbb{N}\cup\{0\}\} = \|f\|_{\log B_{(\alpha,\beta)}}.$$

The proof is now complete.

We know that $B^\alpha = \log B_{(\alpha,0)}$, and the relationship between B^α space and Cesàro means is given in [4] as the following corollary.

Corollary If $f \in H(D)$, then the following statements are equivalent:

i) $f \in B^\alpha, \quad \alpha \in (0,\infty)$;

ii) $\|\sigma_n(f)\|_{B^\alpha} = O(1), \quad n \to \infty$;

iii) $\|\sigma_n'(f)\|_\infty = O(n^\alpha), \quad n \to \infty$.

References

1. K.R.M. Attele, Toeplitz abd Hankel operators on Bergman spaces, *Hokkaido Math. J.*, **211**(1992): 279-293.
2. F. Holland, D. Walsh, Criteria for Membership of Bloch Spaces and its Subspace, *BMOA, Math. Ann.* **273**(1986):317-335, .
3. T. W. Körner, *Fourier Analysis*, Cambrige University Press, London, 1988.
4. H. Wulan, M. Zhan, Cesáro Means and Complex Function Spaces, *Journal of STU*, **20**(1)(2005):3-6.

INFINITE SERIES IN JAPANESE MATHEMATICS OF THE 18TH CENTURY*

MITSUO MORIMOTO

International Christian University
3-10-2 Osawa, Mitaka
Tokyo, 181-8585, Japan
E-mail: mitsuo@icu.ac.jp

During the Edo period, mathematics was practised in Japan almost independently of European influence. Seki Takakazu and his disciple Takebe Katahiro learned from the *Suanxue Qimeng* how to express one-variable polynomials with numerical coefficients on the counting board and developed this idea using the so-called side writing method, by which they could treat one-variable polynomials with polynomial coefficients. (See [4].) This mathematical tool also helped them to investigate various properties of the circle. For example, they were interested in representing arc length s in terms of a given diameter d and a sagitta c. Since, in modern notation, $s/2 = d \arcsin\sqrt{c/d}$, one of their results can be described as follows. Let $g_1(t) = (1 - \sqrt{1-t})/2$ and $g_n(t) = g_1(g_{n-1}(t))$ for $n = 1, 2, 3, \cdots$. Then Takebe recognized

$$\lim_{n\to\infty} 4^n g_n(t) = (\arcsin\sqrt{t})^2, \quad (0 \le t \le 1)$$

Using this formula, Takebe was able to derive the Taylor expansion formula of the inverse trigonometric function $(\arcsin t)^2$, which he found earlier by means of numerical calculation. (See [5].)

Keywords: Japanese Mathematics, the 18th century, Trigonometric functions, Takebe Katahiro.

2000 Mathematics Subject Classification: 01A27, 01A50, 33B10

1. Takebe Katahiro

Takebe Katahiro (1664 – 1739) was one of the greatest Japanese mathematicians of the Edo Period (1603 – 1868). (There is no standard reference book on the history of Japanese mathematics in English. The most informative book in Japanese is [6]. See also [2] and [8].) His master Seki Takakazu (c.1640 – 1708) initiated a theory of polynomial algebra with variable coefficients. This theory was called the side writing method and Seki and Takebe applied it to various problems in the *Hatsubi*

*Abbreviated title "Japanese Mathematics of the 18th Century" or "Infinite Series".

Sampō Endan Genkai (see [7]) and later to the study of circles. For the mathematics of Seki Takakazu see his collected work (see [1]).

The main mathematical works of Takebe Katahiro are as follows:

- the *Kenki Sanpō* (Mathematical Methods to Investigate from Slight Signs, 1683)
- the *Hatsubi Sanpō Endan Genkai* (Colloquial Commentary on Operations in the *Hatsubi Sanpō*, 1685)
  - Seki Takakazu: the *Hatsubi Sanpō* (Methods to Explore Subtle Mathematical Points, 1674)
- the *Sangaku Keimō Genkai Taisei* (Great Colloquial Commentary on the *Suanxue Qimeng*, 1690)
  - Zhu Shijie : the *Suanxue Qimeng* (Introduction to Mathematics, 1299)
- the *Taisei Sankei* (Great Accomplished Mathematical Treatise, 20 volumes, 1710)
- the *Koritsu* (Arc Rate, date unknown)
- the *Tetsujutsu Sankei* (Mathematical Treatise on the Technique of Linkage, 1722), Manuscript of Kokuritsu Kōbunsho-kan Naikaku Bunko.
- the *Fukyū Tetsujutsu* (Fukyū's *Tetsujutsu*, 1722), Manuscript of the University of Tokyo Library.
- the *Enri Kohaijutsu* (Studies on the Circle — Methods to Calculate the Length of Circular Arc, date unknown)
- the *Sanreki Zakkō* (Various Considerations on Mathematics and the Calendar, date unknown)

2. Takebe's Infinite Series Expansion

In the *Tetsujutsu Sankei* Takebe Katahiro described the following infinite series expansion of an inverse trigonometric function:

Proposition 2.1. *[Takebe Katahiro, 1722]*

$$\begin{aligned}&(\arcsin\sqrt{t})^2\\ &= t + \frac{t^2}{3} + \frac{8t^3}{45} + \frac{4t^4}{35} + \frac{128t^5}{1575} + \frac{128t^6}{2079} + \frac{1024t^7}{21021} + \cdots \qquad (1)\\ &= t(1 + \frac{t}{3}(1 + \frac{8t}{15}(1 + \frac{9t}{14}(1 + \frac{32t}{45}(1 + \frac{25t}{33}(1 + \frac{72t}{91}(1 + \cdots))))))).\end{aligned}$$

Whereas the first notation is standard in today's mathematics, the second was standard in 18th century Japan. (The second notation requires less multiplications than the first.) The coefficients in the second formula obey the following rules:

$$\frac{1}{3} \quad \frac{8}{15} \quad \frac{9}{14} \quad \frac{32}{45} \quad \frac{25}{33} \quad \frac{72}{91}$$

$$\frac{1}{1\cdot 3}\ \frac{2\cdot 2^2}{3\cdot 5}\ \frac{3^2}{2\cdot 7}\ \frac{2\cdot 4^2}{5\cdot 9}\ \frac{5^2}{3\cdot 11}\ \frac{2\cdot 6^2}{7\cdot 13}$$

$$\frac{2\cdot 1^2}{2\cdot 3}\ \frac{2\cdot 2^2}{3\cdot 5}\ \frac{2\cdot 3^2}{4\cdot 7}\ \frac{2\cdot 4^2}{5\cdot 9}\ \frac{2\cdot 5^2}{6\cdot 11}\ \frac{2\cdot 6^2}{7\cdot 13}$$

The first rule which relies on parity is given in the *Tetsujutsu Sankei* and the second more general rule is given in the *Enri Kohaijutsu*. Judging from this, the latter book is believed to have appeared later than the former. The manuscript *Enri Kohaijutsu* has been preserved and handed down by Seki's followers as one of the confidential books of his school.

3. Squared Half Back Arc and Sagitta

If the diameter d and the sagitta c are given, Takebe Katahiro and his contemporary Japanese mathematicians can calculate the "squared half back arc" $(s/2)^2$ to any accuracy by the rule of right triangle (i.e., Pythagorian Theorem) and by repeated application of an acceleration method. In his studies of the circle, Takebe's most important result is to give a formula for $(s/2)^2$ in terms of d and c, which we will now describe using modern mathematical terminology. (See Figure 1.)

Similarity of two right triangles implies that $x : c = d : x$ and $x^2 = cd$. For the right triangle with hypotenuse d, we have $\sin\theta = \frac{x}{d} = \sqrt{c/d}$. Since θ is also a central angle, we have $\frac{d}{2}\theta = \frac{s}{4}$. Therefore, the half back arc is given as follows:

$$\frac{s}{2} = d\theta = d \arcsin\sqrt{c/d}.$$

Taking the square of both sides, we obtain

$$\left(\frac{s}{2}\right)^2 = \left(d \arcsin\sqrt{c/d}\right)^2.$$

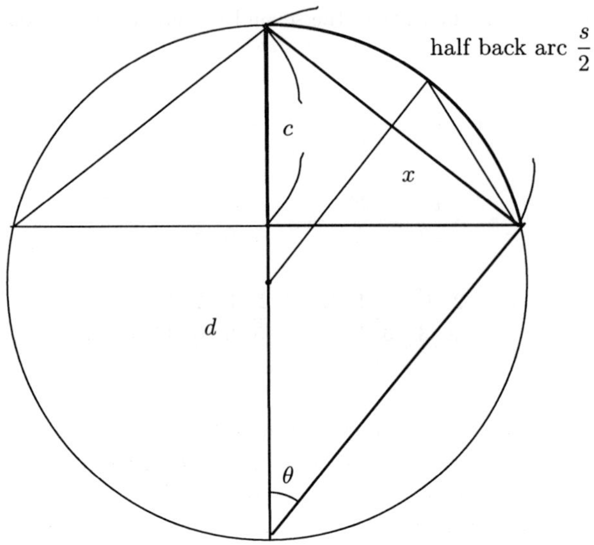

Figure 1

4. Numerical Method to Find the Infinite Series Expansion

Chapter 12 of the *Tetsujutsu Sankei* starts as follows:

> In the search of the form and attribute of the back arc, the true number is hidden if it is close to the half circle and the true number appears if it is close to the side. If it is close to the half circle, it belongs to the latitude and its curve is rapid; if it is close to the side, it belongs to the longitude and its curve is slow. Therefore, taking the sagitta to be extremely small, we should search for the number and seek the procedure.

Now Takebe Katahiro calculates a numerical value of the "squared half back arc" $(s/2)^2$ for $d = 10$ and $c = 10^{-5}$. To do so, he applies the iterated acceleration method, which he employes for the calculation of the π to more than 40 digits in Chapter 11 of the *Tetsujutsu Sankei.* Relying on this numerical value of $(s/2)^2$ he obtains (1) in Chapter 12 of the *Tetsujutsu Sankei.* To show a hint how Takebe finds the formula, we reproduce his calculation using the computer algebra system Mathematica as follows:

```
d=10;c=10^{-5};
N[h=d^2(ArcSin[Sqrt[c/d]])^2,40]
{0.0001000000333333511111225396906666728234777}
```

```
N[c d,40]
{0.0001000000000000000000000000000000000000}
N[h-c d,40]
{3.333335111112253969066667282347769479596 10^{-11}}
N[h-c d -c^2/3,40]
{1.777778920635733333949014436146262542024 10^{-17}}
N[8/45,40]
0.1777777777777777777777777777777777777778
```

Looking at the numerical value of $(s/2)^2$, Takebe assumes its principal part is cd. He then finds the first error $(s/2)^2 - cd$ to be approximately equal to 3.33×10^{-11}, and assumes its principal part to be $c^2/3$. He then finds the second error $(s/2)^2 - cd - c^2/3$ to be approximately equal to 1.77×10^{-17}, and assumes its principal part to be $8c^3/(45d)$. Continueing in this manner, he finally obtains the infinite series expansion formula (1).

5. Approximation Formulas by Interpolation

In the *Katsuyō Sanpō* Seki Takakazu obtained the numerical values of the "back arc" $s = 2d \arcsin\sqrt{c/d}$ for $d = 10$ and $c = 1, 2, 3, 4, 4.5, 5$. On the other hand, Takebe Katahiro gives the numerical values of the "squared back arc" s^2 in the *Koritsu* for the same values d and c. Using the computer algebra system Mathematica we can easily obtain these values:

```
G[c_, d_] = (2*d ArcSin[Sqrt[c/d]])^2
N[G[{1, 2, 3, 4, 45/10, 5}, 10], 25]

{41.40936770181864283627168, 85.98764213286575548425309, \
134.3928914435610127284475, 187.5361547840326075823830, \
216.2749378084305902074557, 246.7401100272339654708623}
```

For Takebe the formula for the "squared back arc" should be exact for these c's. Takebe is not satisfied with (1), which is beautiful and accurate for very small c but becomes less accurate for larger c. He proposes an approximation formula by interpolating these values in the *Koritsu.*

Takebe's interpolation formular has the form

$$G(c,d) \doteqdot K(c,d) := A(c,d)/B(c,d),$$

where

$$\begin{aligned} A(c,d) &= 17243148700cd^4 - 27148244837c^2d^3 + 11453384892c^3d^2 \\ &\quad - 807998619c^4d - 27148244837c^2d^3 - 45408726c^5, \\ B(c,d) &= 4310787175d^3 - 8223990414cd^2 + 4838317774c^2d \\ &\quad - 845423484c^3. \end{aligned}$$

By this fractional approximation formula, the approximate values of the "squared back arc" for the given data are as follows:

```
K[c_, d_] :=
(17243148700 c d^4 - 27148244837 c^2 d^3 + 11453384892 c^3 d^2
- 807998619 c^4 d - 27148244837 c^2 d^3 - 45408726 c^5)/
(4310787175 d^3 - 8223990414 c d^2 + 4838317774 c^2 d
- 845423484 c^3)

N[K[{1, 2, 3, 4, 45/10, 5}, 10], 25]
{41.40936770200963809032453, 85.98764213357462274474809, \
134.3928914446184288161436, 187.5361547834501488901178, \
216.2749378048595665694282, 246.7401100176385622737240}
```

These values are exact up to 8 digits and the graphs of the two functions $G(c, 10)$ and $K(c, 10)$ appear to coincide (Figure 2).

```
Plot[{G[c, 10], K[c, 10]}, {c, 0, 5}]
```

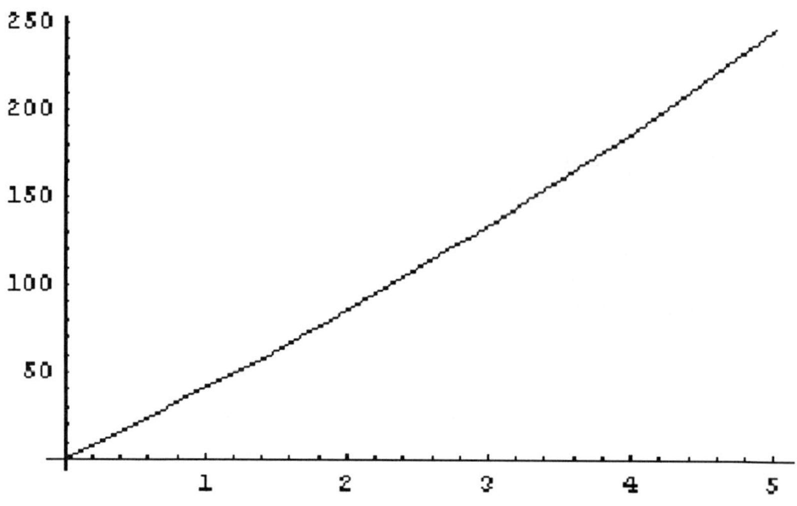

Figure 2

The difference of these two functions is less than 1.5×10^{-8} as seen in Figure 3:

```
Plot[G[c, 10] - K[c, 10], {c, 0, 5}]
```

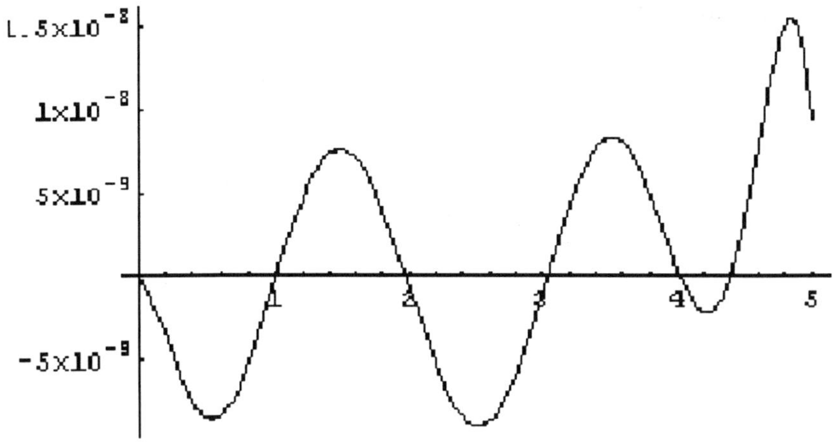

Figure 3

6. Algebraic Method to Find the Infinite Series Expansion

Later in the *Enri Kohaijutsu* Takebe introduces an algebraic method to obtain the infinite series expansion formula (1). This method relies on the following proposition:

Proposition 6.1. *Put*

$$g_1(t) = \frac{1-\sqrt{1-t}}{2}, \quad g_n(t) = g_1(g_{n-1}(t)), \quad n = 1, 2, 3, \cdots .$$

Then we have

$$\lim_{n\to\infty} 4^n g_n(t) = (\arcsin\sqrt{t})^2, \quad (0 \le t \le 1). \tag{2}$$

From Figure 4, the relation $x : c = d : x$ implies $x^2 = cd$. By the rule of right triangle, $x^2 + y^2 = d^2$, and hence $y^2 = d^2 - cd$. It follows that the sagitta c' of the half back arc satisfies $d/2 - c' = y/2$ and is given by $c' = (d - \sqrt{d^2 - cd})/2$. We call c' the A-sagitta. Letting $g(t) = (1 - \sqrt{1-t})/2$, we have $c'/d = g(c/d)$.

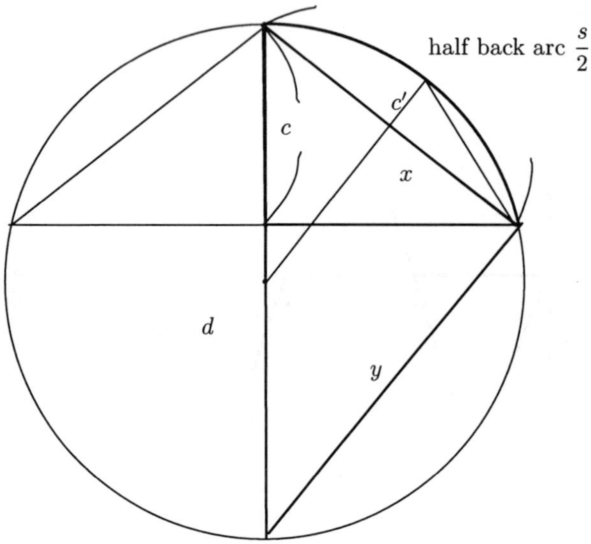

Figure 4

Since the first approximation of $(s/2)^2$ is cd and that of $(s/4)^2$ is $c'd$, the second approximation of $(s/2)^2$ is given by $4c'd = 4d^2g(c/d)$. We call c' the A-sagitta.

Since the second approximation of $(s/4)^2$ is

$$4c''d = 4d^2g(c'/d) = 4d^2g(g(c/d)),$$

the third approximation of $(s/2)^2$ is given by

$$4^2c''d = 4^2d^2g(c'/d) = 4^2d^2g(g(c/d)).$$

We call c'' the B-sagitta.

Iterating this argument, we find $\lim_{n\to\infty} 4^n d^2 g_n(c/d) = (s/2)^2$, which proves (2) geometrically.

7. Counting Board Algebra and Generalized Division

There was no notion of the Cartesian plane in traditional Japanese mathematics, consequently no notion of the graph of a function nor notion of the tangent. This means that there was no notion of differentiation as we have in real variable calculus.

However, Takebe Katahiro and other Japanese mathematicians of the 18th century learned the manipulation of polynomials from Chinese mathematics (The *Suanxue Qimeng* (*Sangaku Keimō* in Japanese), Zhu Shijie, 1299). (For the history of Chinese mathematics, see, for example, [3].) Let

$$P(x) = a + bx + cx^2$$

be a polynomial with numerical coefficients and q be a number. Japanese mathematicians could calculate freely the coefficients a', b', c' of

$$P(x) = a' + b'(x - q) + c'(x - q)^2$$

from the given values a, b, c, and q. In today's terminology, this is nothing but the Taylor expansion of the given polynomial

$$P(x) = P(q) + P'(q)(x - q) + \frac{P''(q)}{2}(x - q)^2.$$

In this sense they knew the derived polynomial $P'(x)$ and the fact that the polynomial function $P(x)$ takes extreme values at $x = q$ when $P'(q) = 0$. (See Chapter 6 of the *Tetsujutsu Sankei.*)

In traditional Chinese and Japanese mathematics, a polynomial with numerical coefficents was represented by the column vector whose components are the coefficients of the polynomial. The polynomial $P(x)$ was represented on a counting board using counting rods as follows:

	Quotient row	shang
a	Reality row	shi
b	Square row	fang
c	Side row	lian
	Corner row	yu

or,

q	Quotient row	shang
a'	Reality row	shi
b'	Square row	fang
c'	Side row	lian
	Corner row	yu

In the sequel, we turn the column vector 90 degree clockwisely and represent it as a row vector. The calculation of a', b' and c' from the old coefficients a, b, c and the number q is done by adding upward from the bottom to the top several times (in the row vector notation, from the right to the left). This algorithm is called generalized division and can be traced back to the *Jiuzhang Suanshu* (Nine Chapters of the Arithmetic Arc) of the Chinese Han Dynasty.

$$\begin{array}{llll}
\hline
 & a & b & c \\
\hline
q) & a & b & c \\
 & (b+cq)q & cq & \\
\hline
 & a+(b+cq)q & b+cq & c \\
 & & cq & \\
\hline
q & a+(b+cq)q & b+2cq & c \\
\hline
\end{array}$$

To calculate a'', b'', c'' we add upward two times:

$$\begin{aligned}
P(x) &= a + bx + cx^2 \\
&= a' + b'(x - q) + c'(x - q)^2 \\
&= a'' + b''(x - q - q') + c''(x^q - q')^2
\end{aligned}$$

	a	b	c
(q)	a	b	c
	$(b+cq)q$	cq	
	$a+(b+cq)q$	$b+cq$	c
		cq	
q	$a+(b+cq)q$	$b+2cq$	c
q	a'	b'	c'
$q+(q')$	a'	b'	c'
	$(b'+c'q')q'$	$c'q'$	
	$a'+(b'+c'q')q'$	$b'+c'q'$	c'
		$c'q'$	
$q+q'$	$a'+(b'+c'q')q'$	$b'+2c'q'$	c'
$q+q'$	a''	b''	c''

If the Reality row becomes null after several operations, the number at the Quotient row $q+q'+\cdots$ gives a solution of the equation $P(x)=0$.

If $P(x)=x^2-2$, this calculation is called the extraction of the square root; if $P(x)=x^3-2$, this is called the extraction of the cubic root. It was in these forms that this algorithm was introduced in the *Jiuzhang Suanshu*. If $P(x)=a+bx$, this algorithm is the usual algorithm of division to find the decimal expansion of $-b/a$. Therefore, this algorithm is called generalized division.

8. The Taylor Expansion Formula in Japanese Mathematics

8.1. *Case of a Quadratic Equation*

Note that $x=g(t)=(1-\sqrt{1-t})/2$ is a solution of the equation

$$P(x)=-t+4x-4x^2=0.$$

If x and t are small, neglecting the terms of higher degree, Takebe finds the equation $P(x)=0$ is approximated by $-4x+t \doteqdot 0$. Therefore, he expands the equation with respect the new variable $x_1=x-\dfrac{t}{4}$. The calculation is done by generalized division. We can recalculate it using the computer algebra system Mathematica as follows:

```
P[x_] = 4x^2 - 4x + t
P1[x1_] := P[x1 + t/4]
Expand[P1[x1]]
```

$$P_1(x_1)=P(x_1+\frac{t}{4})=-\frac{t^2}{4}+(4-2t)x_1-4x_1^2=0.$$

Neglecting the terms of higher degree in x_1 and t, Takebe finds the equation $P_1(x_1)=0$ is approximated by $-\dfrac{t^2}{4}+4x_1 \doteqdot 0$. Therefore, he expands the equation with respect to the new variable $x_2 = x_1 - \frac{t^2}{4 \cdot 4} = x - \frac{t}{4} - \frac{t^2}{16}$. Using generalized division, he finds

$$P_2(x_2) = P_1(x_2 + t^2/16) = -\frac{t^3}{8} - \frac{t^4}{64} + (4 - 2t - \frac{t^2}{2})x_2 - 4x_2^2 = 0.$$

Neglecting the terms of higher degree in x_2 and t, Takebe finds the equation $P_2(x_2) = 0$ is approximated by $-\frac{t^3}{8} + 4x_2 \doteqdot 0$. Therefore, he expands the equation with respect to the new variable $x_3 = x_2 - \frac{t^3}{8 \cdot 4}$. Using generalized division, he finds

$$\begin{aligned} P_3(x_3) &= P_2(x_3 + \frac{t^4}{32}) \\ &= -\frac{5t^4}{64} - \frac{t^5}{64} - \frac{t^6}{256} + (4 - 2t - \frac{t^2}{2} - \frac{t^3}{4})x_3 - 4x_3^2 = 0. \end{aligned}$$

Neglecting the terms of higher degree in x_3 and t, Takebe finds the equation $P_3(x_3) = 0$ is approximated by $-\frac{5t^4}{64} + 4x_3 \doteqdot 0$. Therefore, he expands the equation with respect to the new variable $x_4 = x_3 - \frac{5t^4}{64 \cdot 4}$. Using generalized division, he finds

$$\begin{aligned} P_4(x_4) &= P_3(x_4 + \frac{5t^4}{256}) \\ &= -\frac{7t^5}{128} - \frac{7t^6}{512} - \frac{5t^7}{1024} - \frac{25t^8}{16384} \\ &\quad + (4 - 2t - \frac{t^2}{2} - \frac{t^3}{4} - \frac{5t^4}{32})x_4 - 4x_4^2 = 0. \end{aligned}$$

Neglecting the terms of higher degree in x_4 and t, Takebe finds the equation $P_4(x_4) = 0$ is approximated by $-\frac{7t^4}{128} + 4x_4 \doteqdot 0$. Therefore, he expands the equation with respect to the new variable $x_5 = x_4 - \frac{7t^5}{128 \cdot 4}$. Using generalized division, he finds

$$\begin{aligned} P_5(x_5) &= P_4(x_5 + \frac{7t^5}{512}) \\ &= -\frac{21t^6}{512} - \frac{3t^7}{256} - \frac{81t^8}{16384} - \frac{35t^9}{16384} - \frac{49t^{10}}{65536} \\ &\quad + (4 - 2t - \frac{t^2}{2} - \frac{t^3}{4} - \frac{5t^4}{32} - \frac{7t^5}{64})x_5 - 4x_5^2. \end{aligned}$$

By this repeated application of generalized division, he can find any number of terms in the Taylor expansion of the A-sagitta $g(t)$:

$$g(t) = \frac{t}{4} + \frac{t^2}{16} + \frac{t^3}{32} + \frac{5t^4}{256} + \frac{7t^5}{512} + \cdots, \tag{3}$$

where the coefficients are called, from the lowest order term, the A-original number, the 1st A-difference, the 2nd A-difference, the 3rd A-difference and the 4th A-difference, etc.

8.2. *Iteration*

Since $-t+4g(t)-4g(t)^2=0$, we have $-g(t)+4g(g(t))-4g(g(t))^2=0$. Eliminating x from the system of equations

$$-t+4x-4x^2=0, \quad -x+4y-4y^2=0,$$

we find that $g(g(t))$ satisfies $-t+4(4y-4y^2)-4(4y-4y^2)^2=0$. Therefore, if we employ the equation of degree 4

$$-t+16y-80y^2+128y^3-64y^4=0,$$

we can expand the function $g(g(t))$ by repeated application of generalized division.

The iteration in the *Enri Kohaijutsu* is simpler: Takebe uses the quadratic equation

$$Q(y)=-g(t)+4y-4y^2=0$$

and the expansion formula (3). Namely, put

$$Q(y) \doteqdot -\frac{t}{4}-\frac{t^2}{16}-\frac{t^3}{32}-\frac{5t^4}{256}-\frac{7t^5}{512}-\frac{21t^6}{2048}+4y-4y^2=0.$$

Neglecting the terms of higher order in y and t, Takebe finds the equation $Q(y)=0$ is approximated by $-t/4+4y \doteqdot 0$. Therefore, he expands the equation with respect to the new variable $y_1=y-\dfrac{t}{4\cdot 4}$. Using generalized division, he finds

$$\begin{aligned} Q_1(y_1) &= Q(y_1+\frac{t}{4\cdot 4}) \\ &\doteqdot -\frac{5t^2}{64}-\frac{t^3}{32}-\frac{5t^4}{256}-\frac{7t^5}{512}-\frac{21t^6}{2048}+(4-\frac{t}{2})y_1-4y_1^2=0. \end{aligned}$$

Neglecting the terms of higher order in y_1 and t, Takebe finds the equation $Q_1(y_1)=0$ is approximated by $-\dfrac{5t^2}{64}+4y_1 \doteqdot 0$. Therefore, he expands the equation with respect to the new varianble $y_2=y_1-\dfrac{5t^2}{64\cdot 4}$. Using generalized division, he finds

$$\begin{aligned} Q_2(y_2) &= Q_1(y_2+\frac{5t^2}{64\cdot 4}) \\ &\doteqdot -\frac{21t^3}{512}-\frac{345t^4}{16384}-\frac{7t^5}{512}-\frac{21t^6}{2048}+(4-\frac{t}{2}-\frac{5t^2}{32})y_2-4y_2^2=0. \end{aligned}$$

Neglecting the terms of higher order in y_2 and t, Takebe finds the equation $Q_2(y_2)=0$ is approximated by $-\dfrac{21t^3}{512}+4y_2 \doteqdot 0$. Therefore, he expands the equation with respect to the new variable $y_3 = y_2 - \frac{21t^3}{512 \cdot 4}$. Using generalized division, he finds

$$Q_3(y_3) = Q_2(y_3 + \frac{21t^3}{512 \cdot 4})$$
$$\doteqdot -\frac{429t^4}{16384} - \frac{1001t^5}{65536} - \frac{11193t^6}{1048576} + (4 - \frac{t}{2} - \frac{5t^2}{32} - \frac{21t^3}{256})y_3 - 4y_3^2 = 0.$$

Neglecting the terms of higher order in y_3 and t, Takebe finds the equation $Q_3(y_3) = 0$ was approximated by $-\frac{429t^4}{16384} + 4y_3 \doteqdot 0$. Therefore, he expands the equation with respect to the new variable $y_4 = y_3 - \frac{429t^4}{16384 \cdot 4}$. Using generalized division, he finds

$$Q_4(y_4) = Q_3(y_4 + \frac{429t^4}{16384 \cdot 4})$$
$$\doteqdot -\frac{2431t^5}{131072} - \frac{24531t^6}{2097152} - \frac{9009t^7}{16777216} - \frac{184041t^8}{1073741824}$$
$$+ (4 - \frac{t}{2} - \frac{5t^2}{32} - \frac{21t^3}{256} - \frac{429t^4}{8192})y_4 - 4y_4^2 = 0.$$

Neglecting the terms of higher order in y_4 and t, Takebe finds the equation $Q_4(y_4) = 0$ is approximated by $-\frac{2431t^5}{131072} + 4y_4 \doteqdot 0$. Therefore, he expands the equation with respect to the new variable $y_5 = y_4 - \frac{2431t^5}{131072 \cdot 4}$. Using generalized division, he finds

$$Q_5(y_5) = Q_4(y_5 + \frac{2431t^5}{131072 \cdot 4})$$
$$\doteqdot -\frac{29393t^6}{2097152} - \frac{5291t^7}{4194304} - \frac{592449t^8}{1073741824} - \frac{1042899t^9}{4294967296} - \frac{5909761t^{10}}{68719476736}$$
$$+ (4 - \frac{t}{2} - \frac{5t^2}{32} - \frac{21t^3}{256} - \frac{429t^4}{8192} - \frac{2431t^5}{65536})y_5 - 4y_5^2 = 0.$$

By this repeated application of generalized division, he can find any number of terms in the Taylor expansion of the B-sagitta $g_2(t) = g(g(t))$:

$$g(g(t)) = \frac{t}{16} + \frac{5t^2}{256} + \frac{21t^3}{2048} + \frac{429t^4}{65536} + \cdots,$$

where the coefficients are called the B-original number, the 1st B-difference, the 2nd B-difference, the 3rd B-difference and the 4th B-difference, etc.

9. Taking the Limit

9.1. *Further sagittas*

Taking the diameter $d = 1$ and the sagitta $c = t$, the A-sagitta, B-sagitta, etc. can be given as follows:

A-sagitta	$g_1(t) = g(t)$
B-sagitta	$g_2(t) = g(g(t))$
C-sagitta	$g_3(t) = g(g(g(t)))$
D-sagitta	$g_4(t) = g(g(g(g(t))))$
E-sagitta	$g_5(t) = g(g(g(g(g(t)))))$
F-sagitta	$g_6(t) = g(g(g(g(g(g(t))))))$
G-sagitta	$g_7(t) = g(g(g(g(g(g(g(t)))))))$
H-sagitta	$g_8(t) = g(g(g(g(g(g(g(g(t))))))))$
I-sagitta	$g_9(t) = g(g(g(g(g(g(g(g(g(t)))))))))$
J-sagitta	$g_{10}(t) = g(g(g(g(g(g(g(g(g(g(t))))))))))$

In the *Enri Kohaijutsu*, Takebe does not apply generalized division to obtain the C- and further sagittas.

First, he writes down the rule by which he obtains the expansion of the B-sagitta from that of the A-sagitta. For example, The B-original number is obtained by dividing the A-original number by 4. The 1st B-sagitta is obtained from the A-original number and the 1st A-difference. The 2nd B-difference is obtained by the A-original number, the 1st A-difference, and the 2nd A-difference, etc. Second, he applies this rule to the expansion of the B-sagitta to obtain the expansion of the C-sagitta.

Using Mathematica, we reproduce his calculation of the expansion up to the term of t^{11} (10 terms) for the A-sagitta, the B-sagitta, the C-sagitta, $\cdots$, the J-sagitta.

$$4g_1(t) = t + \frac{t^2}{4} + \frac{t^3}{8} + \frac{5t^4}{64} + \frac{7t^5}{128} + \cdots$$
$$4^2g_2(t) = t + \frac{5t^2}{16} + \frac{21t^3}{128} + \frac{429t^4}{4096} + \frac{2431t^5}{32768} + \cdots$$
$$4^3g_3(t) = t + \frac{21t^2}{64} + \frac{357t^3}{2048} + \frac{29325t^4}{262144} + \frac{666655t^5}{8388608} + \cdots$$
$$4^4g_4(t) = t + \frac{85t^2}{256} + \frac{5797t^3}{32768} + \frac{1907213t^4}{16777216} + \frac{173556383t^5}{2147483648} + \cdots$$
$$4^5g_5(t) = t + \frac{341t^2}{1024} + \frac{93093t^3}{524288} + \frac{122550285t^4}{1073741824} + \frac{44616473759t^5}{549755813888} + \cdots$$

9.2. *Taking the Limit Numerically*

In the *Enri Kohaijutsu* Takebe calculates numerically the following coefficients in the above expansion formulas:

```
N[{1/4, 5/16, 21/64, 85/256, 341/1024}]
{0.25, 0.3125, 0.328125, 0.332031, 0.333008}
```

From this, we find that the coefficient of t^2 in the limit $f(t)$ is equal to $\frac{1}{3}$.

Next, Takebe calculates numerically the ratio of coefficients:

```
N[{1/8/(1/4), 21/128/(5/16), 357/2048/(21/64),
5797/ 32768/(85/256), 93093/524288/(341/1024)}]
{0.5, 0.525, 0.53125, 0.532813, 0.533203}
N[8/15]
0.533333
```

From this, he finds that the coefficient of t^3 in the limit $f(t)$ is equal to $\frac{1}{3} \times \frac{8}{15}$. Continuing in this manner, Takebe gives a different proof to the infinite series expansion formula (1).

10. The *Sanreki Zakkō*

The manuscript of the *Sanreki Zakkō* was held at the "Shōkōkan" in Mito city. The "Shōkōkan" was the library of one branch of the Tokugawa family but was destroyed by fire during the 2nd World War in 1945. Fortunately, the manuscript was copied chemically in 1944 and reproduced recently in [9].

In this book several approximation fractional formula and the addition formula for the trigonometric functions $\sin t$ and $\cos t$ were introduced. Using these formulas, the table for sin was constructed. This is the first occurrence of the trigonometric function in Japanese mathematics.

References

1. Hirayama, Akira; Shimodaira, Kazuo and Hirose, Hideo (ed.), Seki Takakazu Zenshū (The Collected Works of *Seki Takakazu*, in Japanese with English summary), Ōsaka Kyōiku Tosho, 1974.
2. A. Horiuchi, *Les Mathématiques Japonaises à l'Époque d'Edo,* Vrin, 1994.
3. H. C. Martzloff, A History of Chinese Mathematics, Springer 1997. [the French original edition: *Histoire des mathématiques chinoises*, Masson, Paris 1987.]
4. M. Morimoto, A Chinese root of Japanese traditional mathematics, *Proceedings of the 11th International Conference on Finite or Infinite Dimensional Complex Analysis and Applications,* Wasan, (2003):133-141.
5. Morimoto, Mitsuo and Ogawa, Tsukane, Takebe Katahiro no Sūgaku — Tokuni Gyaku-sannkaku Kansū ni kansuru Mittsu no Kōshiki ni tsuite — (The Mathematics of Takebe Katahiro — His Three Formulas for an Inverse Trigonometric Function, in Japanese), Sūgaku, **56**(2004):308 - 319. (English translation will be published by AMS.)
6. Nihon Gakushiin, Nihon Kagakushi Kankōkai (ed.): Meiji-zen Nihon Sūgakushi (History of Japanese Mathematics before the Meiji Restoration, in Japanese), Shinteiban (New Edition)", in 5 volumes, Noma Kagaku Igaku Kenkyū Shiryō kan, 1979. (First edition, Iwanani Shoten, 1954 - 1960.)
7. Ogawa, Tsukane: Seki Takakazu Hatsubi Sanpō, Gendaigoyaku to Kaisetsu (Seki Takakazu's *Hatsubi Sanpō*, Its translation to modern Japanese with annotation, in Japanese), Ōsorasha, 1994.
8. Ogawa, T.: A review of the history of Japanese mathematics, *Revue d'histoire des mathématiques*, **7**(2001):101–119.
9. Satō Ken'ichi, Sr.: Takebe Katahiro no Snreki Zakkō — Nihon hatsu no sankaku kansū hyō (Takebe Katahiro's *Sanreki Zakkō* — Table of trigonometric functions first in Japan, in Japanese), Kenseisha, 1995.

MEROMORPHIC FUNCTIONS THAT SHARE FOUR SMALL FUNCTIONS

QIU GANGDI

Department of Mathematics
Ningde Teachers College
Ningde, Fujian 352100 P. R. China
E-mail: q_g_d@yahoo.com.cn

In this paper, we mainly discuss the uniqueness problem of k order derivatives of meromorphic functions that share four small functions, and consider Nevanlinna's Four Values Theorem when the derivatives of meromorphic functions share some small functions, which generalize and improve some related results of the author.

Keywords: Meromorphic function, small function, uniqueness.

1991 Mathematics Subject Classification: 30D35

1. Introduction and Main Results

In this paper, we use the same symbols as given in Nevanlinna theory of meromorphic functions (see [1]). By $S(r,f)$ we denote any quantity satisfying $S(r,f) = o\{T(r,f)\}$ as $r \to \infty$, possibly outside a set finite linear measure. A meromorphic furction α is said to be a small function of f if and only if $T(r,\alpha) = S(r,f)$

Let $f(z)$ and $g(z)$ be two nonconstant meromorphic functions, and $a(z)$ a small function of $f(z)$ and $g(z)$, if $f(z) - a(z)$ and $g(z) - a(z)$ have the same zeros ignoring (counting) multiplicities, then we say that $f(z)$ and $g(z)$ shares $a(z) IM (CM)$.

In addition, let $S(f = a = g)$ be the set of all common zeros of $f - a$ and $g - a$, $S_{(k,m)}(f = g = a)$ be the set of all points which are zeros of $f - a$ with multiplicities k as well as the zeros of $g - a$ with multiplicities m. Funthermore, we denote $S_E(f = a = g)$ the set of all common zeros of $f - a$ and $g - a$ with the same multiplicities. Denote $N(r, f = a = g)$,

$N_{(k,m)}(r, f = a = g)$ and $N_E(r, f = a = g)$ the reduced counting functions of f and g correspondent to the sets $S(f = a = g)$, $S_{(k,m)}(f = g = a)$ and $S_E(f = a = g)$, respectively.

If $\overline{N}(r, 1/(f-a)) - \overline{N}(r, f = a = g) = S(r, f)$, and $\overline{N}(r, 1/(g-a)) - \overline{N}(r, f = a = g) = S(r, g)$, then we say that f and g share a IM*.

If $\overline{N}(r, 1/(g-a)) - \overline{N}_E(r, f = a = g) = S(r, f)$, and $\overline{N}(r, 1/(g-a)) - \overline{N}_E(r, f = a = g) = S(r, g)$, then we say that f and g share a CM*.

In 1929, R. Nevanlinna proved the following famous Four Values Theorem:

Theorem A (see [2]): Let f and g be two nonconstant functions, $a_j (j = 1, 2, 3, 4)$ be four distinct complex numbers. If f and g share a_j $(j = 1, 2, 3, 4)$ CM, then f is a Mobius transformation of g.

In 1998, the author proved the following theorem when considering derivatives of meromorphic functions that share four small entire functions:

Theorem B (see [3]) Let f and g be two nonconstant meromorphic functions, k be a positive integer, and $a_j (j = 1, 2, 3, 4)$ four distinct small entire functions with respect to f and g. If $f^{(k)}$ and $g^{(k)}$ share $b_j (j = 1, 2, 3, 4)$ CM*, then $f^{(k)} = g^{(k)}$.

In this paper, we generalize and improve the result of theorem B and obtain the following result:

Theorem 1.1. *Let f and g be two nonconstant meromorphic functions, k be a positive integer, and a_1, a_2, a_3 three distinct small functions with respect to f and g. If $f^{(k)}$ and $g^{(k)}$ share ∞, a_1, $a_2 IM^*$, and share $a_3 CM^*$. then one of the following cases does occur:*

(i) $f^{(k)} \equiv g^{(k)}$.

(ii) $(f^{(k)} - a_1)(g^{(k)} - a_1) \equiv (a_1 - a_2)^2$, and ∞, a_1 are two exceptional functions of $f^{(k)}$ such that $a_2 + a_3 \equiv 2a_1$

(iii) $(f^{(k)} - a_2)(g^{(k)} - a_2) \equiv (a_2 - a_3)^2$, and ∞, a_2 are two exceptional functions of $f^{(k)}$ such that $a_1 + a_3 \equiv 2a_2$

(iv) $(f^{(k)} - a_3)(g^{(k)} - a_3) \equiv (a_3 - a_1)^2$, and ∞, a_3 are two exceptional functions of $f^{(k)}$ such that $a_1 + a_2 \equiv 2a_3$

2. Some Lemmas

Lemma 2.1. *Assume the condition same as given in theorem 1.1, then*

$$S(r, f^{(k)}) = S(r, g^{(k)}) \triangleq S(r) \tag{1}$$

Proof. It is easy to see that (1) follows by the conditions of theorem 1 and Nevanlinna's three small functions theorem (P. 47[1]).□

Lemma 2.2. *(see [4,5]) Let f be a nonconstant meromorphic function, b_1 and b_2 be two distinct small functions of f, and*

$$L(f, b_1, b_2) = \begin{vmatrix} f - b_1 & b_1 - b_2 \\ f' - b_1' & b_1' - b_2' \end{vmatrix} \tag{2}$$

then

$$(i) \quad L(f, b_1, b_2) \equiv 0 \tag{3}$$

$$(ii) \quad m[r, \frac{L(f, b_1, b_2)}{(f - b_i)}] = S(r, f), (i = 1, 2) \tag{4}$$

$$(iii) \quad m[r, \frac{L(f, b_1, b_2)}{(f - b_1)(f - b_2)}] = S(r, f) \tag{5}$$

$$(iv) \quad m[r, \frac{L(f, b_i, b_2)}{(f - b_1)(f - b_2)}] = S(r, f) \tag{6}$$

where a is a small function of f

$$(v) \quad \sum_{i=1}^{2} N(r, \frac{1}{f - b_i}) - N(r, \frac{1}{L(f, b_1, b_2)}) \leq \sum_{i=1}^{2} \overline{N}(r, \frac{1}{f - b_i}) + S(r, f) \tag{7}$$

Lemma 2.3. *(see [6]) Let f and g be two nonconstant meromorphic fanctions, a_j $(j = 1,\ 2,\ 3,\ 4)$ be four distinct small functions with respect to f and g. If f and g share a_1, $a_2 CM^*$, and share a_3, $a_4 IM^*$, then*

$$f = \frac{ag + b}{cg + d} \tag{8}$$

where a, b, c, d are small functions with respect to f and g.

3. Proof of Theorem

Let

$$F = \frac{f^{(k)} - a_1}{a_2 - a_1}, G = \frac{g^{(k)} - a_1}{a_2 - a_1}, a = \frac{a_3 - a_1}{a_2 - a_1} \tag{9}$$

then F and G share $0, 1, \infty$ IM*, and share a CM*, $a \neq 0, 1, \infty$
Assume that $f^{(k)} \not\equiv g^{(k)}$, then $F \not\equiv G$.

$$H = \frac{F'}{F(F-1)} \frac{L(G, 0, a)(F - G)}{G(G - a)} - \frac{G'}{G(G-1)} \frac{L(F, 0, a)(F - G)}{F(F - a)} \tag{10}$$

from (4)-(6) of Lemma 2.2 we deduce that

$$m(r,H)=S(r) \tag{11}$$

It is easy to see that the poles of H only possibly comes from the zeros, poles, 1-points and a-points of F.

Combining (7) with (10), we can know that 1 and a are not the poles of H. Next, (10) can be rewritten as

$$\begin{aligned} H &= \left[\left(\tfrac{F'}{F-1}-\tfrac{F'}{F}\right)\left(\tfrac{G'-a'}{G-a}-\tfrac{G'}{G}\right)\left(\tfrac{F'-a}{F-a}-\tfrac{F'}{F}\right)\right](F-G) \\ &= \left[\tfrac{F'}{F}\left(\tfrac{G'}{G-1}-\tfrac{G'-a'}{G-a}\right)-\tfrac{G'}{G}\left(\tfrac{F'}{F-1}-\tfrac{F'-a'}{F-a}\right)\right](F-G) \\ &\quad+\left[\tfrac{F'}{F-1}\tfrac{G'-a'}{G-a}-\tfrac{G'}{G-1}\tfrac{F'-a'}{F-a}\right](F-G) \end{aligned} \tag{12}$$

therefore, 0 is also not a po1e of H. In addition, from (10) we get

$$H=\frac{F-G}{FG}\left[\frac{F'(aG'-a'G)}{(F-1)(G-a)}-\frac{G'(aF'-a'F)}{(G-1)(F-a)}\right]$$

By a simple computation we have

$$\begin{aligned} H &= \tfrac{F-G}{FG}\left[\frac{a(a-1)F'G'+a'FG'(F-1)(G-a)-a'GF'(G-1)(F-a)}{(F-1)(F-a)(G-1)(G-a)}\right] \\ &\triangleq \tfrac{R}{Q} \end{aligned} \tag{13}$$

Now let Z_0 is a pole of f with multiplicities m as well as a pole of g with multiplicities n, from (9) we know that Z_0 is a pole of F with multiplicities $m+k$ as well as a po1e of G with multiplicities $n+k$.

Then it follows that Z_0 is a po1e of R with multiplicities $2m+2n+5k+1+\max(m,n)$ at most, as well as a pole of Q with multiplicities $3(m+n+2k)$

Noting that:

$$\begin{aligned} &3(m+n+2k)-[2m+2n+5k+1+\max(m,n)] \\ &=m+n+k-1-\max(m,n)\geqslant 1 \end{aligned}$$

So Z_0 is not a pole of Q but a zero of H. According to the discussions above, we get

$$N(r,H)=S(r),$$

which leads to

$$T(r,H)=S(r).$$

Hence

$$N(r,F) \leqslant N(r,\tfrac{1}{H}) \leqslant T(r,H) + O(1) = S(r) \tag{14}$$

Therefore F and G snare $0, l, IM^*$, and share ∞, aCM^*.

From lemma 2.3 we get

$$F = \frac{c_1 G + c_2}{c_3 G + c_4} \tag{15}$$

where c_1, c_2, c_3, c_4 are small functions with respect to F and G.

If none of 0, 1, a is the Picard exceptional function of F and G, substituting sharing small functions 0, 1 and a of F and G into the (15), then we deduce that $F \equiv G$, and $f^{(k)} \equiv g^{(k)}$ according to (10), which is a contradiction.

Hence, one of 0, 1 and a must be the Picard exceptionaI function of F and G. From (14) we known that ∞ is a Picard expectional function of F and G, it follows that among 0, 1 and a, there is one and just only one function being an exceptional function of F and G.

We discuss the following three cases:

Case I: 0 and ∞ are two exceptional functions of F and G. Substituting the common 1-points and a-points of F and G in to (15), we have

$$c_1 + c_2 = c_3 + c_4, c_1 a + c_2 = c_3 a^2 + c_4 a \tag{16}$$

Next, from (15) we get

$$F = \frac{c_1\left(G + \frac{c_2}{c_1}\right)}{c_3 G + c_4} \tag{17}$$

So, $\frac{c_2}{c_1} = 0$, or ∞, and it follows that $c_2 = 0$, or $c_1 = 0$

(I.1) If $c_2 = 0$, substituting it into (16), we deduce that $a = 1$, which is a contraction.

(I.2) If $c_1 = 0$, combining (16) with (17),we have

$$F = \frac{c_2}{c_3 G + c_4}, c_4 = -c_3(a+1) \text{ and } c_2 = -c_3 a.$$

Hence

$$F = \frac{-a}{G - (a+1)} \tag{18}$$

which leads to $a + 1 = 0$, i.e, $a = -1$

From (18) we obtain that $FG \equiv 1$, together with (9) we get $(f^{(k)} - a_1)(g^{(k)} - a_1) \equiv (a_2 - a_1)^2$,$a_2 + a_3 \equiv 2a_1$, and ∞, a_1 are two exceptional functions of $f^{(k)}$ and $g^{(k)}$.

Case II: 1 and ∞ are two exceptional functions of F and G, substituting the common zeros and a-points of F and G into (15) we get

$$c_2 = 0, c_1 = c_3 a + c_4 \tag{19}$$

From (15), we have

$$F = \frac{c_1 G}{c_3 \left(G + \frac{c_4}{c_3}\right)} \tag{20}$$

Hence, $-\frac{c_4}{c_3} = 1$ or ∞, and it leads to $c_4 = -c_3$ or $c_3 = 0$
If $c_3 = 0$, then (19) and (20) leads to $F \equiv G$, a contradiction.
If $c_3 = -c_4$, combining (19) with (20) we get

$$F = \frac{(a-1)G}{G-1} \tag{21}$$

Considering that

$$F - 1 = \frac{(a-2)G + 1}{G-1} \tag{22}$$

therefore, $\frac{1}{1-a} = 1$ or ∞, it result in $a = 2$.
Substituting it into (22) we have

$$(F-1)(G-1) \equiv 1$$

Combining (19) with (23) we get $(f^{(k)} - a_2)(g^{(k)} - a_2) \equiv (a_2 - a_3)^2$, $a_3 + a_1 = 2a_2$ and ∞, and a_2 are two exceptional function of $f^{(k)}$ and $f^{(k)}$.

Case III: a and ∞ are two exceptional functions of F and G. Substituting the common zeros and l-pioms of F and G into (15) we obtain

$$c_2 = 0, c_1 = c_3 + c_4, F = \frac{(c_3 + c_4)G}{c_3 G + c_4} \tag{24}$$

Hence,$-\frac{c_4}{c_3} = a$, or ∞, and it leads to $c_3 = 0$, $c_4 = -c_3 a$.
(III.1) If $c_3 = 0$, then $F \equiv G$ is also a contradiction.
(III.2) If $c_4 = -c_3 a$, substituting it into (24) we get

$$F = \frac{(1-a)G}{G-a} \tag{25}$$

Considering that

$$F - a = \frac{(1-2a)G + a}{G-a} \tag{26}$$

Therefore, $\frac{a^2}{2a-1} = a$ or ∞, which result in $a = \frac{1}{2}$, Combining (9) with (26), we have $(f^{(k)} - a_3)(g^{(k)} - a_3) = (a_3 - a_1)^2$, $a_1 + a_3 = 2a_3$, and a_3 and ∞ are two exceptional functions of $f^{(k)}$ and $g^{(k)}$.

Which completes the proof of theorem 1.1.

References

1. W. K. Hayman, *Meromorhic Functions,* Clarendon Press, Oxford, 1964.
2. R. Nevanlinna, *Le theoreme de Picard-Borel eet la theorie Des Fanctions Meromorphes,* Paris, 1929.
3. G. D. Qiu, Four Values Theorem of Derivatives of Meromorphic Functions, *Systems Science and Mathematical Sciences,* **11**(1998):245-248.
4. G. D. Qiu, Uniqueness of Entire Functions that Share Some Small Functions, *Kodai Math, J.* **23**(2003):1-11.
5. Y. H. Li, Entire functions sharing four small functions IM, *Acta Math Sinica,* **41**(1998):249-260(in Chinese).
6. P. Li, Meromorphic Functions that Share Four Small Functions, *Math. Anal. Appl.* **263**(2002):36-42.

INTEGRATION OPERATORS ON THE BMOA TYPE SPACES*

DAN QU
Department of Mathematics
Shantou University
Shantou, Guangdong, 515063 P. R. China
E-mail: g_dqu@stu.edu.cn

Let ϕ be an analytic function on the open unit disk in the complex plane. We study the boundedness of the following operators

$$J_\phi(f)(z) = \int_0^z f(\xi)\phi'(\xi)d\xi, \quad I_\phi(f)(z) = \int_0^z f'(\xi)\phi(\xi)d\xi$$

on Bloch-type spaces B^α and BMOA-type spaces.

Keywords: B^α, $BMOA_\alpha$, $BMOA^\alpha$, integration operator, boundedness.

2000 Mathematics Subject Classification: 47B33

1. Introduction

Let $D = \{z \in C : |z| < 1\}$ denote the open unit disk in the complex plane C and $\partial D = \{z \in C : |z| = 1\}$. For $a \in D$, $0 < r < 1$, let $D(a,r) = \{w \in D : \beta(z,w) < r\}$ denote the Bergman disk, where $\beta(z,w) = \frac{1}{2}\log\frac{1+|\phi_a(z)|}{1-|\phi_a(z)|}$, $\phi_a(z) = \frac{a-z}{1-\bar{a}z}$.

Let $\alpha > 0$, the $\alpha - Bloch$ space B^α is defined to be the space of analytic functions f on D (denoted by H(D)) such that

$$\|f\|_{B^\alpha} := \sup_{z\in D}(1-|z|^2)^\alpha|f'(z)| < +\infty,$$

Note that when $\alpha = 1$, B^α is the well-known Bloch space B. The space

*This work is supported by NSF of Guangdong Province (Grant No. 032028).

$BMOA^\alpha$ is defined to be the space of $f \in H(D)$ with

$$\|f\|^2_{BMOA^\alpha} := \sup_{a\in D} \int_D (1-|z|^2)^{2\alpha-2} |f'(z)|^2 (1-|\phi_a(z)|^2) dA(z) < +\infty.$$

The space $BMOA_\alpha$ is the space of $f \in H(D)$ such that

$$\|f\|^2_{BMOA_\alpha} := \sup_{I\subset \partial D} \frac{|I|^{2\alpha-2}}{|I|} \int_{S(I)} |f'(z)|^2 (1-|z|^2) dA(z) < +\infty,$$

where I is any arc on the unit circle ∂D, $S(I) = \{z \in D : |z| > 1 - |I|, z/|z| \in I\}$, and $|I|$ is the normalized arc length on ∂D. Note that $BMOA^1 = BMOA_1 = BMOA$.

For $\phi \in H(D)$, the operators J_ϕ, I_ϕ are defined respectively by:

$$J_\phi(f)(z) = \int_0^z f(\xi)\phi'(\xi)d\xi, \quad I_\phi(f)(z) = \int_0^z f'(\xi)\phi(\xi)d\xi, \quad f \in H(D).$$

The operator J_ϕ is called the *Cesàro* operator, and I_ϕ is often called the companion operator of J_ϕ. The boundedness of J_ϕ, I_ϕ on analytic function spaces has been studied extensively, see, for example, [1]-[4]. In this article we study the boundedness of the operators J_ϕ and I_ϕ from $B^\alpha(BMOA^\alpha$, $BMOA_\alpha)$ to B^β, and the compactness of I_ϕ from B^α to B^β.

Throughout the article, C denotes positive constant whose value is not necessarily the same at each occurrence.

2. The boundedness and compactness of I_ϕ on B^α

Lemma 2.1. ([4]) *Let $\alpha, \beta > 0$, $\phi \in H(D)$, the operator $I_\phi : B^\alpha \to B^\beta$ is compact if and only if for any bounded sequence $\{f_j\}$ in B^α, which is uniformly convergent to zero on compact subsets of D, $\{f_j\}$ satisfies*

$$\|I_\phi(f_j)\|_{B^\beta} \to 0 \qquad (j \to \infty).$$

Lemma 2.2. ([5]) *Fixed $a \in D$, for $z \in D(a,r)$, $1-|z|^2$ is comparable to $1-|a|^2$, and comparable to $|1-\bar{a}z|$.*

Theorem 2.1. *Let $\alpha, \beta > 0$ and $\phi \in H(D)$. Then the operator I_ϕ is bounded from B^α to B^β if and only if*

$$\sup_{z\in D}(1-|z|^2)^{\beta-\alpha}|\phi(z)| < \infty.$$

Proof. Suppose that $C = \sup_{z\in D}(1-|z|^2)^{\beta-\alpha}|\phi(z)| < \infty$. Then we have

$$\|I_\phi(f)(z)\|_{B^\beta} = \sup_{z\in D}(1-|z|^2)^\beta |f'(z)\phi(z)| \le C\|f\|_{B^\alpha}.$$

To prove the converse, suppose I_ϕ is bounded from B^α to B^β. Firstly, we prove the case $\alpha = 1$. For $a \in D$, let $f_a(z) = \log \frac{1}{1-\bar{a}z}$, then it is clear that $f_a(z)$ is a bounded set in B. So by Lemma 3 and using the subharmonicity of $|\phi(z)|$ and the fact that $\int_{D(a,r)} \frac{1}{(1-|z|^2)^2} dA(z) \le C < \infty$ (C depending only on r), we have

$$\begin{aligned}
&|a|^2(1-|a|^2)^{2(\beta-1)}|\phi(a)|^2 \\
&\le \frac{|a|^2}{(1-|a|^2)^2}\int_{D(a,r)}(1-|z|^2)^{2(\beta-1)}|\phi(z)|^2 dA(z) \\
&\le C\int_{D(a,r)}(1-|z|^2)^{2(\beta-1)}\frac{|a|^2}{|1-\bar{a}z|^2}|\phi(z)|^2 dA(z) \\
&= C\int_{D(a,r)}(1-|z|^2)^{2\beta-2}|(\log\frac{1}{1-\bar{a}z})'|^2|\phi(z)|^2 dA(z) \\
&\le C\sup_{z\in D}(1-|z|^2)^{2\beta}|f_a'(z)|^2|\phi(z)|^2\int_{D(a,r)}\frac{1}{(1-|z|^2)^2}dA(z) \\
&\le C\|I_\phi f_a\|_{B^\beta}^2\int_{D(a,r)}\frac{1}{(1-|z|^2)^2}dA(z) \\
&\le C\|I_\phi\|^2\|f_a\|_B^2 < \infty.
\end{aligned}$$

In the case $\alpha \ne 1$, by putting $f_a(z) = (1-\bar{a}z)^{1-\alpha}$, we can prove that as well, so we omit it.

Remark: (1) R. Yoneda showed that if $\alpha > 1$, then I_ϕ is bounded on B^α if and only if $\phi \in H^\infty$([3]); if $\beta > \alpha > 1$, then $I_\phi : B^\alpha \to B^\beta$ is bounded if and only if $\sup_{z\in D}(1-|z|^2)^{\alpha-\beta+1}|\phi'(z)| < \infty$([2]).

(2) The boundedness of $I_\phi : B^\alpha \to B^\beta$ can also be obtained by the boundedness of $J_\phi : B^\alpha \to B^\beta$ and $M_\phi : B^\alpha \to B^\beta$($M_\phi$ is the multiplication operator), here we give a necessary and sufficient condition of the boundedness.

Theorem 2.2. *Let $\alpha, \beta > 0$, $\phi \in H(D)$. Then I_ϕ: $B^\alpha \to B^\beta$ is compact if and only if*

$$\lim_{|z|\to 1}(1-|z|^2)^{\beta-\alpha}|\phi(z)| = 0.$$

Proof. Suppose that $\lim_{|z|\to 1}(1-|z|^2)^{\beta-\alpha}|\phi(z)| = 0$, then for any $\varepsilon > 0$, there exists $\delta \in (0,1)$ such that $(1-|z|^2)^{\beta-\alpha}|\phi(z) < \varepsilon$ as $|z| > 1-\delta$.

If $\{f_j\}$ is bounded in B^α, and $\{f_j\}$ is uniformly convergent to zero on compact subsets of D, $\{f_j'\}$ uniformly convergent to zero on compact subsets of D, then for the $\varepsilon > 0$ above, there exists $N \in \mathbf{N}$ such that $|f_j'(z)| < \varepsilon$ for all $z \in \{w : |w| \leq 1 - \delta\}$ as $j > N$. Thus

$$\begin{aligned}\|I_\phi f_j\|_{B^\beta} &\leq \sup_{|z|\leq 1-\delta}(1-|z|^2)^\beta|f_j'(z)||\phi(z)| + \sup_{|z|>1-\delta}(1-|z|^2)^\beta|f_j'(z)||\phi(z)| \\ &< C\varepsilon + \sup_{|z|>1-\delta}(1-|z|^2)^{\beta-\alpha}|\phi(z)|\|f_j\|_{B^\alpha} \\ &< (C + \|f_j\|_{B^\alpha})\varepsilon,\end{aligned}$$

when $j > N$, where $C = \sup_{z\in D}(1-|z|^2)^\beta|\phi(z)| < \infty$.

Conversely, suppose $\lim_{|z|\to 1}(1-|z|^2)^{\beta-\alpha}|\phi(z)| \neq 0$, then there exists $\varepsilon_0 > 0$ and $\{z_j\} \subset D$, $|z_j| \to 1$ $(j \to \infty)$ such that $(1-|z_j|^2)^{\beta-\alpha}|\phi(z_j)| > \varepsilon_0$. Let $f_j(z) = \frac{1-|z_j|^2}{(1-\bar{z}_j z)^\alpha}$. It is easy to check that $f_j \in B^\alpha$ is $\{f_j\}$ uniformly convergent to zero on compacts subsets of D, we have

$$\|I_\phi f_j\|_{B^\beta} \geq (1-|z_j|^2)^\beta|f_j'(z_j)||\phi(z_j)| > \varepsilon_0(1-|z_j|^2)^\alpha|f_j'(z_j)| > \varepsilon_0/2.$$

By Lemma 1 this contradicts to the compactness of I_ϕ.

3. The boundedness of J_ϕ and I_ϕ on $BMOA_\alpha$

Lemma 3.1. *For $\alpha > 0$, $BMOA_\alpha \subset B^\alpha$.*

Proof. Suppose $f \in BMOA_\alpha$, Lemma 2, the subhamonicity of $|f'(z)|$ and [3] give

$$\begin{aligned}(1-|a|^2)^{2\alpha}|f'(a)|^2 &\leq \frac{(1-|a|^2)^{2\alpha}}{(1-|a|^2)^2}\int_{D(a,r)}|f'(z)|^2 dA(z) \\ &\leq (1-|a|^2)^{2(\alpha-1)}\int_{D(a,r)}|f'(z)|^2(1-|\phi_a(z)|^2)dA(z) \\ &\leq C(1-|a|^2)^{2(\alpha-1)}\int_D|f'(z)|^2(1-|\phi_a(z)|^2)dA(z) \\ &\leq C\sup_{a\in D}(1-|a|^2)^{2(\alpha-1)}\int_D|f'(z)|^2(1-|\phi_a(z)|^2)dA(z) \\ &\approx C\sup_{I\subset\partial D}\frac{|I|^{2\alpha-2}}{|I|}\int_{S(I)}|f'(z)|^2(1-|z|^2)dA(z) < \infty,\end{aligned}$$

where the equivalence above is a result in [1]. That is $\phi \in B^\alpha$. Hence $BMOA_\alpha \subset B^\alpha$.

Theorem 3.1. *Let $\alpha, \beta > 0$. Then J_ϕ: $BMOA_\alpha \to B^\beta$ is bounded if and only if*

$$\begin{cases} \sup_{z\in D}(1-|z|^2)^\beta|\phi'(z)| < +\infty, & 0<\alpha<1; \\ \sup_{z\in D}(1-|z|^2)^\beta \log\frac{1}{1-|z|^2}|\phi'(z)| < +\infty, & \alpha = 1; \\ \sup_{z\in D}(1-|z|^2)^{\beta-\alpha+1}|\phi'(z)| < +\infty, & \alpha > 1. \end{cases}$$

Proof. By Lemma 3, $BMOA_\alpha \subset B^\alpha$, the necessity can be obtained from the boundedness of J_ϕ: $B^\alpha \to B^\beta$ ([4]).

Suppose J_ϕ: $BMOA_\alpha \to B^\beta$ is bounded. When $\alpha > 1$, it is easy to see that $\{f_a(z) = (1-\bar{a}z)^{1-\alpha}\}$ is a bounded set in $BMOA_\alpha$([3]). Using the subharmonicity of $|\phi'(z)|$ and Lemma 4 yield

$$\begin{aligned} &(1-|a|^2)^{2\beta-2\alpha+2}|\phi'(a)|^2 \\ \le\ & \frac{(1-|a|^2)^{2\beta-2\alpha+2}}{(1-|a|^2)^2}\int_{D(a,r)}|\phi'(z)|^2 dA(z) \\ \approx\ & \int_{D(a,r)}\frac{(1-|z|^2)^{2\beta}|1-\bar{a}z|^{2-2\alpha}|\phi'(z)|^2}{(1-|z|^2)^2}dA(z) \\ \le\ & \sup_{z\in D}(1-|z|^2)^{2\beta}|f_a(z)|^2|\phi'(z)|^2\int_{D(a,r)}\frac{1}{(1-|z|^2)^2}dA(z) \\ =\ & C\|J_\phi f_a\|^2_{B^\beta} \\ \le\ & C\|J_\phi\|^2\|f_a\|^2_{BMOA_\alpha} < \infty, \end{aligned}$$

where $C = \int_{D(a,r)}\frac{1}{(1-|z|^2)^2}dA(z) < +\infty$. Thus $\sup_{z\in D}(1-|z|^2)^{\beta-\alpha+1}|\phi'(z)| < +\infty$.

When $\alpha = 1$, by putting $f_a(z) = \log\frac{1}{1-\bar{a}z} \in BMOA$, we can prove that as well as the case of $\alpha > 1$, so we omit the detail. When $0 < \alpha < 1$, since a non-zero constant C belongs to $BMOA_\alpha$, we have $J_\phi C \in B^\beta$. Hence $\sup_{z\in D}(1-|z|^2)^\beta|\phi'(z)| < +\infty$.

Theorem 3.2.

Let $\alpha, \beta > 0$. Then I_ϕ: $BMOA_\alpha \to B^\beta$ is bounded if and only if

$$\sup_{z\in D}(1-|z|^2)^{\beta-\alpha}|\phi(z)| < +\infty.$$

Proof. Using $BMOA_\alpha \subset B^\alpha$, and choose the same test functions as the function as above, the proof is similar to that of Theorem 1.

4. The boundedness of J_ϕ and I_ϕ on $BMOA^\alpha$

Theorem 4.1. *Let $\alpha, \beta > 0$. Then J_ϕ: $BMOA^\alpha \to B^\beta$ is bounded if and only if*

$$\begin{cases} \sup_{z\in D}(1-|z|^2)^\beta|\phi'(z)| < +\infty, & 0<\alpha<1; \\ \sup_{z\in D}(1-|z|^2)^\beta \log\frac{1}{1-|z|^2}|\phi'(z)| < +\infty, & \alpha = 1; \\ \sup_{z\in D}(1-|z|^2)^{\beta-\alpha+1}|\phi'(z)| < +\infty, & \alpha > 1. \end{cases}$$

Theorem 4.2. *Let $\alpha, \beta > 0$, I_ϕ: $BMOA^\alpha \to B^\beta$ is bounded if and only if*

$$\sup_{z\in D}(1-|z|^2)^{\beta-\alpha}|\phi(z)| < \infty.$$

Note that $BMOA^\alpha \subset B^\alpha$, the proofs of Theorems 5 and 6 are similar to those of Theorems 3 and 4 respectively. The details are omitted.

Remark: In [3], R. Yoneda showed the boundedness of operators J_ϕ, I_ϕ: $BMOA^\alpha \to B^\beta$ as $0 < \alpha < \beta$.

References

1. A. Alema and A. G. Siskakis, Integral operators on H^p, *Complex Variables,* **28** (1995):107-152.
2. R. Yoneda, Multiplication operators, integration operators and companion operators on weighted Bloch spaces, *Hokkaido Math. J.* **34**(2005):135-147.
3. R. Yoneda, Pointwise multipliers from $BMOA^\alpha$ to $BMOA^\beta$, *Complex Variables* **49** (2003):1045-1061.
4. X. J. Zhang, Extend *cesàro* operator on the Dirichlet type spaces and Bloch type spaces of $\mathbf{C^n}$ (in Chinese), *Chinese Ann. Math.* **26A** (2005):139-150.
5. K. H. Zhu, Operator Theory in Function Spaces, *Marcel Dekker*, New York and Basel, 1990.

THE GROWTH OF ANALYTIC FUNCTIONS OF INFINITE ORDER REPRESENTED BY LAPLACE-STIELTJES TRANSFORMATIONS

LINA SHANG
LMIB and Department of Mathematics
Beihang University
Beijing, 100083 P. R. China
E-mail: sln-8108@163.com

ZONGSHENG GAO
LMIB and Department of Mathematics
Beihang University
Beijing, 100083 P. R. China
E-mail: zshgao@buaa.edu.cn

In this paper, we investigate the growth and regular growth of functions of infinite order defined by Laplace-Stieltjes transformations and obtain some necessary and sufficient conditions.

Keywords: Laplace-Stieltjes transformation, order, analytic function.

2000 Mathematics Subject Classification: 30D99; 44A10

1. Introduction

Many problems about the growth and the value distribution of analytic functions defined by Laplace-Stieltjes transformations have been studied and some important results have been obtained in [1],[2]. But the infinite order of Laplace-Stieltjes transformation hasn't been discussed yet. In this paper, we study the analytic functions of infinite order defined by Laplace-Stieltjes transformations in the half right plane and get some results on their growth and regular growth.

Consider the Laplace-Stieltjes transformation

$$F(s) = \int_0^\infty e^{-sx} d\alpha(x) \quad (s = \sigma + it) \tag{1}$$

where $\alpha(x)$ is a complex valued function on $x \geq 0$ and a bounded variation function on arbitrary closed interval [0,X] $(0 < X < +\infty)$. Choose a sequence $\{\lambda_n\}$ such that:

$$0 = \lambda_1 < \lambda_2 < \cdots < \lambda_n \uparrow \infty, \tag{2}$$

$$\overline{\lim_{n\to\infty}} \frac{ln n}{ln \lambda_n} = l < \infty, \tag{3}$$

$$\overline{\lim_{n\to\infty}} (\lambda_{n+1} - \lambda_n) < \infty. \tag{4}$$

If the transformation (1) satisfies

$$\overline{\lim_{n\to\infty}} \frac{ln A_n^*}{\lambda_n} = 0, \tag{5}$$

where

$$A_n^* = \sup_{\lambda_n < x \leq \lambda_{n+1}, -\infty < t < +\infty} \left| \int_{\lambda_n}^x e^{-ity} d\alpha(y) \right|,$$

it follows from [1,Theorem1.3] that the abscissa of uniform convergence of (1) is 0 and the transformation (1) defines an analytic function $F(s)$ in the right-half plane.

Set

$$M(\sigma, F) = \sup_{-\infty < t < +\infty} |F(\sigma + it)|,$$

$$M_u(\sigma, F) = \sup_{0 < x < +\infty, -\infty < t < +\infty} \left| \int_0^x e^{-(\sigma+it)y} d\alpha(y) \right|,$$

$$\mu(\sigma, F) = \max_n \{A_n^* e^{-\lambda_n \sigma}\}, \quad n(\sigma) = \max_k \{k | \ \mu(\sigma, F) = A_k^* e^{-\lambda_k \sigma}\}.$$

Definition:

$$\tau = \overline{\lim_{\sigma\to+0}} \frac{ln^+ ln^+ M(\sigma, F)}{-ln\sigma}, \quad \tau_u = \overline{\lim_{\sigma\to+0}} \frac{ln^+ ln^+ M_u(\sigma, F)}{-ln\sigma},$$

$$\tau_\mu = \overline{\lim_{\sigma\to+0}} \frac{ln^+ ln^+ \mu(\sigma, F)}{-ln\sigma}.$$

We call τ order(R) of $F(s)$ in $\sigma > 0$.

Note: It is obvious that $M(\sigma, F) \leq M_u(\sigma, F)$. Specially, if $\tau = \infty$, we have $\tau_u = \infty$.

2. Main Results

In this paper we use C for a numerical constant. It will not be the same at each occurrence but it is always independent of all variables.

Lemma 1. *Suppose that the function $\varphi(\sigma)$ is continuous on $\sigma > 0$, which satisfies*

$$\overline{\lim_{\sigma\to+0}}\frac{ln^+ln^+\varphi(\sigma)}{-ln\sigma} = \infty.$$

Then there exists a function $\tau(u)$ such that

1) $\tau(u)$ is continuous on $u > 0$ and $\lim_{u\to+\infty}\tau(u) = \infty$;

2) $\lim_{u\to+\infty}\frac{lnU(u')}{lnU(u)} = 1$, *where* $U(u) = u^{\tau(u)}$, $u' = u+\frac{u}{lnU(u)}$ *or* $u-\frac{u}{lnU(u)}$;

3) $\overline{\lim_{\sigma\to+0}}\frac{ln^+ln^+\varphi(\sigma)}{lnU(\frac{1}{\sigma})} = 1.$

Proof When $u' = u + \frac{u}{lnU(u)}$, the lemma is given in [3,Theorem4.3]. Imitating the method of [3,Theorem4.3], we can prove the lemma holds when $u' = u + \frac{cu}{lnU(u)}(c > 0)$. To finish the proof, we need show that the lemma holds when $u' = u - \frac{u}{lnU(u)}$.

For sufficiently large u and $u' = u-\frac{u}{lnU(u)}$, there exists $c > 0$, such that

$$u = u'(1 - \frac{1}{lnU(u)})^{-1} \le u'(1 + \frac{c}{lnU(u)}) \le u'(1 + \frac{c}{lnU(u')}).$$

Consequently,

$$U(u) \le U(u'(1 + \frac{c}{lnU(u')})).$$

Noticing that

$$1 \le \frac{lnU(u)}{lnU(u')} \le \frac{lnU(u'(1 + \frac{c}{lnU(u')}))}{lnU(u')},$$

we derive

$$\lim_{u\to+\infty}\frac{lnU(u)}{lnU(u')} = 1.$$

Hence the lemma.

Lemma 2. *Suppose that $\{\lambda_n\}$ satisfies (2),(3),(4) and the transformation (1) satisfies (5). Then*

$$\overline{\lim_{\sigma\to+0}}\frac{ln^+ln^+M_u(\sigma,F)}{lnU(\frac{1}{\sigma})} = 1 \Longleftrightarrow \overline{\lim_{\sigma\to+0}}\frac{ln^+ln^+\mu(\sigma,F)}{lnU(\frac{1}{\sigma})} = 1,$$

where U is a function satisfying conditions 1), 2) in Lemma 1.

Proof Let

$$I(x;\sigma+it)=\int_0^x e^{-(\sigma+it)y}d\alpha(y),$$

then for $x>\lambda_n$, we have

$$\begin{aligned}\textstyle\int_{\lambda_n}^x e^{-ity}d\alpha(y)&=\textstyle\int_{\lambda_n}^x e^{\sigma y}d_yI(y;\sigma+it)\\&=I(y;\sigma+it)e^{\sigma y}|_{\lambda_n}^x-\sigma\textstyle\int_{\lambda_n}^x e^{\sigma y}I(y;\sigma+it)dy.\end{aligned}$$

By (4), there exists $K>0$ such that for $n\in N$, $\lambda_{n+1}-\lambda_n<K$. Furthermore, for sufficiently small σ, we have $e^{K\sigma}<2$. Hence, when $\lambda_n<x\le\lambda_{n+1}$,

$$|\int_{\lambda_n}^x e^{-ity}d\alpha(y)|\le 2M_u(\sigma,F)e^{x\sigma}\le 2M_u(\sigma,F)e^{\lambda_{n+1}\sigma}\le 4M_u(\sigma,F)e^{\lambda_n\sigma}.$$

Consequently

$$\mu(\sigma,F)\le 4M_u(\sigma,F). \tag{6}$$

On the other hand, notice that

$$\begin{aligned}\textstyle\int_0^x e^{-(\sigma+it)y}d\alpha(y)&=\sum_{k=1}^{n-1}\textstyle\int_{\lambda_k}^{\lambda_{k+1}}e^{-(\sigma+it)y}d\alpha(y)\\&\quad+\textstyle\int_{\lambda_n}^x e^{-(\sigma+it)y}d\alpha(y)\quad(\lambda_n<x\le\lambda_{n+1}).\end{aligned}\tag{7}$$

Let

$$I_k(x;it)=\int_{\lambda_k}^x e^{-ity}d\alpha(y)\quad(\lambda_k<x\le\lambda_{k+1}). \tag{8}$$

Then for $\lambda_k<x\le\lambda_{k+1},\ -\infty<t<+\infty$,

$$|I_k(x;it)|\le A_k^*\le\mu(\sigma,F)e^{\lambda_k\sigma}. \tag{9}$$

By (7) and (8), for $\forall 0<\varepsilon<1, \lambda_n<x\le\lambda_{n+1}$, $\sigma>0$,

$$\begin{aligned}\textstyle\int_0^x e^{-(\sigma+it)y}d\alpha(y)&=\sum_{k=1}^{n-1}\textstyle\int_{\lambda_k}^{\lambda_{k+1}}e^{-\sigma y}d_yI_k(y;it)+\int_{\lambda_n}^x e^{-\sigma y}d_yI_n(y;it)\\&=\sum_{k=1}^{n-1}[e^{-\lambda_{k+1}\sigma}I_k(\lambda_{k+1};it)+\sigma\textstyle\int_{\lambda_k}^{\lambda_{k+1}}e^{-\sigma y}I_k(y;it)dy]\\&\quad+e^{-x\sigma}I_n(x;it)+\sigma\textstyle\int_{\lambda_n}^x e^{-\sigma y}I_k(y;it)dy.\end{aligned}$$

Combine this with (9) and put $\Delta = \frac{lnU(\frac{1}{\sigma})}{1+lnU(\frac{1}{\sigma})}\sigma$ to obtain

$$
\begin{aligned}
|\int_0^x e^{-(\sigma+it)y}d\alpha(y)| &\leq \sum_{k=1}^{n-1}\mu(\Delta,F)e^{\lambda_k\Delta}(e^{-\lambda_{k+1}\sigma}+e^{-\lambda_k\sigma}-e^{-\lambda_{k+1}\sigma}) \\
&\quad +\mu(\Delta,F)e^{\lambda_n\Delta}(e^{-x\sigma}+e^{-\lambda_n\sigma}-e^{-x\sigma}) \\
&= \sum_{k=1}^{n}\mu(\Delta,F)e^{\frac{-\lambda_k\sigma}{1+lnU(\frac{1}{\sigma})}} \leq \mu(\Delta,F)\sum_{k=1}^{+\infty}e^{\frac{-\lambda_k\sigma}{1+lnU(\frac{1}{\sigma})}}.
\end{aligned} \tag{10}
$$

Since (3), for any $\delta > 0$, there exists $N > 0$ such that for $n \geq N$, $\lambda_n \geq n^{\frac{1}{l+\varepsilon}} \geq ln^3 n$. Let $T = \exp\{\frac{2+2lnU(\frac{1}{\sigma})}{\sigma}\}^{\frac{1}{2}}$. Combine with (10) to obtain

$$
\begin{aligned}
M_u(\sigma,F) &\leq \mu(\Delta,F)(N+\sum_{n=N+1}^{[T]}n^{-ln^2 n(\sigma/[1+lnU(\frac{1}{\sigma})])}+\sum_{n=[T]+1}^{\infty}n^{-2}) \\
&\leq \mu(\Delta,F)(C+2\exp((\frac{2+2lnU(\frac{1}{\sigma})}{\sigma})^{\frac{1}{2}}\times 2)).
\end{aligned}
$$

Consequently

$$
ln^+ln^+M_u(\sigma,F) \leq ln^+ln^+\mu(\Delta,F)+\frac{1}{2}lnlnU(\frac{1}{\sigma})+\frac{1}{2}ln\frac{1}{\sigma}+C.
$$

By the property of $U(\frac{1}{\sigma})$, we have

$$
\overline{\lim_{\sigma\to+0}}\frac{ln^+ln^+M_u(\sigma,F)}{lnU(\frac{1}{\sigma})} \leq \overline{\lim_{\sigma\to+0}}\frac{ln^+ln^+\mu(\sigma,F)}{lnU(\frac{1}{\sigma})}.
$$

Combining this with (6), we conclude Lemma 2.

Theorem 1. *Suppose that $\{\lambda_n\}$ satisfies (2),(3),(4) and the transformation (1) satisfies (5) and*

$$
\overline{\lim_{\sigma\to+0}}\frac{ln^+ln^+M(\sigma,F)}{-ln\sigma} = \infty.
$$

Then

$$
\overline{\lim_{\sigma\to+0}}\frac{ln^+ln^+M_u(\sigma,F)}{lnU(\frac{1}{\sigma})} = 1 \iff \overline{\lim_{n\to\infty}}\, t_n = 1, \tag{11}
$$

where

$$
t_n = \begin{cases} ln\lambda_n/lnU(\frac{\lambda_n}{lnA_n^*}) & (A_n^* > 1) \\ 0 & (A_n^* \leq 1) \end{cases}
$$

and U is a function satisfying conditions 1), 2) in Lemma 1.

Proof First we prove that if $\overline{\lim_{n\to\infty}} t_n = 1$, then

$$\overline{\lim_{\sigma\to+0}} \frac{ln^+ ln^+ M_u(\sigma, F)}{lnU(\frac{1}{\sigma})} \leq 1. \tag{12}$$

For any $\eta > 0$, there exists a positive integer $N > 0$ such that for $n > N$ and $A_n^* > 1$,

$$\lambda_n < [U(\frac{\lambda_n}{lnA_n^*})]^{1+\eta}.$$

Let $v = U(u)$ and $u = \varphi(v)$ be two reciprocally inverse functions. Then for $n > N$ and $A_n^* > 1$, we have

$$\varphi(\lambda_n^{\frac{1}{1+\eta}}) < \frac{\lambda_n}{lnA_n^*},$$

and so

$$A_n^* e^{-\lambda_n\sigma} < \exp[\frac{\lambda_n}{\varphi(\lambda_n^{\frac{1}{1+\eta}})} - \lambda_n\sigma]. \tag{13}$$

Evidently (13) holds when $A_n^* \leq 1$. Hence we have (13) for $\forall n > N$. Fix $\sigma > 0$ and take $\overline{\lambda}$ such that

$$\frac{1}{\sigma}(1 + \frac{1}{lnU(\frac{1}{\sigma})}) = \varphi(\overline{\lambda}^{\frac{1}{1+\eta}}).$$

Then $\overline{\lambda} = [U(\frac{1}{\sigma})]^{1+\eta+o(1)}$ $(\sigma \to 0^+)$. By (13), when $\lambda_n > \overline{\lambda}$ and $n > N$,

$$A_n^* e^{-\lambda_n\sigma} < \exp[\lambda_n(\frac{1}{\varphi(\overline{\lambda}^{\frac{1}{1+\eta}})} - \sigma)] = \exp[\frac{-\lambda_n\sigma}{1 + lnU(\frac{1}{\sigma})}] < 1.$$

Then for sufficiently small σ, we have $\lambda_{n(\sigma)} < \overline{\lambda}$. Since([4])

$$ln\mu(\sigma, F) = A + \int_\sigma^a \lambda_{n(t)} dt, \ (0 < \sigma < 1)$$

where $a(> \sigma)$ and A are constants, we have

$$ln\mu(\sigma, F) < U^{1+2\eta+o(1)}(-\sigma), \ (\sigma \to +0).$$

Notice that η can be arbitrarily small to obtain

$$\overline{\lim_{\sigma\to+0}} \frac{ln^+ ln^+ \mu(\sigma, F)}{lnU(\frac{1}{\sigma})} \leq 1.$$

Applying Lemma 2, hence (12) holds.

Now we prove that if $\overline{\lim_{n\to\infty}} t_n = 1$, then we cannot have

$$\overline{\lim_{\sigma\to+0}} \frac{ln^+ln^+M_u(\sigma,F)}{lnU(\frac{1}{\sigma})} = c < 1. \tag{14}$$

Suppose that (14) would hold, choose $\varepsilon > 0$ and $c + 2\varepsilon < 1$. Then there would exist $\sigma_0 (0 < \sigma < 1)$ such that for $0 < \sigma < \sigma_0$,

$$lnM_u(\sigma,F) < [U(\frac{1}{\sigma})]^{c+\varepsilon}$$

and consequently for $n = 0, 1, 2, \cdots$

$$lnA_n^* - \lambda_n\sigma < [U(\frac{1}{\sigma})]^{c+\varepsilon}. \tag{15}$$

On the other hand, there exist arbitrarily large integers n such that

$$\lambda_n > [U(\frac{\lambda_n}{lnA_n^*})]^{1-\varepsilon} \qquad (A_n^* > 1). \tag{16}$$

Take such sufficiently large n and take $\sigma(0 < \sigma < \sigma_0)$ such that

$$[U(\frac{1}{\sigma})]^{c+\varepsilon} = \frac{\lambda_n}{\frac{\lambda_n}{lnA_n^*} lnU(\frac{\lambda_n}{lnA_n^*})}. \tag{17}$$

Combining (15) and (17), we see that for any $\eta > 0$, there are arbitrarily large n and $\sigma \in (0, \sigma_0)$ such that

$$\frac{1}{\sigma} < \frac{\lambda_n}{lnA_n^*}[1 + \frac{1}{lnU(\frac{\lambda_n}{lnA_n^*})}]$$

and consequently

$$U(\frac{1}{\sigma}) < [U(\frac{\lambda_n}{lnA_n^*})]^{1+\eta}. \tag{18}$$

By (17) and (18) we have, for those sufficiently large n for which (16) holds,

$$\lambda_n < [U(\frac{\lambda_n}{lnA_n^*})]^{(1+\eta)(c+\varepsilon)} \frac{\lambda_n}{lnA_n^*} lnU(\frac{\lambda_n}{lnA_n^*}).$$

And since (16), for the other sufficiently large n, $t_n \leq 1 - \varepsilon$. Therefore, $\overline{\lim_{n\to\infty}} t_n < 1$, which is contrary to the hypothesis.

The sufficiency of the condition (11) is proved.

By the prove of sufficiency we can prove necessity easily.

Theorem 2. *Suppose that $\{\lambda_n\}$ satisfies (2),(3),(4) and the transformation (1) satisfies (5) and*

$$\overline{\lim_{\sigma\to+0}} \frac{ln^+ln^+M(\sigma,F)}{-ln\sigma} = \infty.$$

Then

$$\lim_{\sigma\to+0}\frac{ln^+ln^+M_u(\sigma,F)}{lnU(\frac{1}{\sigma})}=1\Longleftrightarrow(i)\overline{\lim_{n\to\infty}}t_n=1,\tag{19}$$

(ii) there exists a non-decreasing positive integer sequence $\{n_v\}$ *satisfying*

$$\lim_{n\to\infty}t_{n_v}=1,\qquad\lim_{v\to\infty}\frac{ln\lambda_{n_{v+1}}}{ln\lambda_{n_v}}=1.\tag{20}$$

where

$$t_n=\begin{cases}ln\lambda_n/lnU(\frac{\lambda_n}{lnA_n^*}) & (A_n^*>1)\\ 0 & (A_n^*\le 1)\end{cases}$$

and U is a function satisfying conditions 1), 2) in Lemma 1.

Proof First, we prove that the conditions are sufficient. From (ii), for any $\varepsilon>0$ and sufficiently large v, we have

$$lnA_{n_v}^*e^{-\lambda_{n_v}\sigma}>\frac{\lambda_{n_v}}{\varphi(\lambda_{n_v}^{\frac{1}{1-\varepsilon}})}-\lambda_{n_v}\sigma,$$

where φ is the verse function of U. Take σ_v such that

$$\frac{1}{\sigma_v}(1-\frac{1}{lnU(\frac{1}{\sigma_v})})=\varphi(\lambda_{n_v}^{\frac{1}{1-\varepsilon}}),$$

then $\{\sigma_v\}$ is a decreasing sequence and $\{\sigma_v\}\to 0$. If $0<\sigma_v<\sigma<\sigma_{v-1}$, we have

$$\begin{aligned}ln^+\mu(\sigma,F)\ge ln^+\mu(\sigma_{v-1},F)&\ge\frac{\lambda_{n_{v-1}}}{\varphi(\lambda_{n_{v-1}}^{\frac{1}{1-\varepsilon}})}-\lambda_{n_{v-1}}\sigma_{v-1}\\&\ge U^{1-\varepsilon}(\frac{1}{\sigma_{v-1}}(1-\frac{1}{lnU(\frac{1}{\sigma_{v-1}})}))\frac{\sigma_{v-1}}{lnU(\frac{1}{\sigma_{v-1}})}.\end{aligned}$$

Noticing that $\lim_{v\to\infty}\frac{ln\lambda_{n_{v+1}}}{ln\lambda_{n_v}}=1$ and

$$\begin{aligned}\lim_{v\to\infty}\frac{ln\lambda_{n_v}}{ln\lambda_{n_{v-1}}}&=\lim_{v\to\infty}\frac{lnU^{1-\varepsilon}(\frac{1}{\sigma_v}(1-\frac{1}{lnU(\frac{1}{\sigma_v})}))}{lnU^{1-\varepsilon}(\frac{1}{\sigma_{v-1}}(1-\frac{1}{lnU(\frac{1}{\sigma_{v-1}})}))}\\&=\lim_{v\to\infty}\frac{lnU(\frac{1}{\sigma_v})}{lnU(\frac{1}{\sigma_{v-1}})}\ge\lim_{v\to\infty}\frac{lnU(\frac{1}{\sigma})}{lnU(\frac{1}{\sigma_{v-1}})}\ge 1,\end{aligned}$$

we have

$$\lim_{v\to\infty}\frac{lnU(\frac{1}{\sigma})}{lnU(\frac{1}{\sigma_{v-1}})}=1.$$

Therefore, by the property of $U(\frac{1}{\sigma})$ we obtain

$$\lim_{\sigma\to 0}\frac{ln^+ln^+\mu(\sigma,F)}{lnU(\frac{1}{\sigma})}=\lim_{v\to\infty}\frac{ln^+ln^+\mu(\sigma,F)}{lnU(\frac{1}{\sigma_{v-1}})}\geq\lim_{v\to\infty}\frac{ln^+ln^+\mu(\sigma_{v-1},F)}{lnU(\frac{1}{\sigma_{v-1}})}\geq 1-\varepsilon.$$

Using (6),

$$\lim_{\sigma\to 0}\frac{ln^+ln^+M_u(\sigma,F)}{lnU(\frac{1}{\sigma})}\geq 1-\varepsilon.$$

Combining this with Theorem 1, we obtain the sufficiency.

Now we prove the necessity of the conditions. The conclusion (i) has been proved in Theorem 1. Here we prove that (ii) is true. Choose $\{\varepsilon_p\}(0<\varepsilon_p<1)$ such that $\varepsilon_p\downarrow 0(p\to\infty)$. Put

$$E_p=\{n:t_n>1-\varepsilon_p\}. \tag{21}$$

From the right hand side of (19), we can see that E_p is an unempty infinite set and $E_{p+1}\subset E_p$ for each p. Arrange the elements of E_p in an increasing order and denote it by $\{n_v^{(p)}\}$.

1) If $\lim_{v\to\infty}\frac{ln\lambda_{n^{(p)}_{v+1}}}{ln\lambda_{n^{(p)}_v}}=1$ for all $p\in N$, there exists $N_p\in E_p(p=1,2,\cdots)$, such that for $n_v^{(p)}\geq N_p$,

$$\frac{ln\lambda_{n^{(p)}_{v+1}}}{ln\lambda_{n^{(p)}_v}}\leq 1+\varepsilon_p. \tag{22}$$

Since $E_{p+1}\subset E_p$, we can suppose $N_{p+1}>N_p$. Put $E_p'=\{n\in E_p:N_p\leq n\leq N_{p+1}\}$. Then the elements of E_p' can satisfy (21) and (22). Finally arrange the elements of $E=\bigcup_p E_p'$ in an increasing order and denote it by $\{n_v\}$. It is easy to see that $\{n_v\}$ satisfy (20). The necessity is proved.

2) If there exists certain $p\in N$ satisfying $\lim_{v\to\infty}\frac{ln\lambda_{n^{(p)}_{v+1}}}{ln\lambda_{n^{(p)}_v}}\neq 1$, then $\overline{\lim_{v\to\infty}}\frac{ln\lambda_{n^{(p)}_{v+1}}}{ln\lambda_{n^{(p)}_v}}>1$. Thus there must exist a subset of $\{n_v^{(p)}\}$ (for convenience we still denote it by $\{n_v^{(p)}\}$), satisfying

$$\frac{ln\lambda_{n^{(p)}_{v+1}}}{ln\lambda_{n^{(p)}_v}}>1+\gamma\quad(v=1,2,\cdots),$$

where γ is a certain constant. Put

$$n_1^{'}=n_1^{(p)},n_2^{'}=n_3^{(p)},\ \cdots,\ n_v^{'}=n_{2v-1}^{(p)},\ \cdots$$

$$n_1^{''}=n_2^{(p)},n_2^{''}=n_4^{(p)},\ \cdots,\ n_v^{''}=n_{2v}^{(p)},\ \cdots$$

Then $\{n_v^{'}\}, \{n_v^{''}\}$ are increasing positive integer sequences which satisfy

$$n_v^{''} < n_{v+1}^{'}, \quad \lambda_{n_v^{''}} > \lambda_{n_v^{'}}^{1+\gamma} \qquad (v = 1, 2, \cdots). \tag{23}$$

Put $\beta = \frac{\varepsilon_p}{2}$. For sufficiently large v, when $n_v^{'} < n < n_v^{''}$, we have $t_n \leq 1 - \varepsilon_p < 1 - \beta$. Consequently

$$lnA_n^* e^{-\lambda_n \sigma} < \frac{\lambda_n}{\varphi(\lambda_n^{\frac{1}{1-\beta}})} - \lambda_n \sigma, \tag{24}$$

where φ is the verse function of U. Choose $\mu > 0$ such that

$$\frac{1+2\mu}{1+\gamma} < 1 - \eta \qquad (\eta > 0). \tag{25}$$

Put $\frac{1}{\sigma_v} = \varphi(\lambda_{n_v^{''}}^{\frac{1}{1+\mu}})$, then

$$\lambda_{n_v^{''}} = U^{1+\mu}(\frac{1}{\sigma_v}). \tag{26}$$

Put $\overline{\lambda} = U^{1-\beta}(\frac{1}{\sigma_v})(0 < \beta < 1)$, then

$$\frac{1}{\sigma_v} = \varphi(\overline{\lambda}^{\frac{1}{1-\beta}}). \tag{27}$$

Furthermore, for above $\mu > 0$, by (i), there exists n_0 such that for $n > n_0$,

$$lnA_n^* e^{-\lambda_n \sigma_v} < \frac{\lambda_n}{\varphi(\lambda_n^{\frac{1}{1+\mu}})} - \lambda_n \sigma_v. \tag{28}$$

Now, for sufficiently large v, we estimate $lnA_n^* e^{-\lambda_n \sigma_v}$ in two cases.
(a) If $\overline{\lambda} > \lambda_{n_v^{'}}$:

(a_1) When $\lambda_n \geq \lambda_{n_v^{''}}$, by (28),(26),

$$lnA_n^* e^{-\lambda_n \sigma_v} < \lambda_n[\frac{1}{\varphi(\lambda_n^{\frac{1}{1+\mu}})} - \sigma_v] \leq \lambda_n[\frac{1}{\varphi(\lambda_{n_v^{''}}^{\frac{1}{1+\mu}})} - \sigma_v] = 0. \tag{29}$$

(a_2) When $\overline{\lambda} \leq \lambda_n < \lambda_{n_v^{''}}$, by (24),(27),

$$lnA_n^* e^{-\lambda_n \sigma_v} < \lambda_n[\frac{1}{\varphi(\lambda_n^{\frac{1}{1-\beta}})} - \sigma_v] \leq \lambda_n[\frac{1}{\varphi(\overline{\lambda}^{\frac{1}{1-\beta}})} - \sigma_v] = 0. \tag{30}$$

By (a_1),(a_2), we can see that for sufficiently large v, $\lambda_{n(\sigma_v)} < \overline{\lambda}$.
(a_3) When $\lambda_{n_0} < \lambda_n < \overline{\lambda}$, by (27),

$$lnA_n^* e^{-\lambda_n \sigma_v} \leq ln\mu(\sigma_v, F) = A + \int_{\sigma_v}^{a} \lambda_{n(t)} dt < U^{1-\frac{\beta}{2}}(\frac{1}{\sigma_v}).$$

(b) If $\overline{\lambda} \le \lambda_{n'_v}$:

(b_1) When $\lambda_n \ge \lambda_{n''_v}$, by (28),(26), we still arrive at (29).

(b_2) When $\lambda_{n'_v} \le \lambda_n < \lambda_{n''_v}$, by (24),(27), we still arrive at (30).

By (b_1),(b_2), we can see that for sufficiently large v, $\lambda_{n(\sigma_v)} < \lambda_{n'_v}$.

(b_3) When $\lambda_{n_0} < \lambda_n < \lambda_{n'_v}$, by (23),(26),(25),

$$lnA_n^* e^{-\lambda_n \sigma_v} \le ln\mu(\sigma_v, F) = A + \int_{\sigma_v}^{a} \lambda_{n(t)} dt < A + \lambda_{n'_v}(a - \sigma_v)$$

$$< A + \lambda_{n''_v}^{\frac{1}{1+\gamma}}(a - \sigma_v) < U^{\frac{1+2\mu}{1+\gamma}}(\frac{1}{\sigma_v}) < U^{1-\eta}(\frac{1}{\sigma_v}).$$

In any case, we can conclude

$$\overline{\lim_{v\to\infty}} \frac{ln^+ ln^+ \mu(\sigma_v, F)}{lnU(\frac{1}{\sigma_v})} \le \max(1-\eta, 1-\frac{\beta}{2}) < 1.$$

Combining with Lemma 2, we have

$$\overline{\lim_{v\to\infty}} \frac{ln^+ ln^+ M_u(\sigma_v, F)}{lnU(\frac{1}{\sigma_v})} < 1,$$

which is contrary to the hypothesis. We obtain the necessity.

The proof is completed.

References

1. J. R. Yu, Borel's line of entire functions represented by Laplace-Stieltjes transformation(in Chinese), *Acta Math Sinica*, **13**(1963):65-104.
2. Z. L. Li, The growth of analytic functions represented by Laplace-Stieltjes transformations in the half right plane, *Wuhan Univ.Nat.Sci.,* **3**(1981):15-27.
3. Q. T. Zhuang, Borel's directions of meromorphic functions(in Chinese), *Sci.Press, Beijing*,1982.
4. X. Luo, D. C. Sun, The transformation related with Laplace-Stieltjes transformation ananlytic in the half right plane, *Acta Math Sinica.*
5. D.V. Widder, *The Laplace transform,* Princeton University Press, Newjersey, 1946.
6. Zongsheng-Gao, Dirichlet Series and random Dirichlet series of infinite order(in Chinese), *J.Sys.Sci.and Math.Scis.*, **20**(2000):187-195.

ON T-DIRECTION OF MEROMORPHIC FUNCTION

MINGCHUN SHU and CAIFENG YI
School of Mathematics and Information Science
Jiangxi Normal University
Nanchang, 330022, P. R. China
E-mail: shuhe7876@sina.com

In this paper, a new singular direction of meromorphic function namely T-direction is investigated. Let a meromorphic function takes the value of its small function, we prove the existence of the corresponding T-direction.

Keywords: Meromorphic function, small function, T-direction, Ahlfors-Shimizu characteristic function.

2000 Mathematics Subject Classification: 30D30

1. Introduction and results

Let $f(z) : C \to \overline{C}$ be a transcendental meromorphic function of order λ, where C is the complex plane and $\overline{C} = C \cup \{\infty\}$. It is well known that $f(z)$ must has at least one Borel direction when $0 < \lambda < +\infty$. But to take into account the $\lambda = 0$ or $\lambda = +\infty$ case, Borel direction has no meaning. Then in 2004 Jianhua Zheng took the Nevanlinna Characteristic as a comparison function to introduce a new singular direction namely T-direction.

Definition[1]: A ray $T : argz = \theta (0 \le \theta < 2\pi)$ is called T-direction for a meromorphic function, if for any $\varepsilon (0 < \varepsilon < \pi)$,

$$\varlimsup_{r\to\infty} \frac{\overline{N}(r, \Omega(\theta-\varepsilon, \theta+\varepsilon), f=a)}{T(r,f)} > 0 \tag{1}$$

for all a on $\overline{C}$ with at most two exceptions, where $\Omega(\theta-\varepsilon, \theta+\varepsilon) = \{z : \theta - \varepsilon < argz < \theta + \varepsilon\}$.

Here $\overline{N}(r,\Omega(\theta-\varepsilon,\theta+\varepsilon),f=a)$ is the counting function of distinct a-points of $f(z)$ in $\Omega(\theta-\varepsilon,\theta+\varepsilon)\cap\{|z|<r\})$ defined by the formula

$$\overline{N}(r,\Omega(\theta-\varepsilon,\theta+\varepsilon),f=a)=\int_0^r\frac{\overline{n}(t,\Omega(\theta-\varepsilon,\theta+\varepsilon),f=a)}{t}dt.$$

The existence of T-direction for a meromorphic function has been proved by Hui Guo, Jianhua Zheng and Tuanwai Ng[2] by using Ahlfors-Shimizu characteristic of the meromorphic function in an angular domain.

Theorem $A^{[2]}$: *If $f(z)$ is a meromorphic function defined on the whole complex plane and satisfies*

$$\overline{\lim_{r\to\infty}}\frac{T(r,f)}{(\log r)^2}=\infty,$$

then $f(z)$ has at least one T-direction.

In this paper, we prove the existence of the corresponding T-direction if $f(z)$ takes the value of its small function, we obtain the following result.

Theorem: *Let $f(z)$ be a transcendental meromorphic function of lower order μ defined on the whole complex plane, $\mu<\infty$ and $f(z)$ satisfies*

$$T(r,f)>(\log r)^p, p>3, \tag{2}$$

then there exists a ray $T: argz=\theta(0\le\theta<2\pi)$, for any small function $\Phi(z)$ of $f(z)$,

$$\overline{\lim_{r\to\infty}}\frac{\overline{N}(r,\Omega(\theta-\varepsilon,\theta+\varepsilon),f=\Phi(z))}{T(r,f)}>0,$$

except at most for two exceptions.

Throughout we follow the standard notation of Nevanlinna theory of meromorphic function. And use $T_0(r,f)$ to denote Ahlfors-Shinizu characteristic of $f(z)$ on the whole complex plane.

$T_0(r,\Omega(\alpha,\beta),f)=\int_0^r\frac{S(t,\Omega(\alpha,\beta),f)}{t}dt$ is the notation of Ahlfors-Shimizu characteristic of $f(z)$ in an angular domain $\Omega(\alpha,\beta)$, where

$$S(t,\Omega(\alpha,\beta),f)=\frac{1}{\pi}\int_\alpha^\beta\int_0^t\left(\frac{|f'(\rho e^{i\phi}|}{1+|f(\rho e^{i\phi})|^2}\right)^2 d\rho d\phi.$$

From [3,Theorem 1.4] we have

$$\left|T(r,f)-T_0(r,f)-\log^+|f(0)|\right|\le\log 2. \tag{3}$$

Thus $T_0(r,f)$ and $T(r,f)$ differ by a bounded term.

2. Lemmas

Lemma 1:*Let $f(z)$ be a meromorphic function satisfying (2) and a_1, a_2, a_3 be three distict points on Riemann sphere, then for any given $\theta \in [0, 2\pi)$ and any small $\varepsilon > 0$, there exists a sequence $\{r_j\}$ such that*

$$T_0(r_j, \Omega(\theta - \frac{\varepsilon}{2}, \theta + \frac{\varepsilon}{2}), f) \leq \sum_{\nu=1}^{3} \overline{N}(r_j, \Omega(\theta - \varepsilon, \theta + \varepsilon), f = a_\nu) + o(T(r_j, f)),$$

where $r_j \to \infty$ when $j \to \infty$, and r_j depends only on $f(z)$ and ε.

This result can be gained from [2, lemma 3] directly.

Lemma 2: *Let $f(z)$ be a meromorphic function satisfying (2) and Φ_1, Φ_2, Φ_3 be three distict meromorphic function satisfying*

$$T(r, \Phi_\nu) = o(T(r, f)), (\nu = 1, 2, 3).$$

then for any given $\theta \in [0, 2\pi)$ and any small $\varepsilon > 0$,there exists a sequence $\{r_j\}$ such that

$$\begin{aligned} &(1 + o(1))T_0(Cr_j, \Omega(\theta - \frac{\varepsilon}{4}, \theta + \frac{\varepsilon}{4}), f) \\ &\quad \leq \sum_{\nu=1}^{3} \overline{N}(r_j, \Omega(\theta - \varepsilon, \theta + \varepsilon), f = \Phi_\nu) + o(T(r_j, f)), \end{aligned}$$

where $r_j \to \infty$ when $j \to \infty$, and r_j depends only on $f(z)$ and ε, $C \in (0, \frac{1}{2}]$ is a constant determined by ε.

Proof: Let

$$g(z) = \frac{f - \Phi_1}{f - \Phi_3} \frac{\Phi_2 - \Phi_3}{\Phi_2 - \Phi_1},$$

it is easy to see $T(r, g) = (1 + o(1))T(r, f)$.

And $f(z)$ satisfies (2), so $g(z)$ satisfies (2), then for $g(z)$, by Lemma 1, for any given $\theta \in [0, 2\pi)$ and any small $\varepsilon > 0$, there exists a sequence $\{r_j\}$ such that

$$\begin{aligned} &T_0(r_j, \Omega(\theta - \frac{\varepsilon}{2}, \theta + \frac{\varepsilon}{2}), g) \\ &\quad \leq \overline{N}(r_j, \Omega(\theta - \varepsilon, \theta + \varepsilon), g = 0) + \overline{N}(r_j, \Omega(\theta - \varepsilon, \theta + \varepsilon), g = 1) \\ &\qquad + \overline{N}(r_j, \Omega(\theta - \varepsilon, \theta + \varepsilon), g) + o(T(r_j, g)), \end{aligned}$$

where $r_j \to \infty$ when $j \to \infty$, and r_j depends only on $g(z)$ and ε.

Notice that

$$\begin{aligned}&\overline{N}(r_j, \Omega(\theta-\varepsilon, \theta+\varepsilon), g=0)\\&\qquad+\overline{N}(r_j, \Omega(\theta-\varepsilon, \theta+\varepsilon), g=1)+\overline{N}(r_j, \Omega(\theta-\varepsilon, \theta+\varepsilon), g)\\&\qquad\leq \sum_{\nu=1}^{3}\overline{N}(r_j, \Omega(\theta-\varepsilon, \theta+\varepsilon), f=\Phi_\nu)+o(T(r_j, f))\end{aligned}$$

Then

$$T_0(r_j, \Omega(\theta-\frac{\varepsilon}{2}, \theta+\frac{\varepsilon}{2}), g) \leq \sum_{\nu=1}^{3}\overline{N}(r_j, \Omega(\theta-\varepsilon, \theta+\varepsilon), f=\Phi_\nu)+o(T(r_j, f)). \tag{4}$$

Construct conformal mapping $\omega=\varphi(z)=\varphi_2\circ\varphi_1(z)$,where $\zeta=\varphi_1(z)=(ze^{-i\theta})^{\frac{\pi}{\varepsilon}}$ is a conformal mapping from $\Omega(\theta-\frac{\varepsilon}{2}, \theta+\frac{\varepsilon}{2})$ and $\Omega'(\theta-\frac{\varepsilon}{4}, \theta+\frac{\varepsilon}{4})$ onto $H_1=\{\zeta:|arg\zeta|<\frac{\pi}{2}\}$ and $H_2=\{\zeta:|arg\zeta|<\frac{\pi}{4}\}$,and $\varphi_2(\zeta)=\frac{\zeta-1}{\zeta+1}$ is from H_1, H_2 onto a unit disk $D=\{|\omega|<1\}$ and the domain between two arcs which is from -1 to 1 in the unit disk.

Let $\tau=r^{\frac{\pi}{\varepsilon}}$ and $R=\frac{\tau-1}{\tau+1}$, where r, τ and $R(<1)$ are positive constants, and

$$\begin{aligned}&\Omega(r):\Omega(\theta-\frac{\varepsilon}{2}, \theta+\frac{\varepsilon}{2})\cap\{z:|z|<r\},\\&\Omega'(r):\Omega'(\theta-\frac{\varepsilon}{4}, \theta+\frac{\varepsilon}{4})\cap\{z:|z|<r\},\\&H_1(\tau):H_1\cap\{\zeta:|\zeta|<\tau\},\\&H_2(\tau):H_2\cap\{\zeta:|\zeta|<\tau\},\\&D(R):D\cap\{\omega:|\omega|<R\}.\end{aligned}$$

Obviously

$$\varphi^{-1}(D(R))=\varphi_1^{-1}\varphi_2^{-1}(D(R))\subset\varphi_1^{-1}(H_1(\tau))=\Omega(r). \tag{5}$$

On the other hand

$$\varphi_2^{-1}(\partial D(R))=\{\zeta=\xi+i\eta:(\xi-\frac{1+\tau^2}{2\tau})^2+\eta^2=(\frac{\tau^2-1}{2\tau})^2\},$$

$$\varphi_2^{-1}(\partial D(R))\cap\partial H_2(\tau)=\{c\tau, \bar{c}\tau, \frac{c'}{\tau}, \frac{\overline{c'}}{\tau}\},$$

where c, c' are complex numbers.

Then we have

$$H_2(|c|\,\tau)-H_2(\frac{|c'|}{\tau})\subseteq\varphi_2^{-1}(D(R)).$$

Combining the above formula and (5), we have

$$\Omega'(|c|\,r) - E \subseteq \varphi^{-1}(D(R)) \subseteq \Omega(r), \tag{6}$$

where $E \subseteq \{z : |z| \leq 1\}$ for sufficiently large r.

Let

$$G(\omega) = \frac{F - \Psi_1}{F - \Psi_3}\frac{\Psi_2 - \Psi_3}{\Psi_2 - \Psi_1},$$

where $G(\omega) = g(\varphi^{-1}(\omega)), F(\omega) = f(\varphi^{-1}(\omega)), \Psi_\nu = \Phi_\nu(\varphi^{-1}(\omega)), (\nu = 1, 2, 3)$.

It is easy to see $T(R,G) = (1 + o(1))T(R,F)$ namely

$$T_0(R,G) = (1 + o(1))T_0(R,F) \tag{7}$$

By (6),

$$S(R,G) \leq S(r, \Omega(\theta - \frac{\varepsilon}{2}, \theta + \frac{\varepsilon}{2}), g)$$

for sufficiently large r, because $S(R,G)$ and $S(r, \Omega(\theta - \frac{\varepsilon}{2}, \theta + \frac{\varepsilon}{2}), g)$ mean the area of $G(|\omega| < R)$ on the Riemann sphere divided by π and the area of $g(\Omega(r))$ on the Riemann sphere divided by π respectively. So

$$T_0(R,G) \leq T_0(r, \Omega(\theta - \frac{\varepsilon}{2}, \theta + \frac{\varepsilon}{2}), g). \tag{8}$$

Similarly we can have

$$T_0(|c|\,r, \Omega(\theta - \frac{\varepsilon}{4}, \theta + \frac{\varepsilon}{4}), f) - A \leq T_0(R,F), (A \text{ is a constant}) \tag{9}$$

Combining (7), (8) and (9),

$$(1 + o(1))T_0(|c|\,r, \Omega(\theta - \frac{\varepsilon}{4}, \theta + \frac{\varepsilon}{4}), f) \leq T_0(r, \Omega(\theta - \frac{\varepsilon}{2}, \theta + \frac{\varepsilon}{2}), g).$$

Take $C = min\{|c|, \frac{1}{2}\}$,by (4), so proved the Lemma.

Lemma 3[4]: Suppose that $T(r)$ is a positive nondecreasing continuous function defined in $[0, \infty)$, and $T(r)$ satisfies

$$\underline{\lim}_{r\to\infty} \frac{\log T(r)}{\log r} = \mu < \infty, \overline{\lim_{r\to\infty}} \frac{T(r)}{(\log r)^2} = \infty,$$

then for any given real number $h > 0$, there exist sequences $\{r_j\}$ and $\{R_j\}$,

$$R_j^{1-o(1)} \leq r_j \leq R_j, (j \to \infty),$$

such that

$$\lim_{j\to\infty} \frac{T(r_j)}{(\log r_j)^2} = \infty,$$

$$T(e^h R_j) \leq e^{h\mu} T(R_j)(1 + o(1)), (j \to \infty).$$

3. Proof of theorem

Proof: Suppose the theorem does not hold. Then for any three distinct meromorphic function Φ_1, Φ_2, Φ_3 satisfying $T(r, \Phi_\nu) = o(T(r, f))$, we have

$$\overline{N}(r, \Omega_\theta, f = \Phi_1) + \overline{N}(r, \Omega_\theta, f = \Phi_2) + \overline{N}(r, \Omega_\theta, f = \Phi_3) = o(T(r, f)). \quad (10)$$

Where $\Omega_\theta = \{z : \theta - \varepsilon_\theta < argz < \theta + \varepsilon_\theta\}$.

A unit disk $D' \subseteq \bigcup_{\theta \in [0,2\pi)} \Omega(\theta - \frac{\varepsilon_\theta}{4}, \theta + \frac{\varepsilon_\theta}{4})$ and D' is a compact closed set, then by finite covering theorem, we can choose finitely many $(\theta - \varepsilon_\theta, \theta + \varepsilon_\theta)$ to cover D'.

$$D' \subseteq \bigcup_{\gamma=1}^{q} \Omega(\theta_\gamma - \frac{\varepsilon_{\theta_\gamma}}{4}, \theta + \frac{\varepsilon_{\theta_\gamma}}{4}).$$

$f(z)$ satisfies (2), then by lemma 2, there exists a common sequence $\{r_j\}$ so that

$$\begin{aligned}(1 + o(1))T_0(Cr_j, \Omega(\theta_\gamma - \frac{\varepsilon_{\theta_\gamma}}{4}, \theta_\gamma + \frac{\varepsilon_{\theta_\gamma}}{4}), f) \\ \leq \sum_{\nu=1}^{3} \overline{N}(r_j, \Omega(\theta_\gamma - \varepsilon_{\theta_\gamma}, \theta_\gamma + \varepsilon_{\theta_\gamma}), f = \Phi_\nu) + o(T(r_j, f)).\end{aligned}$$

Where $r_j \to \infty$ when $j \to \infty$, and $\{r_j\}$ depends only on $f(z)$ and $\varepsilon = min\{\varepsilon_{\theta_1}, \varepsilon_{\theta_2}, \cdots, \varepsilon_{\theta_q}\}$, C is a constant determined by ε.

By (10)

$$(1 + o(1))T_0(Cr_j, \Omega(\theta_\gamma - \frac{\varepsilon_{\theta_\gamma}}{4}, \theta_\gamma + \frac{\varepsilon_{\theta_\gamma}}{4}), f) \leq o(T(r_j, f)), (\gamma = 1, 2, \ldots, q),$$

So $(1 + o(1))T_0(Cr_j, f) \leq o(T(r_j, f))$.

Let $r'_j = Cr_j$, namely $r_j = \frac{r'_j}{C}$, then the formula above can be rewritten as

$$(1 + o(1))T_0(r'_j, f) \leq o(T(\frac{r'_j}{C}, f)). \quad (11)$$

By lemma 3, so there exists a sequence $\{r''_j\}$ such that

$$T(\frac{1}{C}r''_j, f) \leq (1 + o(1))(\frac{1}{C})^\mu T(r''_j, f). \quad (12)$$

Combining (11) and (12)

$$(1 + o(1))T_0(r''_j, f) \leq o(T(r''_j, f)).$$

This contradicts (3). So proved the theorem.

References

1. Jianhua Zheng, On transcendental meromorphic functions with radially distributed values [J], Science in China, Ser. A, Mathematics Vol.47(2004) No.3, 401-416.
2. Hui Guo, Jianhua Zheng and Tuen Wai NG, On a new singular direction of meromorphic functions [J], Bull. Austral. Math. Soc. Vol.69(2004), 277-287.
3. W. K. Hayman, Meromorphic Functions [M], Oxford Clarendon Press, 1964.
4. Yinian Lv, Guanghou Zhang, On Nevanlinna-direction of Algebroids Function [J], Science in China, Ser. A, 3(1983), 215-224.

ON NEVANLINNA TYPE CLASSES*

NATTAKORN SUKANTAMALA
Department of Mathematics
Chiang Mai University
Chiang Mai, 50200 Thailand
E-mail: scmti006@chiangmai.ac.th

ZHIJIAN WU
Department of Mathematics
The University of Alabama
Tuscaloosa, AL 35487
E-mail: zwu@bama.ua.edu

This paper studies the Nevanlinna type classes and operators such as multipliers, nontangential maximal function operator and area operator. These operators are useful in complex and harmonic analysis.

Keywords: Nevanlinna class, multiplier, nontangential maximal function, area operator.

2000 Mathematics Subject Classification: 32A37, 47B35, 47B47

1. Introduction

Let $\mathbb{D}$ be the open unit disk of complex plane, and $\partial\mathbb{D}$ be the boundary of $\mathbb{D}$. For $0 < p \leq \infty$, denote $L^p = L^p(\partial\mathbb{D}, m)$ and $\|\cdot\|_p = \|\cdot\|_{L^p}$, where m is the measure of the normalized arc length on $\partial\mathbb{D}$. The Hardy space H^p for $0 < p < \infty$, consists of all holomorphic functions f defined on $\mathbb{D}$ so that

$$\|f\|_p^p = \sup_{0<r<1} \int_{\partial\mathbb{D}} |f(r\zeta)|^p dm(\zeta) < \infty.$$

*Research was supported in part by National Science Foundation DMS 0200587.

Denote also by H^∞ the space of all bounded holomorphic functions on $\mathbb{D}$ with the norm

$$\|f\|_\infty = \sup_{z\in\mathbb{D}} |f(z)|.$$

The Nevanlinna class N consists of all holomorphic functions f defined on $\mathbb{D}$ so that

$$\sup_{0<r<1} \int_{\partial\mathbb{D}} \log^+ |f(r\zeta)| dm(\zeta) < \infty,$$

or equivalently, the subharmonic function $\log^+ |f(z)| (= \max\{0, \log |f(z)|\})$ has a harmonic majorant in $\mathbb{D}$ (see for example, Chapter 2 in [3]). It is standard that H^p is a subset of N, and N can be viewed as an endpoint of the scalar space H^p at $p = 0$.

In this paper, we introduce the $p-$Nevanlinna type class N_p, and study some basic properties of this scalar space. We characterize the multipliers of N_p, prove the boundedness of the nontangential maximal function operator on N_p for $p > 1$, and characterize the "boundedness" of the area operator on N_p for $p > 1$. Multipliers, nontangential maximal function operator and area operator are useful in complex and harmonic analysis. They relate to, for example, Poisson integral, Littlewood-Paley operator, and tent spaces, etc.

Definitions and preliminaries are gathered in Section 1. Main results are stated and proved in Section 2. Throughout this paper c and C are positive constants which may change from one step to the next. We say two quantities a and b are equivalent if there exists two positive constants c and C such that $ca \le b \le Ca$.

2. Preliminaries

For $f \in L^p$, $p \ge 1$, the harmonic extension of f onto $\mathbb{D}$ is defined by the following Poisson integral

$$f(z) = \int_{\partial\mathbb{D}} \frac{1-|z|^2}{|\zeta - z|^2} f(\zeta)\, dm(\zeta)\,.$$

For any $\zeta \in \partial\mathbb{D}$, define the cone in $\mathbb{D}$ with vertex ζ by

$$\Gamma(\zeta) = \{z \in \mathbb{D} : |\zeta - z| < 1 - |z|^2\}.$$

It is clear that, for $\zeta \in \partial\mathbb{D}$, the cone $\Gamma(\zeta)$ contains the line segment $\{z = r\zeta : 0 \le r < 1\}$, and is nontangential to the unit circle $\partial\mathbb{D}$.

Let h be a subharmonic function on $\mathbb{D}$. The nontangential maximal function of h is defined by

$$h^*(\zeta) = \sup_{z \in \Gamma(\zeta)} |h(z)| .$$

For $f \in L^p$, for convenience, we write f^* for the nontangential maximal function of $f(z)$–the harmonic extension of f onto $\mathbb{D}$.

The following result is standard (See Chapter 1 in [3] for example).

Theorem A *Suppose* $1 < p \leq \infty$. *Then there exists a constant* $C_p > 0$ *such that*

$$\|f^*\|_p \leq C_p \|f\|_p$$

holds for all $f \in L^p$.

Let μ be a nonnegative measure on $\mathbb{D}$. The area operator G_μ acting on a holomorphic function f defined on $\mathbb{D}$ is defined by

$$G_\mu(f)(\zeta) = \int_{\Gamma(\zeta)} |f(z)| \frac{d\mu(z)}{1 - |z|}.$$

A nonnegative measure μ on $\mathbb{D}$ is called a Carleson measure if there exists a constant $C > 0$ such that

$$\mu(S(I)) \leq C|I|$$

holds for all arc I on $\partial\mathbb{D}$. Here $|I|$ denotes the normalized arc length of I and $S(I)$ is the Carleson box based on I defined as $S(I) = \{z = re^{i\theta} \in \mathbb{D} : e^{i\theta} \in I, 0 < 1 - r < |I|\}$.

On Hardy space, the area operator has been studied by Cohn [1]. The following theorem is the main result in [1].

Theorem B *Suppose* μ *is a nonnegative measure on* $\mathbb{D}$ *and* $0 < p < \infty$. *Then* $G_\mu : H^p \to L^p$ *is bounded if and only if* μ *is a Carleson measure.*

On Bergman space $A_p(= \{f$ holomorphic in $\mathbb{D} : \int_{\mathbb{D}} |f(z)|^p dA(z) < \infty\})$, the area operator has been studied systematically in [4]. There the boundedness of the map $G_\mu : A_p \to L^q$ has been characterized.

The Nevanlinna class has been well studied. We cite some standard results in the following. One can refer to [2] and [3] for more details.

- $f \in N$ if and only if $f = \frac{f_1}{f_2}$, where $f_1, f_2 \in H^\infty$ and f_2 is nowhere zero in $\mathbb{D}$.

- Let $f \in N$ and $f \not\equiv 0$. Then
$$f(z) = CB(z)F(z)S_1(z)/S_2(z), \quad \forall z \in \mathbb{D}.$$
Where $|C| = 1$, $B(z)$ is the Blaschke product, $F(z)$ is an outer function, and $S_1(z)$ and $S_2(z)$ are singular functions. Every function of such form is also in N.
- Let $f \in N$. Then the boundary function of f exists, i.e.,
$$f(\zeta) = \lim_{r \to 1^-} f(r\zeta)$$
exists for a.e. $\zeta \in \partial\mathbb{D}$. Moreover $\log|f| \in L^1$.
- $H^p \subset N$ for all $p > 0$.

The last fact above suggests that N can be viewed as an endpoint of the scalar space H^p at $p = 0$ because of the nest property $H^p \subset H^q$ for $p > q$.

3. Main Results

For f defined on $\mathbb{D}$ and $0 < r < 1$, denote $f_r(z) = f(rz)$. Recall that a holomorphic function f on $\mathbb{D}$ is in the Nevanlinna class N if
$$N(f) = \sup_{0<r<1} \|\log^+|f_r|\|_1 < \infty.$$

To introduce $p-$Nevanlinna class, we first establish the following lemma, which gives both lower and upper estimates to the Nevanlinna quantity $N(f)$ defined above.

Lemma 2.1 *Suppose f is holomorphic on $\mathbb{D}$. Then f is in N if and only if*
$$N_1(f) = \sup_{0<r<1} \|\log(1 + |f_r|)\|_1 < \infty.$$
Moreover $N(f) \le N_1(f) \le \log 2 + N(f)$.

Remark 2.1 *Both $N(\cdot)$ and $N_1(\cdot)$ are not norms for N, because for a constant a and $f \in N$, $N_1(af) \neq |a|N_1(f)$ and $N(af) \neq |a|N(f)$ in general. However, it is better to measure $f \in N$ by $N_1(f)$, because the triangle inequality holds for $N_1(\cdot)$ (see Theorem 3 later) but not for $N(\cdot)$ (for example $N(2) = \log 2 > 0 = N(1) + N(1)$).*

Proof. It is clear that the estimate $N(f) \le N_1(f)$ holds.

Let $\zeta \in \partial\mathbb{D}$ and $0 < r < 1$. If $|f(r\zeta)| \le 1$, then
$$\log^+|f(r\zeta)| = \max\{0, \log|f(r\zeta)|\} = 0.$$

Therefore

$$\log(1+|f(r\zeta)|) \le \log 2 = \log 2 + \log^+ |f(r\zeta)|.$$

If $|f(r\zeta)| > 1$, we have that $\log^+ |f(r\zeta)| = \log |f(r\zeta)|$ and $\frac{1+|f(r\zeta)|}{|f(r\zeta)|} < 2$. Consequently,

$$\log(1+|f(r\zeta)|) - \log |f(r\zeta)| = \log \frac{1+|f(r\zeta)|}{|f(r\zeta)|} < \log 2\,.$$

Therefore

$$\log(1+|f(r\zeta)|) < \log 2 + \log |f(r\zeta)| = \log 2 + \log^+ |f(r\zeta)|.$$

Hence for all $\zeta \in \partial\mathbb{D}$ and $0 < r < 1$, we have

$$\log(1+|f(r\zeta)|) \le \log 2 + \log^+ |f(r\zeta)|.$$

Thus, by integrating both sides of above inequality on $\partial\mathbb{D}$, we get

$$N_1(f) \le \log 2 + N(f)\,.$$

The proof is completed. □

For $0 < p < \infty$, define the $p-$Nevanlinna class, N_p, to be the collection of all holomorphic functions f on $\mathbb{D}$ such that

$$N_p(f) = \sup_{0<r<1} \| \log(1+|f_r|)\|_p < \infty.$$

Clearly $N_p(\cdot)$ is not a norm for N_p, and by the proof of Lemma 3 we have that $N_p(f) < \infty$ if and only if $\sup_{0<r<1} \| \log^+ |f_r|\|_p < \infty$. Similar to the reason given in Remark 2.1, it is better to measure $f \in N_p$ by $N_p(f)$ instead of $\sup_{0<r<1} \| \log^+ |f_r|\|_p$.

For any $q > 0$, N_q can be viewed as an endpoint of the scalar space H^p at $p = 0$ because $H^p \subset N_q$ if $p > 0$. The inclusion $H^p \subset N_q$ can be obtained by the estimate

$$[N_q(f)]^q \le C + c\|f\|_p^p, \qquad \forall f \in H^p\,,$$

which is a consequence of the fact that there exists a constant $C \ge 1$ such that

$$1 + x \le Ce^{x^{p/q}}, \qquad \text{for all} \;\; x \ge 0\,.$$

The following nest property can be obtained by using Hölder's inequality:

$$N_p \subset N_q \qquad \text{if} \quad p > q\,.$$

For $p \geq 1$, we have $N_p \subset N_1 = N$. Therefore the boundary function $f(\zeta)$ exists for $f \in N_p$ and $p \geq 1$.

For $0 < p < \infty$, define the class $\tilde{N}_p$ to be the collection of all measure functions f on $\partial\mathbb{D}$ such that

$$\tilde{N}_p(f) = \| \log(1 + |f|)\|_p < \infty .$$

Later, we will compare $N_p(f)$ and $\tilde{N}_p(f)$ when the boundary function of $f \in N_p$ exists.

We need the following lemma, which tells when $\Phi(|f(z)|)$ is subharmonic on $\mathbb{D}$ for any holomorphic f on $\mathbb{D}$.

Lemma 2.2 *Suppose Φ is a C^2 function on $[0, \infty)$. Then $\Phi(|f(z)|)$ is subharmonic on $\mathbb{D}$ for any holomorphic function f on $\mathbb{D}$ if and only if*

$$x\Phi''(x) + \Phi'(x) \geq 0, \quad \textit{for all} \;\; x \in (0, \infty) .$$

Proof. Denote $\partial = \frac{1}{2}\left(\frac{\partial}{\partial x} - i\frac{\partial}{\partial y}\right)$. We know that $\Phi(|f(z)|)$ is subharmonic on $\mathbb{D}$ if and only if $\partial\overline{\partial}\Phi(|f(z)|) \geq 0$ a.e. on $\mathbb{D}$.

For $z \in \mathbb{D}$, direct computation yields

$$\partial|f(z)| = \frac{1}{2}\frac{f'(z)\overline{f(z)}}{|f(z)|}, \quad \overline{\partial}|f(z)| = \frac{1}{2}\frac{f(z)\overline{f'(z)}}{|f(z)|}, \quad \overline{\partial}\partial|f(z)| = \frac{1}{4}\frac{|f'(z)|^2}{|f(z)|}.$$

Therefore

$$\begin{aligned}\partial\overline{\partial}\Phi(|f(z)|) &= \Phi''(|f(z)|) \cdot \partial|f(z)| \cdot \overline{\partial}|f(z)| + \Phi'(|f(z)|)\partial\overline{\partial}|f(z)| \\ &= \frac{|f'(z)|^2}{4|f(z)|}\left[\Phi''(|f(z)|)|f(z)| + \Phi'(|f(z)|)\right].\end{aligned}$$

This is enough to conclude the desired result. □

Corollary 2.3 *Suppose f is holomorphic on $\mathbb{D}$ and $p \geq 1$. Then*

$$(\log(1 + |f(z)|))^p$$

is subharmonic on $\mathbb{D}$.

Proof. We only need to verify that the function $\Phi(x) = [\log(1 + x)]^p$ satisfies $x\Phi''(x) + \Phi'(x) \geq 0$ if $x > 0$. In fact

$$x\Phi''(x) + \Phi'(x) = \frac{p[\log(1 + x)]^{p-2}}{(1 + x)^2}\left[(p - 1)x + \log(1 + x)\right].$$

The conclusion is therefore followed. □

One of the important properties for subharmonic function $h(z)$ defined on $\mathbb{D}$ is that for $0 < r < s < 1$

$$\int_{\partial\mathbb{D}} h(r\zeta)dm(\zeta) \le \int_{\partial\mathbb{D}} h(s\zeta)dm(\zeta).$$

This fact leads to the following corollary.

Corollary 2.4 *Suppose $p \ge 1$ and $f \in N_p$. Then*

$$N_p(f) = \lim_{r\to 1^-} \|\log(1 + |f_r|)\|_p .$$

The following theorem compares all three quantities $N_p(f)$, $\tilde{N}_p(f)$ and $\tilde{N}_p(f^*)$.

Theorem 2.5 *Suppose f is holomorphic on $\mathbb{D}$ and $0 < p < \infty$.*

- *(a) If $p > 0$, then $N_p(f) \le \tilde{N}_p(f^*)$;*
- *(b) If $p \ge 1$, then $\tilde{N}_p(f) \le N_p(f)$;*
- *(c) If $p > 1$, then there exists a constant $C_p > 0$ such that $\tilde{N}_p(f^*) \le C_p\tilde{N}_p(f)$.*

Remark 2.5 *Theorem 2.5 suggests that for $p \ge 1$, if we identify f with its boundary function then $N_p \subset \tilde{N}_p$, and for $p > 1$ all three quantities $N_p(f)$, $\tilde{N}_p(f)$ and $\tilde{N}_p(f^*)$ are equivalent.*

Proof. The estimate $N_p(f) \le \tilde{N}_p(f^*)$ is trivial because $|f(r\zeta)| \le f^*(\zeta)$ for $0 < r < 1$ and $\zeta \in \mathbb{D}$. For $p \ge 1$, since the boundary function $f(\zeta)$ exists for $f \in N_p$, the estimate $\tilde{N}_p(f) \le N_p(f)$ is a consequence of Fatou's lemma and Corollary 2.4.

To prove $\tilde{N}_p(f^*) \le C_p\tilde{N}_p(f)$ for $f \in N_p$ and $p > 1$, we note first that, by (b), the boundary function $f(\zeta)$ satisfies $\log(1 + |f(\zeta)|) \in L^p \subset L^1$. Let $U(z)$ be the harmonic function with boundary function $\log(1+|f(\zeta)|)$, i.e., $U(z)$ is the following Poisson integral

$$U(z) = \int_{\partial\mathbb{D}} \frac{1-|z|^2}{|\zeta - z|^2} \log(1 + |f(\zeta)|)\, dm(\zeta).$$

It is clear that $U(z)$ is a harmonic majorant of the subharmonic function $\log(1 + |f(z)|)$. Therefore for $\zeta \in \mathbb{D}$,

$$\log(1 + f^*(\zeta)) = \sup_{z\in\Gamma(\zeta)} \log(1 + |f(z)|) \le \sup_{z\in\Gamma(\zeta)} U(z) = U^*(\zeta),$$

and then,

$$\left[\log(1+f^*(\zeta))\right]^p \leq \left[U^*(\zeta)\right]^p.$$

Hence by Theorem A, we have a constant $C_p > 0$ such that

$$\tilde{N}_p(f^*) \leq \|U^*\|_p \leq C_p\|U\|_p = C_p\tilde{N}_p(f)\,.$$

The proof is completed. □

The following theorem characterizes functions in N_p for $p > 1$.

Theorem 2.6 *Let f be holomorphic on $\mathbb{D}$ and $p > 1$. Then $f \in N_p$ if and only if $\log^+|f| \in L^p$. Moreover, if $f \in N_p$ and $f \not\equiv 0$. Then*

$$f(z) = CB(z)F(z)S_1(z)/S_2(z)\,.$$

Here $|C| = 1$, $B(z)$ is a Blaschke product, $F(z)$ is an outer function with $\log^+|F| \in L^p$, $S_1(z)$ and $S_2(z)$ are singular functions. Also every function f of this form is in N_p.

Proof Let $p > 1$. Suppose that $f \in N_p$. Then the boundary function $f(\zeta)$ exists. Theorem 2.5 (b) implies that $\log(1 + |f|) \in L^p$. Therefore $\log^+|f| \in L^p$, since $\log^+|f| \leq \log(1 + |f|)$.

Now suppose that $\log^+|f| \in L^p$. Since $\log(1 + |f|) \leq \log 2 + \log^+|f|$ as shown in the proof of Lemma 2.1, we have that $\log(1+|f|) \in L^p$. Theorem 2.5(c) gives that $\log(1 + |f^*|) \in L^p$. Theorem 2.5(a) consequently implies $f \in N_p$.

For the second part, suppose that $f \in N_p$ and $f \not\equiv 0$. Then $f \in N$, and this is enough to conclude that $f(z) = CB(z)F(z)S_1(z)/S_2(z)$ with $|C| = 1$, $B(z)$ is a Blaschke product, $F(z)$ is an outer function, and $S_1(z)$ and $S_2(z)$ are singular functions.

Moreover, for $\zeta \in \partial\mathbb{D}$, we have

$$\log|f(\zeta)| = \log|C|+\log|B(\zeta)|+\log|F(\zeta)|+\log|S_1(\zeta)|-\log|S_2(\zeta)| = \log|F(\zeta)|\,.$$

Therefore $\log^+|f| \in L^p$ is equivalent to $\log^+|F| \in L^p$. This is enough. □

For a function set X, we say a function φ is a multiplier of X if $\varphi X \subset X$, i.e., $\varphi f \in X$, for all $f \in X$. The collection of all multipliers of X is denoted by $M(X)$. We note that if X is a Banach space then, by the Closed Graph Theorem, a multiplier of X defines a bounded linear operator. We note that however N_p is not a normed space.

Theorem 2.7 *Suppose $p > 0$. Then $M(N_p) = N_p$. Moreover if $f,\ g \in N_p$ then*

$$N_p(f+g),\ N_p(fg) \le \max\{2^{1/p-1}, 1\}\,(N_p(f) + N_p(g))\ .$$

Proof. Let $\varphi \in M(N_p)$. Then $\varphi N_p \subset N_p$. This implies $\varphi \in N_p$, since $f \equiv 1$ is in N_p.

Let $f,\ g \in N_p$. We have for $z \in \mathbb{D}$

$$1 + |f(z) + g(z)|,\ 1 + |f(z)g(z)| \le (1 + |f(z)|)(1 + |g(z)|)\,.$$

Therefore

$$\log\left(1 + |f(z) + g(z)|\right), \quad \log\left(1 + |f(z)g(z)|\right) \le \log(1+|f(z)|)+\log(1+|g(z)|)\,,$$

and hence for $0 < r < 1$

$$\begin{aligned}&\|\log(1 + |f_r + g_r|)\|_p\,,\ \|\log(1 + |f_r g_r|)\|_p\\ &\qquad \le \max\{2^{1/p-1}, 1\}\,(\|\log(1 + |f_r|)\|_p + \|\log(1 + |g_r|)\|_p)\ .\end{aligned}$$

Hence $N_p(f+g),\ N_p(gf) \le \max\{2^{1/p-1}, 1\}\,(N_p(g) + N_p(f))$. This implies also $fg \in N_p$. □

The proof of Theorem 2.7 suggests that the following result is also true.

Corollary 2.8 *Suppose $p > 0$. Then $M(\tilde{N}_p) = \tilde{N}_p$. Moreover if $f,\ g \in \tilde{N}_p$ then*

$$\tilde{N}_p(f+g),\ \tilde{N}_p(fg) \le \max\{2^{1/p-1}, 1\}\left(\tilde{N}_p(f) + \tilde{N}_p(g)\right)\ .$$

The following theorem shows that a multiplier defines a "bounded operator" on N_p in the usual sense if and only if it is a bounded holomorphic function.

Theorem 2.9 *Let $p \ge 1$ and φ is holomorphic on $\mathbb{D}$. Then the estimate*

$$N_p(\varphi f) \le A N_p(f)\,, \qquad \forall f \in N_p$$

holds if and only if $\varphi \in H^\infty$.

Proof. Suppose $\varphi \in H^\infty$ and $f \in N_p$. For $z \in \mathbb{D}$, we have clearly

$$1 + |\varphi(z)f(z)| \le 1 + (1 + \|\varphi\|_\infty)\,|f(z)| \le (1 + |f(z)|)^{1+\|\varphi\|_\infty}.$$

Therefore $\log(1 + |\varphi(z)f(z)|) \le (1 + \|\varphi\|_\infty)\log(1 + |f(z)|)$, and hence

$$N_p(\varphi f) \le (1 + \|\varphi\|_\infty)\, N_p(f)\,.$$

(This estimate is true for $p > 1$!)

For the other direction, by Theorem 2.7, we know $\varphi \in N_p \subset N_1$, and therefore the boundary function $\varphi(\zeta)$ exists. Consider $M > 0$ and $E_M = \{\zeta \in \partial\mathbb{D} : |\varphi(\zeta)| > M\}$ with $m(E_M) > 0$. It suffices to show that $M \le 2^A - 1$.

Consider the following harmonic functions on $\mathbb{D}$:

$$h_n(z) = \int_{\partial\mathbb{D}} \frac{1-|z|^2}{|\zeta - z|^2} h_n(\zeta)\, dm(\zeta), \qquad n = 1, 2, \cdots,$$

where $h_n(\zeta)$ equals 0 if $\zeta \in E_M$, and $-\log n$ if $\zeta \in \widetilde{E_M} = \partial\mathbb{D} \setminus E_M$. Clearly

$$-\log n \le h_n(z) \le 0, \qquad \forall z \in \mathbb{D}.$$

Let $H_n(z)$ be the holomorphic function on $\mathbb{D}$ with its real part equals $e^{h_n(z)}$ (In fact, $H_n(z) = \exp(\int_{\partial\mathbb{D}} \frac{\zeta+z}{\zeta-z} h_n(\zeta) dm(\zeta))$). It is easy to see that

$$n^{-1} \le |H_n(z)| \le 1; \ |H_n(\zeta)| = 1 \text{ if } \zeta \in E_M \text{ and } |H_n(\zeta)| = n^{-1} \text{ if } \zeta \in \widetilde{E_M}\,.$$

Hence by Corollary 2.4 and the Bounded Convergence Theorem, we have

$$\begin{aligned} [N_p(H_n)]^p &= \lim_{r \to 1^-} \int_{\partial\mathbb{D}} \left(\log(1 + e^{h_n(r\zeta)})\right)^p dm(\zeta) \\ &= (\log 2)^p m(E_M) + (\log(1+n^{-1}))^p m(\widetilde{E_M})\,. \end{aligned}$$

Together with Theorem 2.5(b) and the assumption $N_p(\varphi H_n) \le A N_p(H_n)$, we obtain

$$\begin{aligned} (\log(1+M))^p m(E_M) &\le [\tilde{N}_p(\varphi H_n)]^p \\ &\le [N_p(\varphi H_n)]^p \\ &\le A^p (N_p(H_n))^p \\ &= A^p \left((\log 2)^p m(E_M) + (\log(1+n^{-1}))^p m(\widetilde{E_M})\right). \end{aligned}$$

This implies $\log(1+M) \le A \log 2$, or equivalently $M \le 2^A - 1$. □

We now consider the area operator G_μ acting on the Nevanlinna type class N_p. We want to characterize the nonnegative measure μ on $\mathbb{D}$ so that $G_\mu(f)$ is finite on every $f \in N_p$, i.e.,

$$\tilde{N}_p(G_\mu(f)) < \infty, \qquad \text{for all } \ f \in N_p\,.$$

We note that for a holomorphic Banach space $(X, \|\cdot\|_X)$, by the Closed Graph Theorem, $\|G_\mu(f)\|_p < \infty$ holds for all $f \in X$ if and only if $\|G_\mu(f)\|_p < C\|f\|_X$ holds for all $f \in X$, i.e., G_μ is bounded on X. However

$N_p(\cdot)$ is not a norm for N_p. We will see in the following theorem that a reasonable form of the estimate for $\tilde{N}_p(G_\mu(f))$ is

$$\tilde{N}_p(G_\mu(f)) \leq C_p N_p(f) + D_{p,\mu},$$

with the positive constant C_p depends only on p, and the positive constant $D_{p,\mu}$ depends only on p and μ. The positive constant $D_{p,\mu}$ is needed because of the fact that $N_p(\cdot)$ is not a norm and $N_p(fg) \leq N_p(f) + N_p(g)$ for $p \geq 1$ as proved in Theorem 2.7.

For convenience, we write

$$G_\mu(\zeta) = G_\mu(1)(\zeta) = \int_{\Gamma(\zeta)} \frac{d\mu(z)}{1-|z|}$$

Theorem 2.10 *Let μ be a nonnegative measure on $\mathbb{D}$ and $1 < p < \infty$. Then*

$$\tilde{N}_p(G_\mu(f)) < \infty, \qquad \forall f \in N_p$$

if and only if

$$\tilde{N}_p(G_\mu(1)) < \infty.$$

Moreover, if $\tilde{N}_p(G_\mu(1)) < \infty$, then there exists a constant $C_p > 0$ such that

$$\tilde{N}_p(G_\mu(f)) \leq C_p N_p(f) + \tilde{N}_p(G_\mu(1)), \qquad \forall f \in N_p.$$

Proof. Suppose first that $\tilde{N}_p(G_\mu(f)) < \infty$ for all $f \in N_p$. Pick $f \equiv 1$ in N_p, we have

$$\tilde{N}_p(G_\mu) = \tilde{N}_p(G_\mu(1)) < \infty.$$

To prove the other direction, we note first that for any $\zeta \in \partial\mathbb{D}$ and $f \in N_p$, we have

$$|f(z)| \leq f^*(\zeta), \qquad \text{for all } z \in \Gamma(\zeta).$$

Therefore

$$G_\mu(f)(\zeta) = \int_{\Gamma(\zeta)} |f(z)| \frac{d\mu}{1-|z|} \leq f^*(\zeta) \int_{\Gamma(\zeta)} \frac{d\mu}{1-|z|} = f^*(\zeta) G_\mu(\zeta).$$

Thus, by the estimate

$$1 + G_\mu(f)(\zeta) \leq 1 + f^*(\zeta) G_\mu(\zeta) \leq (1 + f^*(\zeta))(1 + G_\mu(\zeta)),$$

we have

$$\log(1+G_\mu(f)(\zeta)) \le \log(1+f^*(\zeta)) + \log(1+G_\mu(\zeta))\,.$$

This implies

$$\tilde{N}_p(G_\mu(f)) \le \tilde{N}_p(f^*) + \tilde{N}_p(G_\mu)\,.$$

To complete the proof, we only need to show that $\tilde{N}_p(f^*) \le C_p N_p(f)$ for $p > 1$, which is proved in Theorem 2.5. □

Suppose μ is a Carleson measure on $\mathbb{D}$. By Theorem B, we have $G_\mu(\zeta) = \int_{\Gamma(\zeta)} \frac{d\mu(z)}{1-|z|}$ is in L^p for any $p \in (0,\infty)$. This implies that $\log(1+G_\mu(\zeta))$ is in L^p for any $p > 0$. We have therefore the following corollary.

Corollary 2.11 *Suppose $p > 0$ and μ is a Carleson measure on $\mathbb{D}$. Then $\tilde{N}_p(G_\mu) < \infty$. If in addition $p > 1$, we have*

$$\tilde{N}_p(G_\mu(f)) \le C_p N_p(f) + \tilde{N}_p(G_\mu)$$

holds for all $f \in N_p$.

It will be nice to characterize the nonnegative measure μ on $\mathbb{D}$ so that the estimate

$$\tilde{N}_p(G_\mu(f)) \le A N_p(f)$$

holds for all $f \in N_p$. Similar to the proof of Theorems 2.9, together with the idea in the proof of Theorem 3, we can establish the above estimate if the function $G_\mu = G_\mu(1)$ is bounded. We hope this condition is also necessary.

References

1. W.S. Cohn, Generalized Area Operators on Hardy Spaces, *J. Math. Anal. App.* **216**(1997):112-121.
2. P.L. Duren, *Theory of H^p Spaces*, Academic Press, New York, 1970.
3. J.B. Garnett, *Bounded Analytic Functions*, Academic Press, New York, 1981.
4. Z. Wu, *Area Operator on Bergman Spaces*, to appear.

FREDHOLM MODULE AND CAUCHY INTEGRAL OPERATOR*

JICHENG TAO
Department of Mathematics
China Jiliang University
Hangzhou, 310018 P. R. China
Email: taojc@cjlu.edu.cn

The main goal of this paper is to extend A.Connes's result on Fredholm index to whole Hilbert space. A Fredholm module will be constructed by using singular integral operators with the Cauchy kernel and the relationship between the Fredholm index of the singular integral operators and K-group was set up.

Keywords: Fredholm module, K-group, Cauchy kernel.

2000 Mathematics Subject Classification: 14A22, 19K56, 45E05

1. Introduction

In [1], Connes developed a differential calculus by constructing a Fredholm module. Since then Fredholm module has been steadily evolved into a powerful mathematical framework in studying of noncommutative geometry. Recently, A. Gorokhovsky constructed the Fredholm module in the case of the unital generalized cycles over unital algebras in [6]. T.Schick[7] study the trace on the K-theory of group C^*-algebras by using Fredholm module technique. Further research work in the direction includes [8], [9] etc.

With such a strong theory and applicable background for Fredholm module, it seems desirable to work out other Fredholm module. In this paper, We extend the proposition 2 of A.Connes [1] in page 289 to whole

*This research was partially supported by the National Foundation of China [60473034] and the Education Department Foundation of Zhejiang Province [20040365].

Hilbert space in order to discuss the Fredholm index of the singular integral operators, and then construct a Fredholm module by using the theory of singular integral operators and set up the relationship between the Fredholm index of the singular integral operators and K-group.

The rest of the paper is organized as follows: In Section 2 , we extend a result of A.Connes etc to whole Hilbert space.In Section 3, Fredholm module of Cauchy integral operators will be constructed and K-group index will be obtained.

2. Fredholm Module

Definition 2.1[1] *Let A be an involutive algebras(over C). Then a Ferdholm module over A is given by:*

(1) an involutive representation π of A in a Hilbert space H;
(2) an operator $F = F^*$, $F^2 = I$,on H such that
$[F, \pi a]$ is a compact operator for any $a \in A$.

Such a Fredholm module will be called odd. An even Fredholm module is given by an odd Fredholm module (H, F) as above together with a $\frac{z}{2}$ grading γ,$\gamma = \gamma^*$,$\gamma^2 = I$ of Hilbert space H such that:

a) $\gamma\pi(a) = \pi(a)\gamma, \forall a \in A$
b) $\gamma F = -F\gamma$

The role of such modules in index theory is provided by the following theorem:

Theorem 2.2[1] *Let A be an involutive algebra, (H, F) a Fredholm module over A, and for $q \in N$ let (H_q, F_q) be the Fredholm module over $M_q(A) = A \bigotimes M_q(C)$ given by*

$$H_q = H \bigotimes C^q, F_q = F \bigotimes I, \pi_q = \pi \bigotimes id$$

we extend the action of A on H to a unital action of $\tilde{A}$.

a) Let (H, F) be even, with $\frac{z}{2}$ grading γ, and let $e \in Proj(M_q(\tilde{A})$. Then the operator $\pi_q^-(e)F_q\pi_q^+(e)$ from $\pi_q^+(e)H_q^+$ to $\pi_q^-(e)H_q^+$ is a Fredholm operator. An additive map φ of $K_0(A)$ to Z is determined by

$$\varphi([e]) = Index(\pi_q^-(e)F_q\pi_q^+(e)),$$

where φ is an index map from $K_0(A)$ to Z.

b) Let (H,F) be odd and let $E = (\frac{1+F}{2})$, let $u \in Gl_q(\tilde{A})$. Then the operator $E_q\pi_q(u)E_q$ from E_qH_q to itself is a Fredholm operator. An additive map of $K_1(A)$ to Z is determined by

$$\varphi([u]) = Index(E_q\pi_q(u)E_q),$$

where φ is an index map from $K_1(A)$ to Z, the group of invertible elements in $M_q(\tilde{A})$ is denoted by $Gl_q(\tilde{A})$. For details, see [3], [4].

In fact, we could extend the above result to whole space H by using dual mapping, and then obtained the following relationship between the Fredholm index of the Fredholm operators and K-group in space H:

Theorem 2.3 *Let A be an involutive algebra, (H,F) a Fredholm module over A, and for $q \in N$ let (H_q,F_q) be the Fredholm module over $M_q(A) = A \bigotimes M_q(C)$ given by*

$$H_q = H \bigotimes C^q, F_q = F \bigotimes I, \pi_q = \pi \bigotimes id$$

we extend the action of A on H to the unital action of $\tilde{A}$.

a) Let (H,F) be even, with $\frac{z}{2}$ grading γ, and let $e, v \in Proj(M_q(\tilde{A}))$. Then the operator $S_q := F_q\pi_q^+(e) + F_q\pi_q^-(v) + \kappa_1$ from H_q to itself is a Fredholm operator. An additive map φ of $K_0(A)$ to Z is determined by

$$\varphi([e]) + \varphi([v]) = Index(S_q),$$

where φ is an index map from $K_0(A)$ to Z.

b) Let (H,F) be odd and let $E = (\frac{1+F}{2})$,let $u, v \in Gl_q(\tilde{A})$. Then the operator $S_q := \pi_q(u)E_q + \pi_q(v)(I - E_q) + \kappa_2$ from H_q to itself is a Fredholm operator. An additive map of $K_1(A)$ to Z is determined by

$$\varphi([u]) + \varphi([v]) = Index(S_q),$$

where φ is an index map from $K_1(A)$to Z, κ_1, κ_2 are compact operators from H_q to itself.

Proof: First, we prove b), according to the definition of odd Fredholm module, $E_q = \frac{I+F_q}{2}$ is a project operator in space H_q, in fact, $S_q = E_q\pi_q(u)E_q + (I - E_q)\pi_q(v)(I - E_q) + \kappa'$, where κ' is a compact operator from H_q to itself. Using Theorem 2.1, the operator $E_q\pi_q(u)E_q$ from E_qH_q to itself is a Fredholm operator with the following identity holding:

$$\varphi([u]) = Index(E_q\pi_q(u)E_q).$$

Replace F_q by $-F_q$, hence, $I - E_q = \frac{I+(-F_q)}{2}$, using Theorem 2.1 again, the operator $(I - E_q)\pi_q(v)(I - E_q)$ from $(I - E_q)H_q$ to itself is a Fredholm operator with the following identity holding:

$$\varphi([v]) = Index((I - E_q)\pi_q(v)(I - E_q)).$$

We apply the following index theorem of Fredholm operator:

Let $T \in \zeta(X)$,$S \in \zeta(Y)$, then (1) $T \bigoplus S \in \zeta(X \bigoplus Y)$; (2) $Index(T \bigoplus S) = Index(T) + Index(S)$, where $\zeta(X)$,$\zeta(Y)$ and $\zeta(X \bigoplus Y)$ are included in the set of Fredholm operator in X, Y and $X \bigoplus Y$ respectively;(3)let T be a Fredholm operator from X to Y, U is a compact operator from X to Y, then $Index(T + U) = Index(T)$.

Hence, according to (1),(2), $S_q := \pi_q(u)E_q + \pi_q(v)(I - E_q) + \kappa_2$ from H_q to itself is a Fredholm operator. By using (3),

$$\begin{aligned}\varphi([u]) + \varphi([v]) &:= Index(E_q\pi_q(u)E_q + (I - E_q)\pi_q(v)(I - E_q)) \\ &= Index(\pi_q(u)E_q + \pi_q(v)(I - E_q) + \kappa) \\ &= Index(S_q).\end{aligned}$$

By using the similar method of b), we can prove a).

Remark *Theorem 2.3 extend Theorem 2.2 to the case of whole space H_q, K-group will be characterize by the Fredholm operator index. This theory will be applied to next section, in which we will construct a Fredholm module by using Plemelj-Sokhotzki of singular integral operators with the Cauchy kernel, and discuss relation between the index of singular integral operators and the character of K-group.*

3. Fredholm Module of Cauchy Integral Operators

In this section, we will construct a Fredholm module by using the following singular integral operators with the Cauchy kernel:

$$f(z) := a(z)u(z) + \frac{b(z)}{\pi}P.V. \int_{S^1} \frac{u(s)}{z - s} ds + Tu(z),$$

where S^1 is unit circle, $a, b \in C(S^1)$ and

$$P.V. \int_{S^1} \frac{u(s)}{z - s} ds := \lim_{z \to 0} \int_{|z-s| \geq \varepsilon, z \in s^1} \frac{u(s)}{z - s} ds.$$

Denote

$$(Fu)(z) := \frac{1}{\pi i} P.V. \int_{S^1} \frac{u(s)}{z-s} ds \quad (\forall z \in S^1).$$

According to [5], PP 248-252, F is a linear bounded operator from $L^2(S^1)$ to itself, and $F^2 = I$, $F = F^*$. Let

$$Pu := \frac{1}{2}(u + Fu),$$

then P is a projective operator from $L^2(S^1)$ to itself, and $\forall \varphi \in C(S^1)$, commutator $\varphi \cdot P - P\varphi\cdot$ is a compact operator from $L^2(S^1)$ to itself, where $\varphi\cdot$ is a multiplication operator from $L^2(S^1)$ to itself.

Denote $A := C(S^1)$. It is well known that A is C^*-algebra. Hence, it is an involutive algebra. There exists a natural involutive representation π of C^*-algebra by defining a multiplication operator in $L^2(S^1)$

$$\pi(a)\xi = a\xi, \forall \xi \in L^2(S^1), a \in C(S^1).$$

From the above discuss, we have the following results:

Theorem 3.1 *$(L^2(S^1), F)$ is a Fredholm odd module over $A := C(S^1)$.*

According to Theorem 2.3, we will give the index theorem of singular integral operators:

Theorem 3.2 *Let integral operator $f(z) := a(z)u(z) + \frac{b(z)}{\pi} P.V. \int_{S^1} \frac{u(s)}{z-s} ds + Tu(z)$, $a, b \in C(S^1)$,$(a-ib)(a+ib) \neq 0$, where T is a compact operator from $L^2(S^1)$ to itself. Then the operator $f(z) := a(z)u(z) + \frac{b(z)}{\pi} P.V. \int_{S^1} \frac{u(s)}{z-s} ds + Tu(z)$, from $L^2(S^1)$ to itself is a Fredholm operator. An additive map of $K_1(A)$ to Z is determined by*

$$\varphi([a+ib]) + \varphi([a-ib]) = Index(f(z)).$$

Where φ is an index map from $K_1(A)$ to Z.

Proof: Since $P = \frac{I+F}{2}$,then $f(z) = (a+ib)Pu(z) + (a-ib)(I-P)u(z) + Tu(z)$. Since $(a-ib)(a+ib) \neq 0$,$(a-ib) \in GL(A)$,$(a+ib) \in Gl(A)$, according to Theorem 2.2, $f(z) := a(z)u(z) + \frac{b(z)}{\pi} P.V. \int_{S^1} \frac{u(s)}{z-s} ds + Tu(z)$, from $L^2(S^1)$ to itself is a Fredholm operator. An additive map of $K_1(A)$ to Z is determined by

$$\varphi([a+ib]) + \varphi([a-ib]) = Index(f(z)).$$

Where φ is an index map from $K_1(A)$ to Z.

The above Theorem sets up the relationship between the Fredholm index of the singular integral operators and K-group.According to [5,P 251-252], the following corrollary is obvious.

Corrollary 3.3
Let integral operator $f(z) := a(z)u(z) + \frac{b(z)}{\pi}P.V.\int_{S^1}\frac{u(s)}{z-s}ds + Tu(z)$, $a, b \in C(S^1)$,$(a - ib)(a + ib) \neq 0$, *where* T *is a compact operator from* $L^2(S^1)$ *to itself. Then the operator* $f(z) := a(z)u(z) + \frac{b(z)}{\pi}P.V.\int_{S^1}\frac{u(s)}{z-s}ds + Tu(z)$, *from* $L^2(S^1)$ *to itself is a Fredholm operator. An additive map of* $K_1(A)$ *to* Z *is determined by*

$$\varphi([a + ib]) + \varphi([a - ib]) = \nu_{a-ib} - \nu_{a+ib}.$$

Where φ *is an index map from* $K_1(A)$ *to* Z,ν_{a-ib}, ν_{a+ib} *are the cycle numbers of* $a - ib$, $a + ib$ *winding zero, respectively.*

References

1. A.Connes, *Noncommutative Geometry,* Academic Press, 1994: 288-289.
2. M.Rrdam, F.Larsen and N.J.Lausten, *An Introduction to K-Theory for C-Algebras,* Cambridge University Press. 2000.
3. M.F.Atiyah, Global Theory of Elliptic Operator, *Proc. Internat. Conf. On Functional Analysis and Related Topics (Tokyo,1969): 21-30,* Univ. of Tokyo Press, Tokyo, 1970.
4. G.Kasparov, Topological Invariants of Elliptic Operators,1. K-homology, *Math.USSR Izv.*, **9**(1975): 751-792.
5. G. Q. Zhang, Y. Q. Lin, *Lecture on Functional Analysis,* Beijing University Publication, 1987: 243-252.
6. A. Gorokhovsky, Explicit Formulae for Characteristic Classes in Noncommutative Geometry, *Dissertation for the Degree Doctor of Philosophy in the Graduate School of the Ohio State University,* 1999.
7. T.Schick, The Trace on the K-theory of Group C^*-Algebras, *Duke Math.*, **107**(1)(2001):1-14.
8. E. Getzler, The odd Chern Character in Cyclic Homology and Spectral Flow, *Topology,* **32**(3)(1993): 489-507.
9. A.Connes and H.Moscovici, Hopf algebras, Cyclic Cohomology and the Transverse Index Theorem, *Comm. Math. Phys.* , **198**(1)(1998): 199-246.

ON Ψ-MODULUS AND Ψ-CAPACITIES EQUALITIES IN METRIC MEASURE SPACES*

WU JIONG-QI

Department of Mathematics
Zhangzhou Normal University
Fujian, 363000 P. R. China
E-mail: wujiong@263.net

Suppose X is a metric measure space. As a generalization of the p-modulus and the p-capacities in X, the Ψ-modulus and the Ψ-capacities are defined by means of the Luxemburg norm in the Orlicz space $L^{\Psi}(X)$ with a general Yang function Ψ, instead of the ordinary norm in $L^p(X)$. The relations between the Ψ-modulus and Ψ-capacities are discussed and some related equalities are established.

Keywords: Capacity, modulus, metric measure space, quasiconformal mappings, Orlicz space.

2000 Mathematics Subject Classification: 31C15, 30C62, 46E30

1. Introduction

Many mathematicians studied intensively on modulus of curve families and capacities when they investigated conformal, quasiconformal mappings or related subjects, see also, for example, [1,2,4] and the references therein. In particular, in 1998 of Heinonen and Koskela [2] established the theory of quasiconformal mappings on metric measure spaces by means of p-modulus and p-capacity; they have also posed an open problem on p-capacities, i.e., in what generality is there equality in the following inequality (where U is an open set of X, E and F are two disjoint closed subsets of U, cf. [3.3] and [3.4] below)

$$cap_p(E,F;U) \le cap_{pC}(E,F;U) \le cap_{pL}(E,F;U)? \qquad (1.1)$$

*The project is supported by National Natural Science Fund of China (10271056).

From then many authors have partly answer it in different directions, see [3, 6-8].

At the same time, in 2004 Tuominen [6] extended the concept of the p-modulus to the Ψ-modulus based on the norm of the Orlicz spaces $L^{\Psi}(X)$ with a general Yang function Ψ when he investigated the Orlicz-Soblev spaces, where X is a metric measure space.

In this paper, by introducing a new kind of Ψ-capacities for triple $(E, F; U)$, which is corresponding to Ψ-modulus and different from that in [6, p62] defined as a set function on 2^X with another norm, we study the relations between Ψ-modulus and Ψ-capacities, discuss the question on Ψ-capacities corresponding to the open question of Heinonen and Koskela and establish some related equalities. Our results can be extended to more general types of modulus and capacities.

2. Preliminaries and lemmas

2.1. *Metric measure spaces*

All metric spaces X in this paper are assumed to be rectifiably connected and all measures μ on X are assumed to be locally finite and Borel regular with dense support. A metric space is called rectifiably connected if every pair of two points in it can be joined by a rectifiable curve. We shall denote by $X = (X, d, \mu)$ such a metric measure space [2].

By a curve on a topological space Y we mean either a continuous map γ of an interval $I \subset (-\infty, \infty)$ into Y, or the image $\gamma(I)$ of such a map. We usually abuse notation by writing $\gamma = \gamma(I)$.

The space X is said to be (globally) quasiconvex if there is a constant $C > 0$ so that every pair of points x and y in X can be joined by a curve γ whose length satisfies $l(\gamma) \leq C|x - y|$, here and hereafter we use the Polish distance notation $|x - y|$ in any metric space, and refer to C as the quasiconvexity constant. Moreover, X is locally quasiconvex if every point in X has a neighborhood that is quasiconvex. More generally, X is said to be φ-convex if there is a cover of X by open sets $\{U_\alpha\}$ together with homeomorphisms $\{\varphi_\alpha : [0, \infty) \to [0, \infty)\}$ such that every pair of two points x and y in U_α can be joined by a curve in X whose length does not exceed $\varphi_\alpha(|x - y|)$. X is said to be proper if its closed balls are compact. For a discussion of metric measure spaces and the properties of curves we refer to [2,4,6].

An open set G in X is said to be relatively compact if its closure $\overline{G}$ in X is compact.

Lemma 1([7], P759) *Suppose (i) $A \subset X$ is an open set or the closure of some open set in X; (ii) $\{\alpha_n\}$ is a sequence of locally rectifiable curves in A with $\alpha_n : [0, b_n) \to \overline{G}$ (the arc length parametrization of $\alpha_n, b_n = l(\alpha_n), b_n$ may be finite or infinite), and $\{\alpha_n\}$ converges to a locally rectifiable curve $\alpha : [0, b) \to A$ such that $b = l(\alpha)(b$ may be finite or infinite); (iii) $g : A \to [0, \infty]$ is a lower semi-continuous function and $g_n := \min\{g, n\}$ for all $n \in \mathbf{N}$ (the set of all natural numbers). Then we have*

$$\int_\alpha g ds \leq \liminf_{n\to\infty} \int_{\alpha_n} g_n ds.$$

2.2. *Yang function and Orlicz spaces*

Recall a function $\Psi : [0, \infty) \to [0, \infty]$ be a Yang function if

$$\Psi(s) = \int_0^s \xi(t)dt, \tag{2.1}$$

where $\xi : [0, \infty) \to [0, \infty], x(0) = 0$, is an increasing, left continuous function which is neither identically zero nor identically infinite on $(0, \infty)$. For a general Yang function Ψ the generalized inverse, $\Psi^{-1} : [0, \infty] \to [0, \infty]$ is defined by the formula

$$\Psi^{-1}(t) = \inf\{s : \Psi(s) > t\}, \tag{2.2}$$

where $\inf(\emptyset) = \infty$. The function is a right continuous, increasing substitute for the inverse function and satisfies the inequality

$$\Psi(\Psi^{-1}(t)) \leq t \leq \Psi^{-1}(\Psi(t)) \tag{2.3}$$

for all $t \geq 0$. It is clear that $\Psi^{-1}(t_0) > 0$ for $t_0 > 0$ by the properties of the function ξ in (1.1).

Let Ψ be a Yang function and $\Omega \subset X$ an Borel set. It is known [6] that the set

$$L^\Psi(\Omega) = \{u : \Omega \to [0, \infty] : u \text{ is measurable,}$$

$$\textstyle\int_\Omega \Psi(k|u|)d\mu < \infty \text{ for some } k > 0\}$$

is called Orlicz space, which is a vector space. For a Borel measurable function $v : \Omega \to [0, \infty]$, set

$$F(v) = \inf\{k > 0 : \int_\Omega \Psi(\frac{|v|}{k})d\mu \leq 1\}. \tag{2.4}$$

Lemma 2 ([6],p14) *The Orlicz space $L^{\Psi}(\Omega)$ equipped with Luxemburg norm*

$$||u||_{\Psi,\Omega} := F(u)$$

is a Banach space. (If Ω is the whole space X, we denote $||u||_{\Psi}$ instead of $||u||_{\Psi,X}$.)

The following lemma seems simple but is important for the proof of our main results.

Lemma 3 *Suppose each ball in X has a finite measure. Then there exists a Borel function $f : X \to (0,\infty]$ such that $f \in L^{\Psi}(X)$.*

Proof If the diameter of X, denoted by $diamX$, is finite, then $\mu(X) < \infty$ by the assumption, hence the function $f(x) \equiv c$ is the desired, where $c > 0$ is a constant with $\Psi(c) < \infty$. If $diamX = \infty$, we set $S_n = \{y \in X : n-1 \leq |y - x_0| < n\}, n \in \mathbf{N}$ with some fixed point x_0 in X and define a function $g : X \to (0,\infty]$ as follows

$$g(x) = \frac{1}{2^n(\mu(S_n)+1)} \text{ when } x \in S_n.$$

Set $f = \Psi^{-1}(g)$, where Ψ^{-1} is the generalized inverse of Ψ defined in (1.2), then f is obviously a Borel function and $f > 0$ since $g > 0$. We can easily verify that $f \in L^{\Psi}(X)$. Indeed, by (2.3)

$$\begin{aligned}\textstyle\int_x \Psi(f)d\mu = \int_X \Psi(\Psi^{-1}(g))d\mu \leq \int_X g d\mu = \sum_{n=1}^{\infty} \int_{S_n} g d\mu \\ = \textstyle\sum_{n=1}^{\infty} \frac{1}{2^n(\mu(S_n+1))}\mu(S_n) \leq 1.\end{aligned}$$

3. The concepts of Ψ-modulus and Ψ-capacities and Theorem 1

Hereafter we fix a Yang function $\Psi : [0,\infty) \to [0,\infty]$.

Definition 1[6] *Let Γ be a family of curves in X. Denote by $\Phi(\Gamma)$ the set of all Borel functions $\rho : X \to [0,\infty]$ satisfying $\int_{\gamma} \rho ds \geq 1$ for all locally rectifiable $\gamma \in \Gamma$ (such function ϱ is said to be Γ-admissible). The Ψ-modulus of Γ is defined as*

$$mod_{\Psi}\Gamma = \inf\{||\rho||_{\Psi} : \rho \in \Phi(\Gamma)\}, \tag{3.1}$$

Lemma 4 ([6],p21-22) *The Ψ-modulus has the following properties:(1)mod$_\Psi$ ($\emptyset$) $= 0$; (2) That $\Gamma_1 \subset \Gamma_2$ implies that mod$_\Psi(\Gamma_1) \leq$ mod $_\Psi(\Gamma_2)$; (3) mod $_\Psi(\bigcup_{i=1}^{\infty} \Gamma_i) \leq \sum_{i=1}^{\infty}$ mod $_\Psi(\Gamma_i)$; (4) If Γ and Γ_0 satisfy that each curve $\gamma \in \Gamma$ has a subcurve $\gamma_0 \in \Gamma_0$, then mod$_\Psi(\Gamma) \leq$ mod$_\Psi(\Gamma_0)$.*

In the following we always suppose that X is non-compact because the corresponding conclusions for a compact space are easier to obtain or can be reduced from our correspondent results. Suppose X^* is the Alexandroff compactificaton of X, i.e. $X^* = X \bigcup \{\infty\}, \infty$ is not in X such that any neighborhood of ∞ is of the form $(X \backslash K) \bigcup \{\infty\}$, where K is a compact subset of X. By extending μ to X^* such that $\mu(\{\infty\}) = 0$, we get a new space $X^* = (X^*, \mu)$. It is easy to see that for any point x in X there is a locally rectifiable curve $\gamma : [0, b) \to X$ such that $\gamma(0) = x$ and $\gamma \bigcap (X \backslash K) \neq \emptyset$ for all compact subsets K of X. Then such a γ is denoted by $\gamma_{0\infty}$ and called a curve joining x and ∞. It is easy to see, if X is proper, then each $\gamma_{0\infty}$ is not rectifiable since X is non-compact. A curve in X^* is not locally rectifiable if it contains ∞, hence it may be neglected in our topic. Thus, only the curves contained in X are considered in the following.

Suppose that U is an open subset of X, that Y stands either for U or for X^*. Suppose E and F are disjoint closed subsets of Y (When $Y = X^*$, we suppose also E is a compact subset of X and set $F^* = F \bigcup \{\infty\}$. Then we have $E \bigcap F^* = \emptyset$). By γ_{xy} we denote a rectifiable curve joining two points x and y in X.

Definition 2[4]. *Suppose u is a real-valued function on Y. We say a Borel function $\rho : Y \to [0, \infty]$ is an upper gradient of u in Y if*

$$|u(x) - u(y)| \leq \int_{\gamma_{xy}} \rho ds \tag{3.2}$$

whenever γ_{xy} is contained in Y.

Let $C(S)(LC(S))$ be the family of all continuous (or locally Lipschitz, respectively) real-valued functions defined on a set S. We denote

$$A(E, F; Y) = \{u : Y \to (-\infty, \infty) : u|_E \geq 1, u|_F \leq 0\};$$

$$A_C(E, F; Y) := C(Y) \bigcap A(E, F; Y); A_L(E, F; Y) := LC(Y) \bigcap A(E, F; Y),$$

and

$$A_{C^*}(E, F^*; X^*) = C_0(X^*) \bigcap A_C(E, F;^*),$$

$$A_{L^*}(E, F^*; X^*) := C_0(X^*) \bigcap A_L(E, F; X^*),$$

where $C_0(X^*) := \{u \in C(X^*) : Supp\ u \subset X\}$.

Definition 3 *The triple $(E, F; Y)$ is called a condenser and its Ψ-capacity is defined as*

$$cap_\Psi(E, F; Y) := \inf ||g||_{\Psi, Y}, \tag{3.3}$$

where the infimum is taken over all upper gradients g of all functions in $A(E, F; Y)$; we use the notation $cap_{\Psi C}(E, F; Y)$, $cap_{\Psi L}(E, F; Y)$, $cap_{\Psi C^*}(E, F^*; X^*)$ and $cap_{\Psi L^*}(E, F^*; X^*)$ for the quantity in (2.3) if the infimum is taken over all upper gradients of all functions in $A_C(E, F; Y), A_L(E, F; Y), A_{C^*}(E, F^*; X^*)$ and $A_{L^*}(E, F^*; X^*)$, respectively.

Corresponding to (1.1), we trivially have by Definition 3 that

$$cap_\Psi(E, F; Y) \leq cap_{\Psi C}(E, F; Y) \leq cap_{\Psi L}(E, F; Y); \tag{3.4}$$

$$\begin{aligned} cap_{\Psi C}(E, F; X) &\leq cap_{\Psi C^*}(E, F^*; X^*) \leq cap_{\Psi L}(E, F; X) \\ &\leq cap_{\Psi L^*}(E, F^*; X^*). \end{aligned} \tag{3.5}$$

Just like in [2], the triple $(E, F; Y)$ denotes also the family of all curves in $Y \bigcap X$ joining E and F, and the triple $(E, F^*; X^*)$ denotes the family of all curves in X joining E and F^*.

The following theorem is an analogue of Prop. 2.17 of [2].

Theorem 1 For any open subset U of X we always have

$$cap_\Psi(E, F; U) = mod_\Psi(E, F; U).$$

The proof is similar to that in [2] and hence it is omitted.

4. Main theorems and their proofs

Suppose that E is a compact subset of X, F is a closed subset of X such that F and E are disjoint. Let Ω be the closure of $X \backslash (E \bigcup F)$ in X, i.e. $\Omega \subset X$. For each $x \in X$, denote $\Gamma_{xF} := \{\gamma_{xy} : y \in F\}$. Set $\Gamma_X := \{\gamma \in (E, F; X) : \gamma \textit{ is rectifiable}\}$, $\Gamma_1 := \{\gamma \in \Gamma_X : \gamma \subset \Omega\}$, $\Gamma_{\infty X} := \{\gamma_{x\infty} \subset X : x \in E\}$, $\Gamma_\infty := \{\gamma_{x\infty} \in \Gamma_{\infty X} : \gamma_{x\infty} \subset \Omega \backslash F\}$, and $\Gamma := \Gamma_1 \bigcup \Gamma_\infty$. Then $\Gamma_X \subset (E, F; X)$, and each curve $\gamma \in \Gamma_X$ has a subcurve $\gamma_0 \in \gamma_1$. Thus by Lemma 4, we have $mod_\Psi(E, F, X) = mod_\Psi(\Gamma_X) =$

$mod_\Psi(\Gamma_1) \le mod_\Psi(\Gamma)$ and
$mod_\Psi(\Gamma) = mod_\Psi(\Gamma_1 \bigcup \Gamma_{\infty X}) = mod_\Psi((E,F,X)\bigcup \Gamma_\infty) = mod_\Psi(E,F^*;X^*)$. Hence

$$mod_\Psi(E,F,X) \le mod_\Psi(\Gamma) = mod_\Psi((E,F,X)\bigcup \Gamma_\infty) = mod_\Psi(E,F^*;X^*). \tag{4.1}$$

To describe the sufficient conditions of our theorems, we introduce the following concept.

Definition 4 *The triple $(E,F;X)$ is said to have Property Q, if X has an exhaustive sequence $\{W_n\}$ of relatively compact, open sets with $\overline{W_n} \subset W_{n+1}, n \ge 1, X = \bigcup_{n=1}^\infty W_n$ such that for each $x \in X\backslash\overline{W_n}$ there exists $\gamma \in \Gamma_{xF}\bigcup\Gamma_{x\infty}$ satisfying $\gamma \subset X\backslash\overline{W_n}$.*

Remark If X has an exhaustive sequence $\{W_n\}$ of relatively compact, open sets with $\overline{W_n} \subset W_{n+1}, n \ge 1, X = \bigcup_{n=1}^\infty W_n$ such that each component of $X\backslash\overline{W_n}$ is unbounded and rectifiably connected for every $n \ge 1$, then every triple $(E,F;X)$ has property Q. For example, If X is proper and is globally quasiconvex, then X has property Q.

Theorem 2 *Suppose X is φ-convex and proper, E and F are two disjoint closed sets in X and E is also compact. If the triple $(E,F;X)$ has Property Q, then we have the equality*

$$cap_{\Psi C^*}(E,F^*;X^*) = mod_\Psi(E,F^*;X^*), \tag{4.2}$$

where $F^* = F\bigcup\{\infty\}$. If, moreover, X is locally quasiconvex, (4.2) holds with $cap_{\Psi L^*}$ on the left-side.

Proof Suppose $u \in A_{C^*}(E,F^*;X^*)$. Then $u|_E \ge 1, u|_F \le 0$ and there exists a compact set K such that $E \subset K \subset X$ and $u|_{X^*\backslash K} = 0$. If ρ is an upper gradient of u in X^*, then it is easy to see the inequality $\int_\gamma \rho ds \ge 1$ holds for all $\gamma \in \Gamma = \Gamma_1 \bigcup \Gamma_\infty$, i.e., ρ is Γ-admissible. Then, by definition 1 we have $mod_\Psi(\Gamma) \le ||\rho||_\Psi$. Then by definition 3 (Remark: We have $||g||_{\Psi,Y} = ||g||_{\Psi,X} = ||g||_\Psi$ when $Y = X^*$ in (3.3)) we obtain

$$mod_\Psi(\Gamma) \le cap_{\Psi C^*}(E,F^*,X^*). \tag{4.3}$$

To verify the opposite inequality, one may only consider the case $mod_\Psi(\Gamma) < \infty$. Suppose $\rho \in \Phi(\Gamma)\bigcap L^\Psi(X)$. We may assume that $\rho > 0$ in X. In fact, by Lemma 3 there exists a Borel function $g : X \to (0,\infty]$ with $g \in L^\Psi(X)$. Hence for any $m \in \mathbf{N}$, we have $m^{-1}g + \rho \in \Phi(\Gamma)\bigcap L^\Psi(X)$

and $m^{-1}g+\rho>0$ in X. Since $||\rho||_\Psi \le ||m^{-1}g+\rho||_\Psi \le m^{-1}||g||_\Psi + ||\rho||_\Psi$, one deduces that $||m^{-1}g+\rho||_\Psi$ converges decreasingly to $||\rho||_\Psi$ as $m\to\infty$. Therefore we may use $m^{-1}g+\rho$ instead of ρ to calculate $mod_\Psi(\Gamma)$.

By the Vitali-Caratheodory theorem [5], we may further assume that $\rho>0$ is lower semi-continuous in X because X is proper. Suppose W_n is the exhaustive sequence given in definition 4 with $E\subset W_1, \Omega_n := \Omega\bigcap\overline{W_n}$. Then for every $n\in\mathbf{N}$, there is a constant $\eta_n\in(0,1)$ such that $\rho|_{\Omega_n}\ge\eta_n$ since Ω_n is compact.

Fix an integer $n\in\mathbf{N}$ and consider the function ρ_n such that $\rho_n(x)=\min\{\rho(x),n\}$ when $x\in\Omega_n$ and $\rho_n(x)=0$ when $x\in X\backslash\Omega_n$. Define

$$u_n(x)=\inf\int_{\gamma_x}\rho_n ds, x\in X; u_n(\infty)=0,$$

where the infimum is taken over all $\gamma_x\in\Gamma_{xF}\bigcup\Gamma_{x\infty}$. One can deduce [2,7] that u_n is continuous in X^* and ρ_n is an upper gradient of u_n. Since the triple $(E,F;X)$ has property Q, for each $x\in(X\backslash\overline{W_n})$ there exists $\gamma\in\Gamma_{xF}\bigcup\Gamma_{x\infty}$ such that $\gamma\subset X\backslash\overline{W_n}\subset X\backslash\Omega_n$. Hence we have $u_n(x)=0$, which implies $u_n\in C_0(X^*)$. If, moreover, X is locally quasiconvex, then u_n is locally Lipschitz.

Set $m_n := \inf\{u_n(x) : x\in E\}$. Since the two compact sets E and $(F\bigcap\overline{W_n})\bigcup\partial W_n$ are disjoint and $\rho|_{\Omega_n}\ge\eta_n>0$, we have $m_n>0$. If we set $v_n := u_n/m_n$, then $v_n\in A_{C^*}(E,F^*,X^*)$ (or $v_n\in A_{L^*}(E,F^*,X^*)$ when X is locally quasiconvex), and ρ_n/m_n is an upper gradient of v_n. Hence

$$cap_{\Psi C^*}(E,F^*,X^*)\le||\rho_n/m_n||_\Psi=(m_n)^{-1}||\rho_n||_\Psi\le(m_n)^{-1}||\rho||_\Psi$$

(or $cap_{\Psi L^*}(E,F^*,X^*)\le||\rho_n/m_n||_\Psi\le(m_n)^{-1}||\rho||_\Psi$ when X is locally quasiconvex).

By lemma 1 we can prove (cf. [7],760) that $\limsup_{n\to\infty}m_n\ge1$, and hence

$$cap_{\Psi C^*}(E,F^*;X^*)\le||\rho||_\Psi \text{ (or } cap_{\Psi L^*}(E,F^*;X^*)\le||\rho||_\Psi),$$

which implies that $cap_{\Psi C^*}(E,F^*;X^*)\le mod_\Psi(\Gamma)$ (or $cap_{\Psi L^*}(E,F^*;X^*)\le mod_\Psi(\Gamma)$) since $\rho\in\Phi(\Gamma)\bigcap L^\Psi(X)$ is arbitrary. Then the proof is complete.

Theorem 3 *Suppose X is φ-convex and proper, E and F are two disjoint closed sets in X and E is also compact. If $(E,F;X)$ has Property Q and $mod_\Psi(\Gamma_\infty)=0$, then*

$$cap_{\Psi C}(E,F;X)=mod_\Psi(E,F;X), \tag{4.4}$$

where $\Gamma_\infty := \{\Gamma_{x\infty} \subset \Omega\backslash F : x \in E\}$. *If, moreover,* X *is locally quasiconvex, then (4.4) holds with* $cap_{\Psi L}$ *on the left-side.*

Proof By (3.5), (4.1) and (4.2),

$$\begin{aligned} cap_{\Psi C}(E,F,X) &\le cap_{\Psi C^*}(E,F^*,X^*) = mod_\Psi(E,F^*,X^*) \\ &= mod_\Psi((E,F;X)\bigcup\Gamma_\infty). \end{aligned} \tag{4.5}$$

By Lemma 4 we have

$$mod_\Psi(G) \le mod_\Psi(E,F;X) + mod_\Psi(\Gamma_\infty).$$

From this inequality and (4.5) it follows that

$$cap_{\Psi C}(E,F;X) \le mod_\Psi(E,F;X) + mod_\Psi(\Gamma_\infty).$$

By Theorem 1 and (3.4) we see that (4.4) holds when $mod_\Psi(\Gamma_\infty) = 0$. If, moreover, X is locally quasiconvex, a similar procedure deduces that (4.4) holds with $cap_{\Psi L}$ on the left-side.

References

1. L. Ahlfors & A. Beurling, Conformal invariants and function-theoretic null-sets, *Acta Math.,* **83**(1950):101-129.
2. J. Heinonen & P. Koskela, Quasiconformal mappings in metric space with controlled geometry, *Acta Math.,* **181**(1998): 1-61.
3. S. Kallunki& N. Shanmugalingam, Modulus and continuous capacity, *Ann. Acad. Sci. Fenn. Math.,* **26** (2002):455-464.
4. Kinnunen & O. Martio, Nonlinear potential theory on metric spaces, *Illinois J. Math.,* **46**(2002): 857-883.
5. W. Rudin, *Real and complex analysis,* New York: McGraw-Hill, 1974
6. H. Tuominen, Orlicz-Sobolev spaces on metric measure spaces, *Ann. Acad. Sci. Fenn. Dissertations* **135**(2004).
7. J. Q. Wu , Modulus and capacity equalities in metric measure spaces, *J. Xiamen univ.* **43**(6) (2004):757-761.
8. Z. Wu & A. Fang, The modulus and capacity on metric measure spaces, *Chinese Math. Ann.,* **22**A(1) (2001): 65-70.

A CRITERION OF BLOCH FUNCTIONS AND LITTLE BLOCH FUNCTIONS*

PENGCHENG WU
School of Mathematics and Computer Science
Guizhou Normal University
Guiyang, 550001 P.R. China
E-mail: wupc@mail.gznu.edu.cn

In this note, we give a characterization of Bloch function and little Bloch function, which extends results of Aulaskari-Lappan, Minda, and Aulaskari-Wulan.

Keywords: Bloch function, little Bloch function.

2000 Mathematics Subject Classification: Primary 30D50

An analytic function f in the unit disk $\triangle$ is said to be a Bloch function if

$$\sup_{z\in\triangle}(1-|z|^2)|f'(z)| < \infty.$$

An analytic function f in the unit disk $\triangle$ is called a little Bloch function if

$$\lim_{|z|\to 1}(1-|z|^2)|f'(z)| = 0.$$

Aulaskari and Lappan [1] and Minda [3] gave an alternative characterization of Bloch functions. They proved the following

Theorem ALM. *A function f analytic in the disk $\triangle$ is not a Bloch function if and if there exist a sequence $z_n \subset \triangle$with $|z_n| \to 1$, and a sequence $\{\rho\}$ of positive numbers satisfying $\frac{\rho_n}{(1-|z_n|)} \to 0$, such that the se-*

*Research supported by the Fred and Barbara Kort Sino-Iarael Post Doctoral Fellowship Program at Bar-Ilan University, the German-Israel Foundation for Scientific Research and Development, G.I.F. Grant No. G-643-117.6/1999.

quence $\{f(z_n+\rho_n\xi)-f(z_n)\}$ *converges locally uniformly to a nonconstant analytic function in* C.

For little Bloch functions, Aulaskari and Wulan [2] proved the following

Theorem AW. *A function f analytic on the unit disk* $\triangle$ *is not a little Bloch function if and only if there exist a constant* $R>0$, *a sequence* $z_n \subset \triangle$ *with* $|z_n|\to 1$, *and a sequence* ρ_n *of positive numbers satisfying* $\frac{\rho_n}{(1-|z_n|)}<\frac{1}{2R}$, *such that the sequence* $\{f(z_n+\rho_n\xi)-f(z_n)\}$ *converges locally uniformly to a nonconstant analytic function in* $|\xi|<R$.

In this paper, we investigate the possibility of introducing a sliding scale involving a parameter α into these theorems in analogy to Pang's generalization[4] of Zalcman's Lemma[5].our fist result is the following.

Theorem 1. *Let f be an analytic function in the disk* $\triangle$, *and* α *give real number with* $0\le\alpha<1$. *Then f is not a Bloch function if and only if there exist a sequence* $z_n\subset\triangle$ *with* $|z_n|\to 1$, *and a sequence* $\{\rho_n\}$ *of positive numbers satisfyingand a sequence* $\{\rho_n\}$ *of positive numbers satisfying* $\frac{\rho_n^{1-\alpha}}{(1-|z_n|)}\to 0$, *such that the sequence* $\{\frac{f(z_n+\rho_n\xi)-f(z_n)}{\rho_n^\alpha}\}$ *converges locally uniformly to* $a\xi$ *in* C, *where a is a constant with* $|a|=1$.

Proof. Assume that f is not a Bloch function. Then there exists a sequence $\{z_n^*\}\subset\triangle$ with $|z_n^*|\to 1$ such that

$$(1-|z^a st_n|^2)|f'(z_n^*)|\to\infty,\ \ as\ n\to\infty. \tag{1}$$

Without loss of generality, we can assume $|z_n^*|>\frac{1}{2}$ for all $n\in N$. Let

$$r_n=\frac{2|z_n^*|}{1+|z_n^*|},\ n\in N.$$

Then $|z_n^*|<r_n<1$ and $r_n\to 1$ as $n\to\infty$. Choose$z_n\subset\triangle$ such that

$$\begin{aligned} M_n &= \sup_{|z|\le r_n}(1-\tfrac{|z|^2}{r_n^2})^{\frac{1}{1-\alpha}}(|f'(z)|)^{\frac{1}{1-\alpha}} \\ &= (1-\tfrac{|z_n|^2}{r_n^2})^{\frac{1}{1-\alpha}}(|f'(z_n)|)^{\frac{1}{1-\alpha}} \end{aligned} \tag{2}$$

Since $|z_n^*|<r_n$, Thus

$$M_n\ge(\tfrac{1}{4})^{\frac{1}{1-\alpha}}(1-|z_n^*|^2)^{\frac{1}{1-\alpha}}(|f'(z_n^*)|)^{\frac{1}{1-\alpha}}\to\infty. \tag{3}$$

Now set

$$\rho_n=\frac{1}{M_n}\left(1-\frac{|z_n|^2}{r_n^2}\right)^{\frac{1}{1-\alpha}}=\left(\frac{1}{|f'(z_n)|}\right)^{\frac{1}{1-\alpha}}. \tag{4}$$

Then, by (2) and (3), we have

$$\frac{\rho_n^{1-\alpha}}{r_n - |z_n|} \to 0. \tag{5}$$

In particular,

$$\frac{\rho_n^{1-\alpha}}{1 - |z_n|} \to 0. \tag{6}$$

$$\frac{\rho_n}{r_n - |z_n|} \to 0. \tag{7}$$

and

$$\frac{\rho_n}{1 - |z_n|} \to 0. \tag{8}$$

Therefore, the functions

$$g_n(\xi) = \frac{f(z_n) + \rho_n\xi - f(z_n)}{\rho_n^\alpha}$$

are defined on the disk $|\xi| < \frac{r_n - |z_n|}{\rho_n} := R_n \to \infty$. Hence, according to (2) and (5), for each fixed $R > 0$, when $|\xi| \leq R \leq R_n$, we have

$$\begin{aligned} |g_n'(\xi)| &= \rho_n^{1-\alpha}|f'(z_n + \rho_n\xi)| \\ &\leq \tfrac{r_n + |z_n|}{r_n + |z_n| + \rho_n R} \cdot \tfrac{r_n - |z_n|}{r_n - |z_n| - \rho_n R}. \end{aligned} \tag{9}$$

Since the last term of the inequality (9) tend to 1, g_n' is a normal family on C. Taking s subsequence and renumbering, we may assume that g_n' converges locally uniformly on compacta to an entire function G, and $|G| \leq 1$. By Liouville's Theorem, G is a constant; and since $|g_n'(0)| = 1$, we have $G \equiv a$, where $|a| = 1$. It follows from

$$g_n(\xi) - g_n(0) = \int_0^\xi g_n'(\xi)d\xi \to \ a\xi, \quad as\ n \to \infty.$$

and $g_n(0) = 0$ that $\{g_n(\xi)\}$ tends to $a\xi$ locally uniformly on any compact set in C.

Conversely, suppose that there exist z_n, ρ_n with $|z_n| < 1$, $|z_n| \to 1$, $\frac{\rho^{1-\alpha}}{(1-|z_n|)} \to 0$ such that $g_n(\xi) = \{f(z_n + \rho\xi) - f(z_n)\}/\rho_n^\alpha \to g(\xi)$ locally uniformly in C. Evidently,

$$|g_n'(\xi)| \leq \frac{\rho_n^{1-\alpha}}{(1 - |z_n|)} \cdot \frac{1}{(1 - \frac{\rho_n}{1-z_n}|\xi|)} \cdot (1 - |z_n + \rho_n\xi|^2)|z_n + \rho_n\xi|.$$

If f is a Bloch function, the product of the last two terms on the right is bounded; the second term on the right tends to 1 by (8) and hence is also bounded; and the first term on the right tends to by assumption as n tends to infinity. So $g'(\xi) = 0$ for all $\xi \in C$, and therefore g is constant. This completes the proof of Theorem 1.

Example 1. Let $f(z) = \frac{1}{1-z}$; then f is analytic in the unit disk, and it is evident that f is not a Bloch function. Fix $0 \leq \alpha < 1$, and set $z_n = 1 - \frac{1}{n}, \rho_n = n^{-\frac{2}{1-\alpha}}$, so that $n\rho_n^{1-\alpha} \to 0, n^2\rho_n^{1-\alpha} = 1$, and $n\rho_n \to 0$. then

$$\frac{f(z_n + \rho z) - f(z_n)}{\rho_n^\alpha} = \frac{z}{1 - n\rho_n z} \to z,$$

locally uniformly in C.

Remark 1. *Theorem 1 does not hold for $\alpha > 1$. Indeed, let $\{z_n\}$ be an arbitrary sequence in the unit disk with $|z_n| \to 1$, and ρ_n an arbitrary sequence of positive numbers with $\rho_n \to$. Then with f as in the previous example, we have*

$$\frac{f(z_n + \rho z) - f(z_n)}{\rho_n^\alpha} = \frac{\rho_n^{1-\alpha} z}{(1 - z_n - \rho_n z)(1 - z_n)} \to \infty, \ \mathit{for}\ z \neq 0.$$

Thus Theorem 1 does not hold if $\alpha > 1$.

For the little Bloch functions, we have

Theorem 2. *Let f be an analytic function in the unit disk $\triangle$, and α a given real number with $0 < \alpha < 1$. Then f is not a little Bloch function if and only if there exist a sequence $z_n \subset \triangle$ with $|z_n| \to 1$, a constant $K > 0$, and a sequence $\{\rho_n\}$ of pisitive numbers satisfying $\frac{\rho_n^{1-\alpha}}{1-|z_n|} < K$,* such that the sequence $\{\frac{f(z_n+\rho_n\xi)-f(z_n)}{\rho_n^\alpha}\}$ converges locally uniformly to $\alpha\xi$ in C, where α id a constant with $|\alpha| = 1$.

Proof. The proof follow the pattern in [2], and the proof of Theorem 1.

Example 2. Let $f(z) = \log \frac{1}{1-z}$; since $(1 - |z|^2)|f'(z)| = \frac{1-|z|}{|1-z|}(1 + |z|)$, it is evident that f is a Bloch function, but not a little Bloch function. Fix $0 < a < 1$, and set $z_n = 1 - \frac{1}{n}, \rho_n = n^{\frac{-1}{1-\alpha}}$, so that $n\rho_n \to 0$ as $n \to \infty$. Then

$$\frac{f(z_n + \rho_n z) - f(z_n)}{\rho_n^\alpha} = n^{\frac{\alpha}{1-\alpha}} \log\left(\frac{1}{1 - \frac{z}{n^{\frac{\alpha}{(1-\alpha)}}}}\right)$$

tends to z as $n \to \infty$.

Remark 2. *Theorem 2 does not hold if* $\alpha = 0$. *Let* $\{z_n\}$ *be an arbitrary sequence in the unit disk with* $|z_n| \to 1$, *and* $\{\rho_n\}$ *an arbitrary sequence of positive numbers with* $\rho_n \to 0$, *and satisfying* $\frac{\rho_n}{1-|z_n|} \to 0$. *Then with* f *as above, we have*

$$f(z_n + \rho_n z) - f(z_n) = \log \frac{1}{1 - \frac{\rho_n z}{1-z_n}} \to 0$$

as $n \to \infty$. *Thus Theorem 2 does not hold when* $\alpha = 0$.

Indeed, when $\alpha = 0$, (17) above show that $\frac{\rho_n}{(1-|z_n|)}$ need not tend to 0 , so the functions g_n may not be defined on arbitrarily large disks about the origin. In this case, Theorem AW insures that the sequence $\{g_n\}$ converges locally uniformly on some disk of fixed radius about the origin.

Acknowledgment: This work was done during the author's stay at Department of Mathematics and Statistics of Bar-Ilan University in a postdoctoral fellowship. Thank Professor L.Zalcman for his helpful suggestions and discussions. My thanks are also due to the staff of Department of Mathematics and Statistics for creating a friendly environment for my research. Thanks for referee's comments and pointing out the misprints in original manuscript.

References

1. R.Aulaskari and P.Lappan, A criterion for a rotation automorphic function to be normal, *Bull. Inst. Math. Acad. Sinica* **15**(1987):73-79.
2. R.Aulaskari and H.wulan, A version of the Lohwater-Pommerenke theorem for strongly normal functions, *Comp. Meth. Funct. Theory,* **2**(2001): 99-105.
3. D. Minda , Bloch and normal functions on general planar regions, Holomorphic functions and moduli, Vol, I (Berkeley, CA,1986), *Math. Sci. Res. Publ. 10,* Springer, New York-Berlin, 1988: 101-110.
4. X.C.Pang, Bloch's principle and normal criterion, *Sci. China Ser. A* **32**(1989):781-491.
5. L.Zalcman, A heruistic principle in complex function theory, *Amer. Math. Monthly,* **82**(1975):813-817.

SOME RESULTS OF UNIQUENESS FOR ALGEBROID FUNCTIONS*

ZU-XING XUAN
LMIB and Department of Mathematics
Beihang University
Beijing 100083, P. R. China
E-mail: xuanzuxing@ss.buaa.edu.cn

ZONGSHENG GAO
LMIB and Department of Mathematics
Beihang University
Beijing 100083, P. R. China
E-mail: zshgao@buaa.edu.cn

In this paper, we investigate the uniqueness theory of algebroid functions and extend some uniqueness theorems of meromorphic functions (H.X. Yi [**14**], R. Nevanlinna [**15,16**]) to algebroid functions.

Keywords: Algebroid function; uniqueness theorem; multiple values; deficient values; characteristic function.

2000 Mathematics Subject Classification: 32C20, 30D45

The value distribution theory of meromorphic functions due to R. Nevanlinna (see [**1**] for standard reference) was extended to the corresponding theory of algebroid functions by H. Selberg [**2,3**], E. Ullrich [**4**] and G. Valiron [**5**] around 1930. For the uniqueness theory of algebroid functions, G. Valiron [**5**], N.Baganas [**6**], Y.Z. He [**7**] and others have ever done a lot of works. In this paper, after the $4k+1$ values theorem [**7**] discussed, we will investigate the uniqueness theory of algebroid functions dealing with

*The project is supported by NSFC (Grant No. 10271011).

multiple values and deficient values. Moreover, just as the results about meromorphic functions which were deduced by H.X. Yi [**14**] and R. Nevanlinna [**15,16**], we will get some uniqueness theorems of algebroid functions.

Let $A_k(z), ..., A_0(z)$ be analytic functions with no common zeros in the complex plane, then the following equivalent equation

$$A_k(z)W^k + A_{k-1}(z)W^{k-1} + \cdots + A_0(z) = 0 \tag{1}$$

defines a k-valued algebroid function $W(z)$ [**12**].

Let $W(z)$ be a k-valued algebroid function and $a \in \overline{C}$ be any complex number. $\overline{E}_{t)}(W = a)$ denotes the set of zeros of $W(z) - a$ whose multiplicity is equal to or less than t. $\overline{n}_{t)}(r, W = a)$ denotes the number of distinct zeros of $W(z) - a$ in $|z| \leq r$ whose multiplicity is equal to or less than t(ignoring multiplicity but including zeros which are branch points). Similarly, we define the functions $\overline{n}_{(t+1}(r, W = a)$, $\overline{N}_{t)}(r, W = a)$ and $\overline{N}_{(t+1}(r, W = a)$.

Definition 1 *Assume that $W(z)$ is a k-valued algebroid function and $M(z)$ is a s-valued algebroid function. We call $b \in \overline{C}$ is a t-common value of $W(z)$ and $M(z)$ if $\overline{E}_{t)}(W = b) = \overline{E}_{t)}(M = b)$. We call the ∞-common value of $W(z)$ and $M(z)$ the common value.*

Definition 2[9] *Define the operations of $W(z) = \{(w_j(z), b)\}_{j=1}^{k}$ and $M(z) = \{(m_j(z), b)\}_{j=1}^{s}$ as follows*

$-W(z) := \{(-w_j(z), b)\}_{j=1}^{k}$,

$(W \pm M)(z) := \{(w \pm m)_j(z), b\}_{j=1}^{ks} = \{(w_j(z) \pm m_t(z), b); j = 1, 2, \ldots, k; t = 1, 2, \ldots, s\}$,

$W^{-1}(z) := \{(\frac{1}{w_j(z)}, b)\}_{j=1}^{k}$,

$(W \cdot M)(z) := \{((w \cdot m)_j(z), b)\}_{j=1}^{ks} = \{(w_j(z)m_t(z), b); j = 1, 2, \ldots, k; t = 1, 2, \ldots, s\}$.

Lemma 1 ([**12**, Theorem 2.22]) *Let $W(z)$ be a k-valued algebroid function and $\{a_j\}_{j=1}^{q} \subset \overline{C}$ be q distinct complex numbers (finite or infinite). Then*

$$(q - 2k)T(r, W) < \sum_{j=1}^{p} \overline{N}(r, W = a_j) + S(r, W),$$

where $S(r, W) = O\{\log(rT(r, W))\}$, outside a possible exceptional set of finite linear measure.

Lemma 2[9] *$(W \pm M)(z)$ and $(W \cdot M)(z)$ defined in Definition 2 are both ks-valued algebroid functions. $-W(z)$ and $W^{-1}(z)$ are still k-valued algebroid functions.*

Lemma 3[10] *Let $W(z)$ and $M(z)$ be k-valued and s-valued algebroid functions respectively. If 0 is not the pole of $W(z)$ and $M(z)$, then*

$$T(r, W \pm M) \leq T(r, W) + T(r, M) + \log 2,$$

$$T(r, W \cdot M) \leq T(r, W) + T(r, M).$$

Let $W(z)$(k-valued) and $M(z)$(ν-valued) be irreducible, non-constant algebroid functions and a be any complex number. Denote $\overline{E}(a, W) \bigcap \overline{E}(a, M)$ by $\overline{E}_0(a)$. $\overline{n}_0(r, a)$ denotes the number of $\overline{E}_0(a)$ in $|z| \leq r$. Put (see [**12**])

$$\overline{N}_0(r, a) = \frac{\nu + k}{2\nu k} \int_0^r \frac{\overline{n}_0(t, a) - \overline{n}_0(0, a)}{t} dt + \frac{\nu + k}{2\nu k} \overline{n}_0(0, a) \log r,$$

and

$$\overline{N}_{12}(r, a) = \overline{N}(r, W = a) + \overline{N}(r, M = a) - 2\overline{N}_0(r, a).$$

We have the next results for algebroid functions.

Theorem 1 *Let $W(z)$ and $M(z)$ be two k-valued, irreducible and non-constant algebroid functions. If*

$$\limsup_{r \to \infty} \frac{\sum_{a \in C} \overline{N}_0(r, a)}{T(r, W) + T(r, M)} > k,$$

then $W(z) \equiv M(z)$.

Proof. If $W(z) \not\equiv M(z)$. Observe

$$\sum_{a \in C} \overline{n}_0(r, a) \leq n(r, W - M = 0).$$

Hence

$$\sum_{a \in C} \overline{N}_0(r, a) \leq kN(r, W - M = 0).$$

We can assume that 0 is not the pole of $W_1(z)$ and $W_2(z)$. Otherwise, we can multiply by a proper factor z^n. By the first fundamental inequality ([**12**, Theorem 2.17, pp.78]) and Lemma 3, we have

$$\begin{aligned} \textstyle\sum_{a \in C} \overline{N}_0(r, a) \leq kN(r, W - M = 0) &\leq kT(r, W - M = 0) + O(1) \\ &\leq k\{T(r, W) + T(r, M)\} + O(1). \end{aligned}$$

So

$$\limsup_{r\to\infty}\frac{\sum_{a\in C}\overline{N}_0(r,a)}{T(r,W)+T(r,M)}\le k$$

which contradicts $\limsup_{r\to\infty}\frac{\sum_{a\in C}\overline{N}_0(r,a)}{T(r,W)+T(r,M)}>k$.

Remark 1 *In Theorem 3.1*[7]*, if $a_j (j=1,2,\ldots,4k+1)$ are all finite, then*

$$\sum_{a\in C}\overline{N}_0(r,a)\ge\sum_{j=1}^{4k+1}\overline{N}(r,W=a_j)=\sum_{j=1}^{4k+1}\overline{N}(r,M=a_j).$$

By lemma 1, letting $q=4k+1$, we have

$$(4k+1-2k)T(r,W)<\sum_{j=1}^{p}\overline{N}(r,W=a_j)+S(r,W).$$

So we obtain that

$$\limsup_{r\to\infty}\frac{\sum_{a\in C}\overline{N}_0(r,a)}{T(r,W)+T(r,M)}\ge\frac{2k+1}{2}>k.$$

$W(z)\equiv M(z)$ follows by Theorem 1.

This indicates that Theorem 3.1 ($4k+1$ values Theorem) in [**7**] is a special case of Theorem 1 above.

Theorem 2 *Let $W_1(z)$ and $W_2(z)$ be two k-valued and irreducible algebroid functions. Put*

$$S_j=\{c+a_j,c+a_j\omega,\cdots,c+a_j\omega^{n-1}\}\ \ (j=1,2,\cdots,m),$$

where $\omega=\exp(\frac{2\pi i}{n}), a_j\ne 0\ \ (j=1,2,\cdots,m), S_{j_1}\bigcap S_{j_2}=\emptyset\ (j_1\ne j_2)$ and $m>\frac{2}{n}k+2k^{2n-1}$. If

$$\overline{E}_{W_1}(S_j)=\overline{E}_{W_2}(S_j)\ (j=1,2,\cdots,m),$$

then

$$(W_1-c)^n\equiv(W_2-c)^n,$$

where $\overline{E}_{W_i}(S_j)=\bigcup_{a\in S_j}\{z|W_i(z)-a=0\}$ (ignoring multiplicity), $i=1,2; j=1,2,\ldots,m$.

Proof. If $(W_1 - c)^n \not\equiv (W_2 - c)^n$. By Lemma 1 and Lemma 3 we have

$$\begin{aligned}(nm-2k)T(r,W_1) &< \sum_{j=1}^{m}\sum_{t=0}^{n-1}\overline{N}(r,(W_1-c)=a_j\omega^t)+S(r,W_1)\\ &\le k^{2n-1}N(r,(W_1-c)^n-(W_2-c)^n=0)+S(r,W_1)\\ &\le nk^{2n-1}\{T(r,W_1)+T(r,W_2)\}+S(r,W_1).\end{aligned}$$

Similarly,

$$(nm-2k)T(r,W_2) < nk^{2n-1}\{T(r,W_1)+T(r,W_2)\}+S(r,W_2).$$

So

$$\begin{aligned}(nm-2k)\{T(r,W_1)+T(r,W_2)\} &< 2nk^{2n-1}\{T(r,W_1)+T(r,W_2)\}\\ &\quad + S(r,W_1)+S(r,W_2). \qquad (2)\end{aligned}$$

Using (2) we get

$$n(m-2k^{2n-1}-\frac{2}{n}k)\{T(r,W_1)+T(r,W_2)\} < S(r,W_1)+S(r,W_2).$$

This contradicts $m > 2k^{2n-1}+\frac{2}{n}k$. So $(W_1-c)^n \equiv (W_2-c)^n$.

Remark 2. *Using* Theorem *2 we get Theorem 3.1 (4k+1 values Theorem) in* [**7**] *when* $n = 1$ *and* $m = 4k+1$.

Now we give a definition and extend some properties dealing with multiple values of meromorphic functions [**14**, Lemma3.6 and Theorem3.9] to algebroid functions.

Definition 3. *Let $W(z)$ be a k-valued algebroid function defined in the complex plane and its order be a finite positive number λ. If*

$$\limsup_{r\to\infty}\frac{\log^+\overline{n}_{t)}(r,W=a)}{\log r} < \lambda$$

for any complex number a and integer t, then we call 'a' a $t-$Borel exceptional value of $W(z)$.

Lemma 4.[10] *Let $W(z)$ be a nonconstant, irreducible and k-valued algebroid function. Let $\{a_j\}_{j=1}^q \subset \overline{C}$ be q distinct complex numbers and $\{t_j\}_{j=1}^q \subset N$ be q positive integers. Then*

$$\begin{aligned}(q-2k)T(r,W) &< \textstyle\sum_{j=1}^q \frac{t_j}{t_j+1}\overline{N}_{t_j)}(r,W=a_j)\\ &\quad + \textstyle\sum_{j=1}^q \frac{1}{t_j+1}N(r,W=a_j)+S(r,W),\end{aligned}$$

$$(q-2k-\sum_{j=1}^{q}\frac{1}{t_j+1})T(r,W)<\sum_{j=1}^{q}\frac{t_j}{t_j+1}\overline{N}_{t_j)}(r,W=a_j)+S(r,W),$$

where $S(r,W)=O\{\log(rT(r,W))\}$, *outside a possible exceptional set of finite linear measure.*

Lemma 5. *a is a $t-$ Borel exceptional value of $W(z)$ if and only if*

$$\limsup_{r\to\infty}\frac{\log^+\overline{N}_{t)}(r,W=a)}{\log r}<\lambda, \tag{3}$$

where λ is the order of $W(z)$.

Proof. If a is a $t-$ *Borel* exceptional value of $W(z)$. By the definition, there exists a $\tau\in(0,\lambda)$ such that

$$\overline{n}_{t)}(r,W=a)<r^{\tau}, when \quad r>r_0.$$

So we have

$$\begin{aligned}\overline{N}_{t)}(r,W=a)&=\overline{N}_{t)}(r_0,W=a)+\frac{1}{k}\int_{r_0}^{r}\frac{\overline{n}_{t)}(t,W=a)}{t}dt\\&\le\overline{N}_{t)}(r_0,W=a)+\frac{1}{k}\cdot r^{\tau}\cdot\log\frac{r}{r_0}.\end{aligned}$$

Then

$$\limsup_{r\to\infty}\frac{\log^+\overline{N}_{t)}(r,W=a)}{\log r}\le\tau<\lambda.$$

Contrarily, if the inequality (3) above is right, then we have

$$\overline{N}_{t)}(r,W=a)<r^{\tau}, if \quad r>r_0, \qquad where \quad 0<\tau<\lambda.$$

Then

$$\begin{aligned}\overline{n}_{t)}(r,W=a)&=\frac{\overline{n}_{t)}(r,W=a)}{\log 2}\int_{r}^{2r}\frac{dt}{t}\\&\le\frac{1}{\log 2}\int_{r}^{2r}\frac{\overline{n}_{t)}(t,W=a)}{t}dt\\&\le k\frac{1}{\log 2}\overline{N}_{t)}(2r,W=a)<k\frac{(2r)^{\tau}}{\log 2}.\end{aligned}$$

So

$$\limsup_{r\to\infty}\frac{\log^+\overline{n}_{t)}(r,W=a)}{\log r}\le\tau<\lambda.$$

Theorem 3. *Let $W(z)$ be a k-valued algebroid function defined in the complex plane and its order be a finite positive number. Let $\{a_j\}_{j=1}^q \subset \overline{C}$ be q distinct complex numbers and $\{t_j\}_{j=1}^q \subset N$ be q positive integers. If a_j is $t_j -$ Borel exceptional value of $W(z)$ respectively, then*

$$\sum_{j=1}^{q}(1-\frac{1}{t_j+1}) \leq 2k.$$

Proof. When $q \leq 2k$, we can easily get the conclusion. Next we assume that $q \geq 2k+1$. By Lemma 4 we have

$$(q-2k-\sum_{j=1}^{q}\frac{1}{t_j+1})T(r,W) < \sum_{j=1}^{q}\frac{t_j}{t_j+1}\overline{N}_{t_j)}(r,W=a)+S(r,W).$$

If $q-2k-\sum_{j=1}^{q}\frac{1}{t_j+1} > 0$, noticing that a_j is $t_j - Borel$ exceptional value of $W(z)$ respectively, by Lemma 5, we know that the order of the right formula above is smaller than λ. However, the order of the left formula above is larger than λ. This is a contradiction. So $q-2k-\sum_{j=1}^{q}\frac{1}{t_j+1} \leq 0$. We get the conclusion.

Corollary. *The number of $t-$Borel exceptional value is less than $[\frac{2k(t+1)}{t}]$. Specially, the number of $1-$ Borel points of $W(z)$ is less than $4k$ and the number of $2-$ Borel points of $W(z)$ is less than $3k$ and the number of $3-$ Borel points of $W(z)$ is less than $[\frac{8k}{3}]$.*

Next we will consider the relationship of two characteristic functions when there are valent values. It extends one of H.X. Yi's results in [**14**] of meromorphic functions to algebroid functions.

Theorem 4. *Let $W_1(z)$ and $W_2(z)$ be two k-valued and irreducible algebroid functions. Let $\{a_j\}_{j=1}^q \subset \overline{C}$ be q distinct complex numbers (finite or infinite) and $\{t_j\}_{j=1}^q$ be q positive integers or ∞ satisfying that*

$$1 \leq t_q \leq t_{q-1} \leq \cdots \leq t_2 \leq t_1 \leq \infty. \tag{4}$$

Furthermore,

$$\overline{E}_{t_j)}(W_1=a_j)=\overline{E}_{t_j)}(W_2=a_j), j=1,2,\cdots,q.$$

Denote

$$A_i=\frac{\delta(a_1,W_i)+\delta(a_2,W_i)+\cdots+\delta(a_{2k},W_i)}{t_{2k+1}+1}+\sum_{j=2k+1}^{q}\frac{\delta(a_j,W_i)}{t_j+1} \quad (i=1,2.).$$

If

$$A_i + \frac{kt_{2k+1}}{t_{2k+1}+1} + \sum_{j=2k+1}^{q} \frac{t_j}{t_j+1} - 2k > 0 \quad (i = 1, 2), \tag{5}$$

then

$$T(r, W_1) = O(T(r, W_2)) \ (r \notin E),$$

$$T(r, W_2) = O(T(r, W_1)) \ (r \notin E).$$

Proof. We assume that $a_j (j = 1, 2, \cdots, q)$ are q distinct finite values. Otherwise, we can make a transform. We can assume that 0 is not the pole of $W_1(z)$ and $W_2(z)$. Otherwise, we can multiply by a proper factor z^n. Using Lemma 4 we have

$$(q - 2k)T(r, W_1) < \textstyle\sum_{j=1}^{q} \frac{t_j}{t_j+1} \overline{N}_{t_j)}(r, W_1 = a_j)$$
$$+ \textstyle\sum_{j=1}^{q} \frac{1}{t_j+1} N(r, W_1 = a_j) + S(r, W_1).$$

We get

$$\begin{aligned}(q - 2k)T(r, W_1) &< \frac{t_{2k+1}}{t_{2k+1}+1} \sum_{j=1}^{q} \overline{N}_{t_j)}(r, W_1 = a_j) \\ &+ \sum_{j=1}^{2k} \left(\frac{t_j}{t_j+1} - \frac{t_{2k+1}}{t_{2k+1}+1}\right) N(r, W_1 = a_j) \\ &+ \sum_{j=1}^{q} \frac{1}{t_j+1} N(r, W_1 = a_j) + S(r, W_1) \\ &= \frac{t_{2k+1}}{t_{2k+1}+1} \sum_{j=1}^{q} \overline{N}_{t_j)}(r, W_1 = a_j) \\ &+ \frac{1}{t_{2k+1}+1} \sum_{j=1}^{2k} N(r, W_1 = a_j) \\ &+ \sum_{j=2k+1}^{q} \frac{1}{t_j+1} N(r, W_1 = a_j) + S(r, W_1) \\ &< \frac{t_{2k+1}}{t_{2k+1}+1} \sum_{j=1}^{q} \overline{N}_{t_j)}(r, W_1 = a_j)\end{aligned}$$

$$+ \frac{1}{t_{2k+1}+1}\sum_{j=1}^{2k}(1-\delta(a_j,W_1))T(r,W_1)$$

$$+ \sum_{j=2k+1}^{q}\frac{1}{t_j+1}(1-\delta(a_j,W_1))T(r,W_1)+S(r,W_1).$$

Then

$$(A_1+\frac{2kt_{2k+1}}{t_{2k+1}+1}+\sum_{j=2k+1}^{q}\frac{t_j}{t_j+1}-2k)T(r,W_1)$$
$$<\frac{t_{2k+1}}{t_{2k+1}+1}\sum_{J=1}^{q}\overline{N}_{t_j)}(r,W_1=a_j)+S(r,W_1).$$

Furthermore,

$$\overline{N}_{t_j)}(r,W_1=a_j)\le kN(r,W_1-W_2=0)\le kT(r,W_1-W_2=0)+O(1)$$
$$\le k\{T(r,W_1)+T(r,W_2)\}+O(1)$$

So

$$(A_1+\frac{kt_{2k+1}}{t_{2k+1}+1}+\sum_{j=2k+1}^{q}\frac{t_j}{t_j+1}-2k)T(r,W_1)$$
$$<\frac{kt_{2k+1}}{t_{2k+1}+1}T(r,W_2)+S(r,W_1).$$

Combining (4) and (5) we obtain $T(r,W_1)=O(T(r,W_2))\quad(r\notin E)$. Similarly, we have

$$T(r,W_2)=O(T(r,W_1))\quad(r\notin E).$$

Now we generalize one of R.Nevanlinna's theorems for meromorphic functions in [**15**] to algebroid functions.

Theorem 5. *Let $W(z)$ and $M(z)$ be two distinct k-valued, irreducible, transcendental, non-constant algebroid functions and a be any complex number. Put*

$$\lambda(a)=1-\limsup_{r\to\infty}\frac{\overline{N}_{12}(r,a)}{T(r,W)+T(r,M)}.$$

Then

(i) $0\le\lambda(a)\le 1$,

(ii) The set of 'a' which satisfies that $\lambda(a)>0$ is at most numerable,

(iii) $\sum_a \lambda(a) \leq 4k$.

Proof. We can assume that $\lambda(\infty) = 0$. Otherwise, by Lemma 2, we can make a transform. First we have

$$0 \leq 2\overline{N}_0(r,a) \leq \overline{N}(r, W = a) + \overline{N}(r, M = a).$$

It follows that

$$\begin{aligned} 0 \leq \overline{N}_{12}(r,a) &\leq \overline{N}(r, W = a) + \overline{N}(r, M = a) \\ &\leq T(r,W) + T(r,M) + O(1). \end{aligned}$$

Thus $0 \leq \lambda(a) \leq 1$. (i) follows.

By Lemma 1, for any q distinct finite complex numbers a_j $(j = 1, 2, \ldots, q)$, there is

$$\begin{aligned} (q-2k)\{T(r,W) + T(r,M)\} &< \sum_{j=1}^{q} \{\overline{N}(r, W = a_j) + \overline{N}(r, M = a_j)\} \\ &\quad + S(r,W) + S(r,M) \\ &= \sum_{j=1}^{q} \overline{N}_{12}(r,a_j) + 2\sum_{j=1}^{q} \overline{N}_0(r,a_j) \\ &\quad + S(r,W) + S(r,M). \end{aligned} \tag{6}$$

Note that $W(z) \not\equiv M(z)$, we have

$$\sum_{j=1}^{q} \overline{N}_0(r,a_j) \leq kN(r, W - M = 0) \leq k\{T(r,W) + T(r,M)\} + O(1). \tag{7}$$

From (6) and (7) we get

$$(q-4k)\{T(r,W) + T(r,M)\} < \sum_{j=1}^{q} \overline{N}_{12}(r,a_j) + S(r,W) + S(r,M).$$

By the definition of $\lambda(a)$ we have $\sum_{j=1}^{q} \lambda(a_j) \leq 4k$. (iii) follows.

(iii) implies $\sharp A_p \leq 4kp$, where $A_p = \{a | a \in \overline{C}, \frac{1}{p} \leq \lambda(a) < \frac{1}{p-1}\}$. So $A = \{a | a \in \overline{C}, \lambda(a) > 0\} = \bigcup_{p=1}^{\infty} A_p$ is at most numerable. (ii) follows.

R. Nevalinna proved in [**16**] the 4-*value Theorem* for meromorphic functions. We extend it to algebroid functions.

Theorem 6. *Let $W_1(z)$ and $W_2(z)$ be two k-valued, irreducible algebroid functions and $\overline{E}(a_j, W_1) = \overline{E}(a_j, W_2) (j = 1, 2, \cdots, 4k)$. We have*

(i) $\lim_{r\to\infty} \frac{T(r,W_1)}{T(r,W_2)} = 1$,

(ii) $\lim_{r\to\infty}\sum_{j=1}^{4k}\frac{\overline{N}(r,W_i=a_j)}{T(r,W_i)}=2k, i=1,2,$
(iii)

$$\lim_{r\to\infty}\frac{\overline{N}(r,W_i=a)}{T(r,W_i)}=1,$$

for any $a\neq a_j$, $r\notin E$, *where* E *is an exceptional set of finite linear measure.*

Proof. Letting $q=4k$ in Lemma 1 we have

$$(4k-2k)T(r,W_i)=2kT(r,W_i)<(1+o(1))\sum_{j=1}^{4k}\overline{N}(r,W_i=a_j)\ (i=1,2).$$

Furthermore,

$$\sum_{j=1}^{4k}\overline{N}(r,W_i=a_j)\leq kN(r,W_1-W_2=0)\leq k\{T(r,W_1)+T(r,W_2)\}\\ +O(1)\ (i=1,2).$$

Therefore

$$T(r,W_1)\leq(1+o(1))T(r,W_2),\qquad T(r,W_2)\leq(1+o(1))T(r,W_1).$$

(i) follows. By (i)

$$\begin{aligned}2kT(r,W_i)&\leq(1+o(1))\sum_{j=1}^{4k}\overline{N}(r,W_i=a_j)\\&\leq(1+o(1))k\{T(r,W_1)+T(r,W_2)\}\\&=(2k+o(1))T(r,W_i),i=1,2.\end{aligned}$$

(ii) follows.

If $a\neq a_j$, applying Lemma 1 to $W_i(z)$ and $\{a_j\}\bigcup\{a\}$ implies that

$$\begin{aligned}(2k+1)(1+o(1))T(r,W_i)&\leq\sum_{j=1}^{4k}\overline{N}(r,W_i=a_j)+N(r,W_i=a)\\&\leq 2k(1+o(1))T(r,W_i)+\overline{N}(r,W_i=a).\end{aligned}$$

So

$$T(r,W_i)\leq(1+o(1))\overline{N}(r,W_i=a),(i=1,2).$$

This is (iii).

References

1. W.K. Hayman, *Meromorphic Functions,* Clarendon Press, Oxford, 1964.
2. H. Selberg, Über eine Eigenschaft der logarithmischen Ableitung einer meromorphen oder algebroide Funktion endlicher Ordnung, in: Avh.Norske Vid.Akad.Oslo I. 14 (1929).
3. H. Selberg, Über die Wertverteilung der algebroiden Funktionen, *Math.Z.* **31** (1930):709-728.
4. E. Ullrich, Über den Einfluss der verzweigtheit einer Algebroide auf ihre Wertverteilung, *J.reine ang. Math.* **169** (1931):198-220.
5. G. Valiron, Sur quelques propriétés des fonctions algébroïdes, *C.R.Acad.Sc.Paris.* **189** (1929):824-826.
6. N. Baganas, Sur les valeurs algébriques d'une fonctions algébroïdes et les integrales psendo-abélinnes, *Ann.Ec.Norm.Sup.3 Serie.* **66** (1949):161-208.
7. Y.Z. He, Y.Z. Li, Some results on algebroid functions, *Complex Variables,* **43**(2001):299-313.
8. M.L. Fang, Uniqueness theorems for algebroid functions, *Acta. Math.Sinica.* **36** (1993):217-222.
9. D.C. Sun, Z.S. Gao, The operations on algebroid functions, *Science in China.*
10. Z.S. Gao, D.C. Sun, Theorems for algebroid functions, *Acta Math.Sinica.* To appear.
11. H.X. Yi, Multiple Values and Uniqueness Theorems of Meromorphic Functions, *Chin, Ann.of Math.* **10A** (1989): 421-427.
12. Y.Z. He, X.Z. Xiao, *Algebroid functions and ordinary differential equations,* Sci. Press, Beijing, 1988.
13. L. Yang, Precise estimate of total deficiency of meromorphic derivatives, *J.D'Qnal.Marh.* **55** (1990):287-296.
14. C.C. Yang, H.X. Yi, *Uniqueness Theory of Meromorphic Functions,* Kluwer Academic Publishers, Boston, 2003.
15. R. Nevanlinna, *Le Théorème de Picard-Borel etlathéorie des fonctions méromorphes,* Paris, 1929.
16. R. Nevanlinna, Einige Eindeutigkeitssätze in der Theorie der meromorphen Funktionen, *Acta Math.* **48** (1926): 367-391.

ON NON-EXISTENCE OF TEICHMÜLLER EXTREMAL*

GUOWU YAO
Department of Mathematical Sciences
Tsinghua University
Beijing, 100084, P. R. China
E-mail: gwyao@math.tsinghua.edu.cn

Let Δ be the unit disk in the complex plane. In this paper, we give a simple proof that there exists a point $[\mu]$ in the universal Teichmüller space $T(\Delta)$ which contains infinitely many extremal Beltrami coefficients but admits no extremal Teichmüller Beltrami coefficients.

Keywords: Teichmüller space, Teichmüller extremal, Beltrami coefficient.

2000 Mathematics Subject Classification: Primary 30C75; Secondary 30C62

1. Introduction

Let $\mathfrak{D}$ be a domain in the complex plane $\mathbb{C}$ with at least two boundary points and let $M(\mathfrak{D})$ be the open unit ball of $L^\infty(\mathfrak{D})$. Every element $\mu \in M(\mathfrak{D})$ can be regarded as an element in $L^\infty(\mathbb{C})$ by putting μ equal to zero in the outside of $\mathfrak{D}$. Every $\mu \in M(\mathfrak{D})$ induces a global quasiconformal self-mapping f of the plane which solves the Beltrami equation [1],

$$f_{\bar{z}}(z) = \mu(z) f_z(z), \tag{1}$$

and f is defined uniquely up to postcomposition by a complex affine map of the plane. Conversely, any quasiconformal mapping f defined on $\mathfrak{D}$ has a Beltrami coefficient $\mu(z) = f_{\bar{z}}(z)/f_z(z)$ in $M(\mathfrak{D})$.

Two Beltrami coefficients $\mu, \nu \in M(\mathfrak{D})$ are equivalent if they induce quasiconformal mappings f and g by (1) such that there is a conformal map

*The research was supported by a Foundation for the Author of National Excellent Doctoral Dissertation and the National Natural Science Foundation of China (Grant No. 10401036).

c from $f(\mathfrak{D})$ to $g(\mathfrak{D})$ and an isotopy through quasiconformal mappings h_t, $0 \le t \le 1$, from $\mathfrak{D}$ to $\mathfrak{D}$ which extend continuously to the boundary of $\mathfrak{D}$ such that

1. $h_0(z)$ is identically equal to z on $\mathfrak{D}$,
2. h_1 is identically to $g^{-1} \circ c \circ f$, and
3. $h_t(p) = g^{-1} \circ c \circ f(p)$ for any $p \in \partial\mathfrak{D}$.

The equivalence relation partitions $M(\mathfrak{D})$ into equivalence classes and the space of equivalence classes is by definition the Teichmüller space $T(\mathfrak{D})$ of $\mathfrak{D}$.

Given $\mu \in M(\mathfrak{D})$, we denote by $[\mu]$ the set of all elements $\nu \in M(\mathfrak{D})$ equivalent to μ, and set

$$k_0([\mu]) = \inf\{\|\nu\|_\infty : \nu \in [\mu]\}.$$

We say that μ is extremal (in $[\mu]$) if $\|\mu\|_\infty = k_0([\mu])$, μ is uniquely extremal if $\|\nu\|_\infty > k_0([\mu])$ for any other $\nu \in [\mu]$; the alternative is that μ is non-uniquely extremal.

Let $A(\mathfrak{D})$ be the space of functions $\phi(z)$ holomorphic in Δ with norm

$$\|\phi\| = \iint_{\mathfrak{D}} |\phi(z)| dxdy < \infty.$$

Two elements μ and ν in $L^\infty(\mathfrak{D})$ are infinitesimally equivalent, which is denoted by $\mu \approx \nu$, if $\iint_{\mathfrak{D}} \mu\phi dxdy = \iint_{\Delta} \nu\phi dxdy$ for all $\phi \in A(\Delta)$. Denote by $N(\mathfrak{D})$ the set of all the elements in $L^\infty(\mathfrak{D})$ which are infinitesimally equivalent to zero. Then $B(\mathfrak{D}) = L^\infty(\mathfrak{D})/N(\mathfrak{D})$ is the tangent space of the space $T(\mathfrak{D})$ at the zero point.

Given $\mu \in L^\infty(\mathfrak{D})$, we denote by $[\mu]_B$ the set of all elements $\nu \in L^\infty(\mathfrak{D})$ infinitesimally equivalent to μ, and set

$$\|\mu\| = \inf\{\|\nu\|_\infty : \nu \in [\mu]_B\}.$$

We say that μ is extremal (in $[\mu]_B$) if $\|\mu\|_\infty = \|\mu\|$, uniquely extremal if $\|\nu\|_\infty > \|\mu\|$ for any other $\nu \in [\mu]_B$.

A quasiconformal mapping $f(z)$ on $\mathfrak{D}$ is called a Teichmüller mapping if its Beltrami coefficient has the form

$$\mu(z) = \frac{f_{\bar{z}}(z)}{f_z(z)} = k\frac{\overline{\varphi(z)}}{|\varphi(z)|},$$

where $k \in [0, 1)$ is a constant and $\varphi \not\equiv 0$ is a holomorphic function on $\mathfrak{D}$. A Beltrami coefficient μ with the above form is called a Teichmüller Beltrami coefficient, or TBC for short. If in addition, μ is extremal in its class, we call it a Teichmüller extremal.

Let Δ be the unit disk. For a given point $[\mu]$ ($[\mu]_B$) in $T(\Delta)$ ($B(\Delta)$), there are two cases for the extremal Beltrami coefficients among $[\mu]$ ($[\mu]_B$). One is that there is a unique extremal Beltrami coefficient in $[\mu]$ ($[\mu]_B$) which may be TBC or not. Even there exist uniquely extremal Beltrami coefficients with non-constant modulus [2].

The other is that there are more than one extremal Beltrami coefficient in $[\mu]$ or $[\mu]_B$. In this setting, in fact there are infinitely many extremal Beltrami coefficients in $[\mu]$ or $[\mu]_B$ (see [7,3,6]). It is natural to ask whether there always exists an extremal TBC in $[\mu]$ or $[\mu]_B$. The following two theorems give the negative answers to the problem.

Theorem 1. *There exists a point $[\mu]$ in $T(\Delta)$ that contains more than one extremal Beltrami coefficient but admits no Teichmüller extremals.*

Theorem 2. *There exists a point $[\mu]_B$ in $B(\Delta)$ that contains more than one extremal Beltrami coefficient but admits no Teichmüller extremals.*

In fact, the generalized forms of the above theorems were obtained in [8] by the author. Even there exists an extremal Beltrmai coefficient μ such that $[\mu]$ (or $[\mu]_B$) contains infinitely many extremals but none of which is of constant modulus [9]. We present Theorems 1 and 2 with simpler proofs here.

2. Delta inequalities

For $\mu \in L^\infty(\Delta)$, $\phi \in A(\Delta)$, let

$$\Lambda_\mu[\phi] = \iint_\Delta \mu(z)\phi(z)dxdy, \quad \lambda_\mu[\phi] = Re\Lambda_\mu[\phi].$$

As is well known (for example, see [5]), a Beltrami coefficient μ is extremal if and only if it has a so-called Hamilton sequence, namely, a sequence $\{\phi_n \in A(\Delta) : \|\phi_n\| = 1, n \in \mathbb{N}\}$, such that

$$\lim_{n\to\infty} \Lambda_\mu[\phi_n] = \lim_{n\to\infty} \iint_\Delta \mu\phi_n(z)dxdy = \|\mu\|_\infty. \tag{2}$$

Now, we introduce Reich's Delta Inequality and Infinitesimal Delta Inequality on the unit disk Δ. Their generalized forms play important roles in the joint work [2] of Božin, Lakic, Marković and Mateljević.

Given $\mu \in M(\Delta)$, let $f = f^\mu$ be the uniquely determined quasiconformal mapping of Δ onto itself with Beltrami coefficients μ and normalized to fix 1, -1 and i.

Suppose that μ and ν are two equivalent Beltrami coefficients in $T(\Delta)$. Let $\widetilde{\mu}$ and $\widetilde{\nu}$ be the Beltrami coefficients of the quasiconformal mappings f^{-1} and g^{-1}, respectively, where $f = f^{\mu}$ and $g = f^{\nu}$.

Delta Inequality. *If μ and ν are equivalent Beltrami coefficients in $T(\Delta)$ with*

$$\|\nu\|_\infty \le k = \|\mu\|_\infty < 1,$$

then

$$\iint_\Delta \left|\frac{\widetilde{\mu}(f) - \widetilde{\nu}(f)}{1 - \overline{\widetilde{\mu}(f)}\widetilde{\nu}(f)}\right|^2 |\varphi| \le C(k\|\varphi\| - Re \iint_\Delta \mu\varphi), \tag{3}$$

for all φ in $A(\Delta)$. The constant C depends only on $k = \|\mu\|_\infty$.

Infinitesimal Delta Inequality. *There exists a universal constant C such that for every pair of infinitesimally equivalent Beltrami coefficients μ and ν with*

$$\|\nu\|_\infty \le \|\mu\|_\infty < \infty,$$

we have

$$\iint_\Delta |\mu - \nu|^2 |\varphi| \le C\|\mu\|_\infty(\|\mu\|_\infty \|\varphi\| - Re \iint_\Delta \mu\varphi), \tag{4}$$

for all φ in $A(\Delta)$. The constant C is independent of μ and ν.

3. Proof of Theorem 1

Let $\mathfrak{J} \subsetneqq \Delta$ be a Jordan domain such that $\Delta \backslash \overline{\mathfrak{J}}$ is connected. Let μ be a Beltrami coefficient in $M(\Delta)$.

Lemma 1. *Let μ and ν be two equivalent Beltrami coefficients in $T(\Delta)$. In addition, suppose $\mu(z) = \nu(z)$ for almost every $z \in \Delta \backslash \overline{\mathfrak{J}}$. Then, $f^{\mu}(z) = f^{\nu}(z)$ for all z in $\Delta \backslash \mathfrak{J}$ and hence $\widetilde{\mu}(w) = \widetilde{\nu}(w)$ for almost all w in $f(\Delta \backslash \mathfrak{J})$, where $\widetilde{\mu}$ and $\widetilde{\nu}$ are the Beltrami coefficients of the quasiconformal mappings $(f^{\mu})^{-1}$ and $(f^{\nu})^{-1}$ respectively.*

Proof. For the sake of convenience, let $f = f^{\mu}$ and $g = f^{\nu}$. Let $\mu_{g \circ f^{-1}}(w)$ denote the Beltrami coefficient of $g \circ f^{-1}$. By a simple computation, we have

$$\mu_{g \circ f^{-1}} \circ f(z) = \frac{1}{\tau} \frac{\mu(z) - \nu(z)}{1 - \overline{\mu(z)}\nu(z)},$$

where $\tau = \overline{f_z}/f_z$.

Thus, $\mu_{g\circ f^{-1}}(w) = 0$ for almost all $w \in f(\Delta\backslash\overline{\mathfrak{J}})$ and hence $\Psi = g\circ f^{-1}$ is conformal on $\Delta\backslash\overline{\mathfrak{J}}$. Since $\Psi|_{S^1} = g\circ f^{-1}|_{S^1} = id$, we conclude that $\Psi = id$ in $f(\Delta\backslash\mathfrak{J})$. Thus, $g|_{\Delta\backslash\overline{\mathfrak{J}}} = f|_{\Delta\backslash\overline{\mathfrak{J}}}$. By the continuity of quasiconformal mappings, it follows that $g|_{\Delta\backslash\mathfrak{J}} = f|_{\Delta\backslash\mathfrak{J}}$. In addition, it is evident that $\widetilde{\mu}(w) = \widetilde{\nu}(w)$ for almost all w in $f(\Delta\backslash\mathfrak{J})$. □

Lemma 2. *Let $\mu = k\overline{\varphi}/|\varphi|$ be a TBC in $M(\Delta)$, where φ is holomorphic in Δ and $k \in (0,1)$ is a constant. Define*

$$\nu = \begin{cases} \mu(z) & z \in \Delta\backslash\overline{\mathfrak{J}}, \\ 0, & z \in \overline{\mathfrak{J}}. \end{cases}$$

Then $\nu \notin [\mu]$ in $T(\Delta)$.

Proof. Suppose to the contrary. Then by virtue of Lemma 1, we have $f^{\mu}(z) = f^{\nu}(z)$ for all z in $\Delta\backslash\mathfrak{J}$, and hence $f^{\mu}|_{\partial\mathfrak{J}} = f^{\nu}|_{\partial\mathfrak{J}}$. This shows that $[0]$ and $[k\overline{\varphi}/|\varphi|]$ (φ restricted to $\mathfrak{J}$) are identical in the Teichmüller space $T(\mathfrak{J})$ of $\mathfrak{J}$. Note that the restriction of φ on $\mathfrak{J}$ is integrable and hence $k\overline{\varphi}/|\varphi|$ is uniquely extremal in $[k\overline{\varphi}/|\varphi|]$ in $T(\mathfrak{J})$. The contradiction indicates $\nu \notin [\mu]$ in $T(\Delta)$. □

The following Reich's Construction Theorem plays a most important role in the proof of our main results.

Theorem 3.1. *Let A be a compact subset of Δ containing at least two points and such that $\Delta\backslash A$ is doubly connected. There exists a function $\mu \in L^{\infty}(\Delta)$ and a sequence $\varphi_n \in A(\Delta)$ $(n = 1, 2, \ldots)$ satisfying the following conditions* (5) – (7):

$$|\alpha(z)| = \begin{cases} 0, & z \in A, \\ 1, & \text{for a.a. } z \in \Delta\backslash A, \end{cases} \tag{5}$$

$$\lim_{n\to\infty} \{\|\varphi_n\| - \lambda_{\alpha}[\varphi_n]\} = 0, \tag{6}$$

and

$$\lim_{n\to\infty} |\varphi_n(z)| = \infty \quad \text{a.e. in } \Delta\backslash A. \tag{7}$$

Proof of Theorem 1. Let $0 < r < 1$. Choose $A = \overline{\Delta_r} = \{z : |z| \leq r\}$. Let $\alpha(z)$ and the sequence $\varphi_n \in A(\Delta)$ be constructed by Reich's Construction

Theorem. Put $\mu(z) = k\alpha(z)$, where $k \in (0,1)$ is a constant. Then $\|\mu\|_\infty = k$. Let $\widetilde{\mu}(w)$ denote the Beltrami coefficient of $(f^\mu)^{-1}$. By (6), we have

$$\lim_{n\to\infty}\{k\|\varphi_n\| - \lambda_\mu[\varphi_n]\} = \lim_{n\to\infty}\{k\|\varphi_n\| - \iint_\Delta \mu(z)\varphi_n(z)\} = 0.$$

Thus, by equation (7) and Fatou's lemma,

$$k - Re\iint_\Delta \mu(z)\frac{\varphi_n(z)}{\|\varphi_n\|} \to 0,\ n \to \infty,$$

which shows that $\mu(z)$ is extremal in $[\mu]$.

Since $\mu(z) = 0$ for z in A, obviously μ is extremal instead of being uniquely extremal. We claim that $[\mu]$ contains no Teichmüller extremals.

First step: For any given ν extremal in $[\mu]$, let $\widetilde{\nu}(w)$ denote the Beltrami coefficient of $(f^\nu)^{-1}$. We will prove that $\widetilde{\nu}(f(z)) = \widetilde{\mu}(f(z))$ for almost every $z \in U_r$. Suppose to the contrary. Then there would exist $\epsilon > 0$ and a compact subset S of U_r with positive Lebesgue measure such that $|\frac{\widetilde{\mu}(f)-\widetilde{\nu}(f)}{1-\overline{\widetilde{\mu}(f)}\widetilde{\nu}(f)}| \geq \epsilon > 0$ on S. Then, by the Delta Inequality (3) there exists a positive constant C depending only on k such that

$$\frac{\epsilon^2}{4}\iint_S |\varphi_n| \leq C(k\|\varphi_n\| - Re\iint_\Delta \mu\varphi_n) = Ck(\|\varphi_n\| - \lambda_\alpha[\varphi_n]).$$

The left of the above inequality has a positive lower bound by (7) and Fatou's lemma while the right tends to 0 as $n \to \infty$ by (6). This contradiction induces our claim. Thus we have proved that for any extremal Beltrami coefficient ν in $[\mu]$, $\widetilde{\nu}(w) = \widetilde{\mu}(w)$ for almost all $w \in f(U_r)$. Applying Lemma 1 to the target unit disk, we find that $\mu(z) = \nu(z)$ for almost all z in U_r.

Second step: Now, suppose $[\mu]$ contains a Teichmüller extremal $\nu = k\overline{\varphi}/|\varphi|$ with φ holomorphic in Δ. On the other hand, by the first step, we see that $\mu(z) = \nu(z)$ for almost all $z \in U_r$. But Lemma 2 tells $\mu \notin [\nu]$. This contradiction gives Theorem 1.

4. Proof of Theorem 2

Lemma 3. *Let μ, $\nu \in M(\Delta)$ be given as in Lemma 2. Then $\nu \notin [\mu]_B$ in $B(\Delta)$.*

Proof. Suppose to the contrary. Then $[0]_B$ and $[k\overline{\varphi}/|\varphi|]_B$ (φ restricted to $\mathfrak{J}$) represent a point in the space $B(\mathfrak{J})$ of $\mathfrak{J}$. Note that $k\overline{\varphi}/|\varphi|$ is uniquely extremal in $[k\overline{\varphi}/|\varphi|]_B$ in $B(\mathfrak{J})$. The contradiction indicates $\nu \notin [\mu]_B$ in $B(\Delta)$. □

Proof of Theorem 2: Let μ be constructed as in the proof of Theorem 1. We only need to show $[\mu]_B$ satisfies the requirement of Theorem 2. We use some same denotations as proving Theorem 1 for simplicity.

Obviously, μ is extremal in $[\mu]_B$ instead of being uniquely extremal by the Equivalence Theorem of [2]. Furthermore, we claim that, for any ν extremal in $[\mu]_B$, $\mu(z) = \nu(z)$ for almost every z in $U_r = \Delta\backslash A$. Suppose to the contrary. Then there would exist $\epsilon > 0$ and a compact subset S of U_r with positive Lebesgue measure such that $|\mu(z) - \nu(z)| \geq \epsilon > 0$ on S. Then, by the Infinitesimal Delta Inequality (4) there exists a universal constant C such that

$$\epsilon^2 \iint_S |\varphi_n| \leq Ck(k\|\varphi_n\| - Re \iint_\Delta \mu\varphi_n) = Ck^2(\|\varphi_n\| - \lambda_\alpha[\varphi_n]).$$

The left of the above inequality has a positive lower bound by (7) and Fatou's lemma while the right tends to 0 as $n \to \infty$ by (6). This contradiction confirms our claim.

Now, suppose $[\mu]$ contains a Teichmüller extremal $\nu = k\overline{\varphi}/|\varphi|$ with φ holomorphic in Δ. From the above argument, we see that $\mu(z) = \nu(z)$ for almost all $z \in U_r$. Thus, we derive the desired contradiction with Lemma 3, proving Theorem 2.

References

1. L. V. Ahlfors and L. Bers, Riemann's mapping theorem for variable metrics, *Ann. Math.*, **72** (1960): 385-404.
2. V. Božin, N. Lakic, V. Marković and M. Mateljević, Unique extremality, *J. Anal. Math.*, **75** (1998):299-338.
3. C. J. Earle and N. Lakic, Variability sets on Riemann surfaces and forgetful maps between Teichmüller spaces, *Ann. Acad. Sci. Fenn. Math.*, **27** (2002): 307-324.
4. E. Reich, The unique extremality counterexample,*J. Anal. Math.*, **75** (1998): 339-347.
5. E. Reich and K. Strebel, *Extremal quasiconformal mappings with given boundary values,* Contributions to Analysis, A Collection of Papers Dedicated to Lipman Bers, Academic Press, New York, 1974, pp. 375-392.
6. Y. Shen, On Teichmüller geometry, *Complex Var. Theory Appl.*, **44** (2001): 73-83.
7. K. Strebel, Point shift differentials and extremal quasiconformal mappings, *Ann. Acad. Sci. Fenn. Math.*, **23** (1998): 475-494.
8. G. W. Yao, Is there always an extremal Teichmüller mappings, *J. Anal. Math.* **94** (2004):363-375.
9. G. W. Yao, *Existence of extremal Beltrami coefficients with non-constant modulus,* preprint.

TWO MEROMORPHIC FUNCTIONS SHARING FOUR SMALL FUNCTIONS IN THE SENSE OF $E_{K)}(\beta, F) = E_{K)}(\beta, G)$*

YAO WEIHONG

Department of Mathematics
Shanghai Jiao Tong University
Shanghai 200240, P. R. China
E-mail: whyao@sjtu.edu.cn

In this paper, we proved a result that if two meromorphic functions $f(z)$ and $g(z)$ share four small functions $a_j(z)$ $(j = 1, \cdots, 4)$ in the sense of $E_{k)}(a_j, f) = E_{k)}(a_j, g)$, $(j = 1, \cdots, 4)$ $(k \geq 11)$, then f is a quasi-Möbius transformation of g; i.e., there exist four small functions $\alpha_i (i = 1, \cdots, 4)$ of f and g such that $f = (\alpha_1 g + \alpha_2)/(\alpha_3 g + \alpha_4)$, where $\alpha_1\alpha_4 - \alpha_2\alpha_3 \not\equiv 0$.

Keywords: Meromorphic function, small function, $E_{k)}(\beta, f) = E_{k)}(\beta, g)$, quasi-Möbius transformation.

2000 Mathematics Subject Classification: 30D35

1. Introduction and Main Result

In this paper the term "meromorphic function (entire function)" will mean a meromorphic function (entire function) in C. We will use the standard notations of Nevanlinna theory: $T(r, f)$, $S(r, f)$, $m(r, \beta, f)$, $N(r, \beta, f), \overline{N}(r, \beta, f)$, CM*, IM*, ..., and we assume that the reader is familiar with the basic results in Nevanlinna theory as found in [4].

Let $f(z)$ be a non-constant meromorphic function in the complex plane and let $S(f)$ be the set of meromorphic functions $\beta(z)$ in the complex plane which satisfy

$$T(r, \beta) = S(r, f),$$

*Project supported by the NSFC (No. 10271077).

where $S(r,f)$ is any quantity satisfying

$$S(r,f) = o(T(r,f))$$

for $r \to \infty$, $r \notin E$, $\text{mes}E < +\infty$. Such a meromorphic function $\beta(z)$ is said to be a small function of $f(z)$.

For a non-constant meromorphic function f, a small function $\beta \in S(f) \cup \{\infty\}$ and a positive integer k (or $+\infty$), we write $E_{k)}(\beta, f)$ for the set of zeros of $f(z) - \beta$ with multiplicity $\leq k$ (counting multiplicity); we write $\overline{E}_{k)}(\beta, f)$ for the set of zeros of $f(z) - \beta$ with multiplicity $\leq k$ (each zero counted only once).

If two non-constant meromorphic function f and g and a small function $\beta \in S(f) \cap S(g) \cup \{\infty\}$ satisfy

$$E_{+\infty)}(\beta, f) = E_{+\infty)}(\beta, g),$$

then we say that f and g share β CM. If f and g satisfy

$$\overline{E}_{+\infty)}(\beta, f) = \overline{E}_{+\infty)}(\beta, g)$$

then we say that f and g share β IM.

In 1929, Nevanlinna proved the following well-know result which is the so called Nevanlinna four-values theorem.

Theorem A. (See [1]) *Let f and g be two non-constant meromorphic functions. If f and g share four distinct values CM, then f is a Möbius transformation of g.*

In 1995, Li and Yang proved the following result.

Theorem B. (See [2]) *Let f and g be two non-constant meromorphic functions, and let a_j $(j = 1, \cdots, 4)$ be distinct small functions of f and g. If f and g share a_j $(j = 1, \cdots, 4)$ CM*, then f is a quasi-Möbius transformation of g.*

In this paper, we obtain a result related to the above theorems as follows.

Theorem 1. *Let f and g be two non-constant meromorphic functions, and let a_j $(j = 1, \cdots, 4)$ be distinct small functions of f and g. If f and g satisfy $E_{k)}(a_j, f) = E_{k)}(a_j, g)$, $(j = 1, \cdots, 4)$, $k(\geq 11)$ is a positive integer, then $f(z)$ is a quasi-Möbius transformation of $g(z)$.*

Note: Without loss of generality, we suppose that $a_1(z) = \infty$, $a_2(z) = 0$, $a_3(z) = 1$, $a_4(z) = a(z)$ $(\not\equiv \infty, 0, 1)$.

From Theorem 1, we can get the following result. That is a improvement of Theorem A.

Corollary 1. *Let f and g be two non-constant meromorphic functions, and let a_j $(j = 1, \cdots, 4)$ be four distinct values. If f and g satisfy $E_{k)}(a_j, f) = E_{k)}(a_j, g)$, $(j = 1, \cdots, 4)$, $k(\geq 11)$ is a positive integer, then $f(z)$ is a Möbius transformation of $g(z)$.*

2. Some Lemmas and Notations

In the rest of this section, we assume that f and g are distinct non-constant meromorphic functions sharing $a_1 = \infty$, $a_2 = 0$, $a_3 = 1$ and $a_4 = a(z)$ in the sense of $E_{k)}(a_j, f) = E_{k)}(a_j, g)$ for $j = 1, \cdots, 4$, where $k(\geq 11)$ is a positive integer.

Lemma 0 (See [3]). *Suppose that f, $a(z)$ and $b(z)$ are all non-∞ meromorphic functions $a(z)$ and $b(z)$ are distinct small functions of f. Set*

$$L(f, a, b) := \begin{vmatrix} f & f' & 1 \\ a & a' & 1 \\ b & b' & 1 \end{vmatrix}.$$

Then

$$L(f, a, b) \not\equiv 0,$$

and

$$m(r, \frac{L(f, a, b) f^k}{(f-a)(f-b)}) = S(r, f) \quad (k = 0, 1).$$

Lemma 1 (See [8]). *Let f be non-constant meromorphic functions, and a_j $(j = 1, \cdots, q)$ be distinct small functions of f. Then we have the second main theorem*

$$(q - 2 - \varepsilon)T(r, f) \leq \sum_{j=1}^{q} \overline{N}(r, a_j, f) + S(r, f), \tag{1}$$

for all $\varepsilon > 0$.

Lemma 2 *Let f and g be two non-constant meromorphic functions, and let a_j $(j = 1, \cdots, 4)$ be distinct small functions of f and g, and $k(\geq 3)$*

be a positive integer. If $f(z) \not\equiv g(z)$ and satisfy $E_{k)}(a_j, f) = E_{k)}(a_j, g)$, $(j = 1, 2, 3, 4)$, then we have

(i)

$$S(r) = S(r, f) = S(r, g). \tag{2}$$

(ii)

$$\sum_{j=1}^{4} \overline{N}_{(k+1}(r, a_j, f) \le \frac{(2+\varepsilon)}{k} T(r, f) + S(r). \tag{3}$$

$$\sum_{j=1}^{4} \overline{N}_{(k+1}(r, a_j, g) \le \frac{(2+\varepsilon)}{k} T(r, g) + S(r). \tag{4}$$

$$[2-\varepsilon-\frac{(2+\varepsilon)}{k}]\{T(r,f)+T(r,g)\} \le \sum_{j=1}^{4}\{\overline{N}_{k)}(r,a_j,f)+\overline{N}_{k)}(r,a_j,g)\}+S(r). \tag{5}$$

Proof. From Lemma 1, we know that

$$\begin{aligned}(2-\varepsilon)T(r,f) &\le \textstyle\sum_{j=1}^{4} \overline{N}(r, a_j, f) + S(r, f)\\ &= \textstyle\sum_{j=1}^{4} \overline{N}_{k)}(r, a_j, f) + \sum_{j=1}^{4} \overline{N}_{(k+1}(r, a_j, f) + S(r, f)\\ &\le \textstyle\frac{k}{k+1} \sum_{j=1}^{4} \overline{N}_{k)}(r, a_j, f) + \frac{1}{k+1} \sum_{j=1}^{4} \overline{N}_{k)}(r, a_j, f)\\ &+ \textstyle\frac{1}{k+1} \sum_{j=1}^{4} \overline{N}_{(k+1}(r, a_j, f) + S(r, f)\\ &\le \textstyle\frac{k}{k+1} \sum_{j=1}^{4} \overline{N}_{k)}(r, a_j, f) + \frac{4}{k+1} T(r, f)\\ &\le \textstyle\frac{k+4}{k+1} T(r, f) + \frac{k}{k+1} T(r, g).\end{aligned}$$

Noting that $k \ge 3$, and we let $\varepsilon < 1/4$, then we have

$$T(r, f) \le \frac{k}{(1-\varepsilon)k - 2 - \varepsilon} T(r, g) + S(r, f). \tag{6}$$

Similarly, we have

$$T(r, g) \le \frac{k}{(1-\varepsilon)k - 2 - \varepsilon} T(r, f) + S(r, f). \tag{7}$$

From (6) and (7), we get (2).

(ii) From Lemma 1, we deduce that

$$(2-\varepsilon)T(r,f)+k\sum_{j=1}^{4}\overline{N}_{(k+1}(r,a_j,f)$$

$$\leq\sum_{j=1}^{4}\overline{N}(r,a_j,f)+k\sum_{j=1}^{4}\overline{N}_{(k+1}(r,a_j,f)+S(r).$$

$$\leq 4T(r,f)+S(r). \tag{8}$$

So we have

$$\sum_{j=1}^{4}\overline{N}_{(k+1}(r,a_j,f)\leq\frac{(2+\varepsilon)}{k}T(r,f)+S(r). \tag{9}$$

Similarly, we have

$$\sum_{j=1}^{4}\overline{N}_{(k+1}(r,a_j,g)\leq\frac{(2+\varepsilon)}{k}T(r,g)+S(r). \tag{10}$$

From Lemma 1 and (9), we have

$$(2-\varepsilon)T(r,f)\leq\sum_{j=1}^{4}\overline{N}_{k)}(r,a_j,f)+\sum_{j=1}^{4}\overline{N}_{(k+1}(r,a_j,f)+S(r)$$

$$\leq\sum_{j=1}^{4}\overline{N}_{k)}(r,a_j,f)+\frac{(2+\varepsilon)}{k}T(r,f)+S(r).$$

That is

$$[(2-\varepsilon)-\frac{(2+\varepsilon)}{k}]T(r,f)\leq\sum_{j=1}^{4}\overline{N}_{k)}(r,a_j,f)+S(r).$$

Similarly, we have

$$[(2-\varepsilon)-\frac{(2+\varepsilon)}{k}]T(r,g)\leq\sum_{j=1}^{4}\overline{N}_{k)}(r,a_j,g)+S(r).$$

Hence

$$[(2-\varepsilon)-\frac{(2+\varepsilon)}{k}]\{T(r,f)+T(r,g)\}\leq\sum_{j=1}^{4}\{\overline{N}_{k)}(r,a_j,f)+\overline{N}_{k)}(r,a_j,g)\}+S(r).$$

3. Proof of Theorem 1

In what follows we assume that f and g are distinct and satisfy the assumptions of Theorem 1, the set $S_0 = \{z|\ a(z) = 0,\ \ 1,\ \ \infty\}$.

Set

$$H_1 = \frac{\begin{vmatrix} f(f-1) & f' \\ a(a-1) & a' \end{vmatrix}}{f(f-1)(f-a)} - \frac{\begin{vmatrix} g(g-1) & g' \\ a(a-1) & a' \end{vmatrix}}{g(g-1)(g-a)}.$$

From property of determinant and the lemma of the logarithmic derivative, we have

$$m(r, H_1) = S(r, f).$$

If $H_1 \not\equiv 0$, then from Lemma 2 we have

$$\begin{aligned}
N(r, H_1) &\leq \textstyle\sum_{j=2}^{4} \overline{N}_{(k+1}(r, a_j, f) + \sum_{j=2}^{4} \overline{N}_{(k+1}(r, a_j, g) + S(r) \\
&\leq \tfrac{1}{k+1} \textstyle\sum_{j=2}^{4} \{\overline{N}_{k)}(r, a_j, f) + \overline{N}_{k)}(r, a_j, g)\} \\
&\quad + \tfrac{1}{k+1} \textstyle\sum_{j=2}^{4} \{N_{(k+1}(r, a_j, f) + N_{(k+1}(r, a_j, g)\} \\
&\quad - \tfrac{2}{k+1} \textstyle\sum_{j=2}^{4} \overline{N}_{k)}(r, a_j, f) + S(r) \\
&\leq \tfrac{1}{k+1} \textstyle\sum_{j=2}^{4} \{N(r, a_j, f) + N(r, a_j, g)\} \\
&\quad - \tfrac{2}{k+1} \textstyle\sum_{j=2}^{4} \overline{N}_{k)}(r, a_j, f) + S(r) \\
&\leq \tfrac{3}{k+1} \{T(r, f) + T(r, g)\} - \tfrac{2}{k+1} \textstyle\sum_{j=1}^{4} \overline{N}_{k)}(r, a_j, f) \\
&\quad + \tfrac{2}{k+1} \overline{N}_{k)}(r, a_1, f) + S(r).
\end{aligned}$$

Noting that

$$\overline{N}_{k)}(r, a_1, f) \leq N(r, \frac{1}{H_1}) \leq N(r, H_1) + S(r),$$

so we have

$$\begin{aligned}
(1 - \tfrac{2}{k+1}) \overline{N}_{k)}(r, a_1, f) &\leq \tfrac{3}{k+1} \{T(r, f) + T(r, g)\} \\
&\quad - \tfrac{2}{k+1} \textstyle\sum_{j=1}^{4} \overline{N}_{k)}(r, a_j, f) + S(r).
\end{aligned}$$

Similarly, we have

$$\begin{aligned}
(1 - \tfrac{2}{k+1}) \overline{N}_{k)}(r, a_1, g) &\leq \tfrac{3}{k+1} \{T(r, f) + T(r, g)\} \\
&\quad - \tfrac{2}{k+1} \textstyle\sum_{j=1}^{4} \overline{N}_{k)}(r, a_j, g) + S(r)
\end{aligned} \tag{11}$$

Hence, noting Lemma 2, we have

$$(1-\frac{2}{k+1})\{\overline{N}_{k)}(r,a_1,f)+\overline{N}_{k)}(r,a_1,g)\}$$

$$\leq \frac{6}{k+1}\{T(r,f)+T(r,g)\}-\frac{2}{k+1}\sum_{j=1}^{4}\{\overline{N}_{k)}(r,a_j,f)+\overline{N}_{k)}(r,a_j,g)+S(r)$$

$$\leq \frac{6}{k+1}\{T(r,f)+T(r,g)\}-\frac{2}{k+1}[(2-\varepsilon)-\frac{(2+\varepsilon)}{k}]\{T(r,f)+T(r,g)\}+S(r)$$

$$\leq \frac{2k+4+2(k+1)\varepsilon}{k(k+1)}\{T(r,f)+T(r,g)\}+S(r). \tag{12}$$

So we have

$$\overline{N}_{k)}(r,a_1,f)+\overline{N}_{k)}(r,a_1,g)\leq \frac{2k+4+2(k+1)\varepsilon}{k(k-1)}\{T(r,f)+T(r,g)\}+S(r). \tag{13}$$

Similarly, we set

$$H_2=\frac{L(f,1,a)f}{(f-1)(f-a)}-\frac{L(g,1,a)g}{(g-1)(g-a)},$$

$$H_3=\frac{L(f,0,a)(f-1)}{f(f-a)}-\frac{L(g,0,a)(g-1)}{g(g-a)},$$

$$H_4=\frac{f'(f-a)}{f(f-1)}-\frac{g'(g-a)}{g(g-1)}.$$

And if $H_j\not\equiv 0\ \ (j=2,3,4)$, noting Lemma 0 we also have

$$\overline{N}_{k)}(r,a_j,f)+\overline{N}_{k)}(r,a_j,g)\leq \frac{2k+4+2(k+1)\varepsilon}{k(k-1)}\{T(r,f)+T(r,g)\}+S(r). \tag{14}$$

From (5), i.e.,

$$[2-\varepsilon-\frac{(2+\varepsilon)}{k}]\{T(r,f)+T(r,g)\}\leq \sum_{j=1}^{4}\{\overline{N}_{k)}(r,a_j,f)+\overline{N}_{k)}(r,a_j,g)\}+S(r).$$

we know that there exist at least two of the four $\overline{N}_{k)}(r,\frac{1}{f-a_j})+\overline{N}_{k)}(r,\frac{1}{g-a_j})$, $(j=1,2,3,4)$, without loss of generality, we suppose $j=3,4$, such that

$$\overline{N}_{k)}(r,a_3,f)+\overline{N}_{k)}(r,a_3,g)\geq \tfrac{1}{3}\{[1-\varepsilon-\tfrac{(2+\varepsilon)}{k}]+o(1)\}\{T(r,f)+T(r,g)\}$$

$$(r\in R^+, r\notin E), \tag{15}$$

and

$$\overline{N}_{k)}(r,a_4,f)+\overline{N}_{k)}(r,a_4,g)\geq\tfrac{1}{3}\{[1-\varepsilon-\tfrac{(2+\varepsilon)}{k}]+o(1)\}\{T(r,f)+T(r,g)\}$$

$$(r\in R^+,r\notin E). \tag{16}$$

If $H_3\not\equiv 0$, then from (14) and (15) we have

$$\frac{1}{3}[1-\varepsilon-\frac{(2+\varepsilon)}{k}]\leq\frac{2k+4+2(k+1)\varepsilon}{k(k-1)}.$$

Let $\varepsilon<1/4$, this contradicts $k\geq 11$. So we have $H_3\equiv 0$. Similarly, we have $H_4\equiv 0$. Thus

$$\frac{L(f,0,a)(f-1)}{f(f-a)}\equiv\frac{L(g,0,a)(g-1)}{g(g-a)}, \tag{17}$$

and

$$\frac{f'(f-a)}{f(f-1)}\equiv\frac{g'(g-a)}{g(g-1)}. \tag{18}$$

Let z_a be a zero of $f-a$, but not a zero of $g-a$, and $z_a\notin S_0$. Then z_a is a pole of the left side of (17). So we have $g(z_a)=0$, or $g(z_a)=\infty$. From (17), (18), we have

$$\frac{L(f,0,a)f'}{f^2}\equiv\frac{L(g,0,a)g'}{g^2}. \tag{19}$$

The left side of (19) is analytic at z_a but z_a is a pole of the right side with multiplicity 2. This is a contradiction. So we have $g(z_a)-a(z_a)=0$. By symmetry and comparing the residues of the two sides of (17) at z_a, we know that f and g share a CM*. Similarly, we can get that f and g share 1 CM*.

Developing (17) and (18) , and eliminating f'/f and g'/g , we have

$$a(a-1)\frac{f'-a'}{f-a}-(a-1)\frac{f'}{f-1}=a(a-1)\frac{g'-a'}{g-a}-(a-1)\frac{g'}{g-1}. \tag{20}$$

From (20) , it's obviously that f and g share ∞ CM*.

Then from (17), we know that f and g share 0 CM*.

Hence, from Theorem B we get f is a quasi-Möbius transformation of g.

This completes the proof of Theorem 1 .

Acknowledgement The author wish to thank Prof. H.X. Yi for his valuable advice.

References

1. R. Nevanlinna, *Le théorème de Picard-Borel et la théorie des fonctions meromorphes,* Gauthier-Villars, Paris, 1929.
2. P. Li and C. C. Yang, On Two Meromorphic Functions that Share Pairs of Small Functions, *Complex Variables,* **32**(1997): 177-190.
3. Y. H. Li , Entire Functions Sharing Four Small Functions IM, *Acta Math. Sinica* , **41**(1998): 249-260 (in Chinese).
4. H. X. Yi and C. C. Yang, *Uniqueness Theory of Meromorphic Functions (in Chinese)*, Science Press, Beijing, 1995.
5. Q. C. Zhang, Meromorphic Functions Sharing Values or Sets, *Doctoral Thesis,* Shandong University, 1999.
6. W. K. Hayman, *Meromorphic Functions,* Clarendon Press, Oxford, 1964.
7. E. Mues, Meromorphic Functions Shareing Four Values, *Complex Variables Theory Appl.*, **12**(1989):169–179.
8. K. Yamanoi, The second main theorem for small functions and related problems, *Acta Math.* **192**(2)(2004):225–294.

MULTIPLICATION OPERATORS IN THE α-BLOCH SPACES*

SHANLI YE

Department of Mathematics
Fujian Normal University
Fuzhou 350007, P. R. China
E-mail: shanliye@fjnu.edu.cn

In this paper we discuss the operator of multiplication M_ϕ between the Bloch-type spaces on the unit disc. Some sufficient and necessary conditions are given for which M_ϕ is a compact operator from B^α to B^β.

Keywords: Pointwise multipliers; Bloch space; compactness

2000 Mathematics Subject Classification: 30D05, 30H05, 47B38

1. Introduction

Let D be the open unit disk in the complex plane $\mathbf{C}$, and $H(D)$ denote the set of all analytic functions on D. For $\alpha > 0$, a function $f \in H(D)$ is said to belong to the α-Bloch space B^α if

$$\|f\|_{B^\alpha} = \sup\{(1-|z|^2)^\alpha |f'(z)| : z \in D\} < +\infty$$

and to the little α-Bloch space B_0^α if

$$\lim_{|z|\to 1}(1-|z|^2)^\alpha |f'(z)| = 0.$$

Especially, when $\alpha = 1$ B^1 and B_0^1 are the classic Bloch space and little Bloch space. It is well known that B^α is a Banach space under the norm

$$\|f\|_\alpha = |f(0)| + \|f\|_{B^\alpha},$$

and that B_0^α is a closed subspace of B^α. See [1, 10].

*The research was supported in part by the Foundation of Fujian Educational Committee (JA04171).

Let X and Y be two analytic function spaces. We say a function ϕ is a pointwise multiplier from X to Y, if $\phi f \in Y$ for all $f \in X$. Let $M(X,Y)$ denote the space of all pointwise multipliers from X to Y, and $M(X) = M(X,X)$. By M_ϕ we denote the operator of multiplication by ϕ: $M_\phi f = \phi f, f \in X$. An application of the closed graph theorem shows that if $\phi \in M(X)$, then M_ϕ is a bounded linear transformation from X to Y. Hence it has a finite norm $\|M_\phi\|$.

In [2], L. Brown and A. L. Shields have characterized the classic pointwise multipliers spaces $M(B^1)$ and $M(B_0^1)$. Later, for all $\alpha > 0$, the pointwise multipliers spaces $M(B^\alpha)$ and $M(B_0^\alpha)$ have been studied in [10] by K. H. Zhu. Recently, between B^α and B^β, Lou and Ye have characterized the pointwise multipliers spaces $M(B^\alpha, B^\beta)$ in [3] and [9], respectively. These results are the following the Theorem A.

Theorem A
(1) $M(B^\alpha) = M(B_0^\alpha) = H^\infty$ *if* $\alpha > 1$;
(2) $M(B^\alpha) = B^\alpha$, $M(B_0^\alpha) = B_0^\alpha$ *if* $0 < \alpha < 1$;
(3) $M(B^\alpha) = M(B_0^\alpha) = \{\phi \in H^\infty : \sup_z (1-|z|^2)\ln\frac{1}{1-|z|^2}|f'(z)| < +\infty\}$ *if* $\alpha = 1$;
(4) If $0 < \beta < \alpha$, *then* $M(B^\alpha, B^\beta) = \{\phi : \phi = 0\}$;
(5) If $1 < \alpha < \beta$, *then* $M(B^\alpha, B^\beta) = B^\beta$, $M(B_0^\alpha, B_0^\beta) = B_0^\beta$;
(6) If $\beta > 1$, *then* $M(B^1, B^\beta) = M(B_0^1, B_0^\beta) = \{\phi \in H(D) : \sup_z (1-|z|^2)^\beta \ln\frac{1}{1-|z|^2}|f'(z)| < +\infty\}$.

Recently the composition operator to be compact has been studied in many papers such as [4], [5] and [7]. In this paper, we will discuss the conditions for the pointwise multiplier operator M_ϕ to be a compact operator from B^α to B^β. In this paper, C denotes the constant depending only on the index α, β. The C may differ at different places.

2. Some lemmas

Lemma 2.1. *For* $\alpha > 0$, $f \in B^\alpha$, *then*

(1) $|f(z)| \le (1 + \frac{2}{1-\alpha})\|f\|_\alpha$, *where* $\alpha < 1$;

(2) $|f(z)| \le \frac{1}{\ln 2}\ln\frac{2}{1-|z|^2}\|f\|_\alpha$, *where* $\alpha = 1$;

(3) $|f(z)| \le (1 + \frac{1}{(\alpha-1)(1-|z|)^{\alpha-1}})\|f\|_\alpha$, *where* $\alpha > 1$; *Especially, if* $|z| > r_* = 1 - \sqrt[(\alpha-1)]{\frac{1}{\alpha-1}}$, *then* $|f(z)| < \frac{2}{(\alpha-1)(1-|z|)^{\alpha-1}}\|f\|_\alpha$.

The proof follows from a direct calculation. We omit the details.

Lemma 2.2. *If $f \in B^\alpha$, $\alpha > 0$ and $0 < t < 1$, then $\|f_t\|_\alpha \leq \|f\|_\alpha$, where $f_t(z) = f(tz)$.*

The proof follows from a direct calculation. The details are omitted here.

Lemma 2.3. *Let M_ϕ be a bounded operator from B^α to B^β, then M_ϕ is compact if and only if for any bounded sequence $\{f_n\}$ in B^α which converges to 0 uniformly on compact subset of D, we have $\|M_\phi(f_n)\|_\beta \to 0$ as $n \to \infty$.*

Proof: The result can be proved by using Montel theorem, Lemma 2.1 and Lemma 2.2. The details are omitted here.

Lemma 2.4. *If $\phi \in H^\infty(D)$, then $(1 - |z|^2)|\phi'(z)| \leq \|\phi\|_\infty$.*

Proof: Using the following equation

$$\frac{1}{2\pi}\int_0^{2\pi} \frac{d\theta}{|e^{i\theta} - z|^2} = \frac{1}{1 - |z|^2}$$

and Cauchy integral formula, we can easily prove it; the details are omitted here.

Lemma 2.5.

If $0 < \alpha < 1$, and let $\{f_n\}$ be a bounded sequence in B^α which converges to 0 uniformly on compact subsets of D, then $\lim_{n\to\infty} \sup_D |f_n(z)| = 0$.

Proof: See [8, Lemma 3.2].

3. The main theorems

Theorem 3.1. *Let $\alpha > 0$, then M_ϕ is a compact operator from B^α to B^α if and only if $\phi = 0$.*

Proof: Suppose M_ϕ is compact in B^α, certainly M_ϕ is a bounded operator. We break the proof of the theorem down into three steps.

Step 1. Suppose $0 < \alpha < 1$. We know $\phi \in B^\alpha$ by Theorem A. Now assume that $\lim_{|z|\to 1} |\phi(z)| = 0$ fails. Then there exists a subsequence $\{z_n\} \subset D$ and an $\epsilon_0 > 0$ such that $|z_n| \to 1 (n \to \infty)$ and $|\phi(z_n)| \geq \epsilon_0$ for all n.

Let $z_n = r_n e^{i\theta_n}$, we take

$$f_n(z) = \frac{1 - r_n^2}{(1 - \overline{z_n} z)^\alpha}.$$

We get $\|f_n\|_\alpha \le 1 + 2^{\alpha+1}|\alpha|$ by a directly calculation. Then $\{f_n\}$ be a bounded sequence in B^α which converges to 0 uniformly on compact subsets of D.

On the other hand, we have

$$\begin{aligned}\|M_\phi f_n\|_\alpha &\ge (1-|z_n|^2)^\alpha |f_n'(z_n)\phi(z_n) + f_n(z_n)\phi'(z_n)| \\ &\ge (1-|z_n|^2)^\alpha \frac{\alpha(1-r_n^2)r_n}{(1-|z_n|^2)^{\alpha+1}}|\phi(z_n)| \\ &-(1-|z_n|^2)^\alpha \tfrac{1-r_n^2}{(1-|z_n|^2)^\alpha}|\phi'(z_n)| \\ &\ge \alpha r_n|\phi(z_n)| - \|\phi\|_\alpha (1-r_n)^{1-\alpha}.\end{aligned}$$

So we get $\|M_\phi(f_n)\|_\alpha \ge \alpha\epsilon_0$ as $n \to \infty$. This contradicts the compactness of M_ϕ by Lemma 2.3. This shows that $\lim_{|z|\to 1}|\phi(z)| = 0$ holds. Hence $\phi = 0$.

Step 2. Suppose $\alpha > 1$. We know $\phi \in H^\infty$ by Theorem A. Let $k = \sup_D |\phi(z)| < +\infty$. By virtue of the maximum principle, there exists a subsequence $\{z_n\} \subset D$ such that $\lim_{n\to\infty}|z_n| = 1$ and $\lim_{n\to\infty}|\phi(z_n)| = k$.

Let $z_n = r_n e^{i\theta_n}$, we take

$$f_n(z) = \frac{1-r_n^2}{(1-\overline{z_n}z)^\alpha}.$$

We get $\|f_n\|_\alpha \le 1 + 2^{\alpha+1}|\alpha|$ by a directly calculation. Then $\{f_n\}$ be a bounded sequence in B^α which converges to 0 uniformly on compact subsets of D.

On the other hand, using Lemma 2.4, we have

$$\begin{aligned}\|M_\phi f_n\|_\alpha &\ge (1-|z_n|^2)^\alpha |f_n'(z_n)\phi(z_n) + f_n(z_n)\phi'(z_n)| \\ &\ge (1-|z_n|^2)^\alpha \frac{\alpha(1-r_n^2)r_n}{(1-|z_n|^2)^{\alpha+1}}|\phi(z_n)| \\ &-(1-|z_n|^2)^\alpha \tfrac{1-r_n^2}{(1-|z_n|^2)^\alpha}|\phi'(z_n)| \\ &\ge \alpha r_n|\phi(z_n)| - (1-|z_n|^2)|\phi'(z_n)| \\ &\ge \alpha r_n|\phi(z_n)| - k\end{aligned}$$

Then $\liminf_{n\to\infty}\|M_\phi f_n\|_\alpha \ge (\alpha-1)k$. However, by Lemma 2.3, we know $\lim_{n\to\infty}\|M_\phi f_n\|_\alpha = 0$, since $\alpha - 1 > 0$. So $k = 0$. Hence $\phi = 0$.

Step 3. Suppose $\alpha = 1$. See [6, Corollary 1].

Completing the proof of Theorem 3.1.

Theorem 3.2. *Suppose $\beta > \alpha$ and $0 < \alpha < 1$, then M_ϕ is a compact operator from B^α to B^β if and only if $\phi \in B^\beta$.*

Proof: Assume $\phi \in B^\beta$. By Theorem A, we get M_ϕ is a bounded operator from B^α to B^β. Let $\{f_n\}$ be a bounded sequence in B^α which converges to 0 uniformly on compact subsets of D and $M = \sup_n \|f_n\|_\alpha < +\infty$. Note that $0 < \alpha < 1$. Given $\epsilon > 0$, there is $0 < r < 1$ such that $|z| > r$ implies

$$(1 - |z|^2)^{\beta-\alpha} < \frac{\epsilon}{\|\phi\|_\beta M}, \tag{1}$$

$$(1 - |z|^2)^{1-\alpha} < \frac{\epsilon}{\|\phi\|_\beta M}, \tag{2}$$

and

$$(1 - |z|^2)^{1-\alpha} \ln \frac{2}{1 - |z|^2} < \frac{\epsilon}{\|\phi\|_\beta M}. \tag{3}$$

We know that the sequence $\{f_n\}$ converges to uniformly on $|z| \leq r$, which follows that the $\{f_n'\}$converges to uniformly on $|z| \leq r$. It follows that for large enough n,

$$\sup_{|z|\leq r} |f_n'(z)| < \frac{\epsilon}{\|\phi\|_\beta}. \tag{4}$$

On the other hand, by Lemma 2.5, for large enough n we have

$$\sup_{z\in D} |f_n(z)| < \frac{\epsilon}{\|\phi\|_\beta}. \tag{5}$$

So, by (5),

$$\begin{aligned}
\|M_\phi f_n\|_{B^\beta} &= \sup_z (1 - |z|^2)^\beta |(\phi f_n)'(z)| \\
&\leq \sup_z (1 - |z|^2)^\beta |\phi'(z)||f_n(z)| + \sup_z (1 - |z|^2)^\beta |\phi(z)||f_n'(z)| \\
&\leq \epsilon + + \sup_z (1 - |z|^2)^\beta |\phi(z)||f_n'(z)| \\
&\triangleq \epsilon + I_1.
\end{aligned}$$

Now the remain is broken into three steps.

First, suppose $\beta < 1$. By (2), (4) and Lemma 2.1, for large enough n we have

$$\begin{aligned} I_1 &= \sup_{|z|\le r}(1-|z|^2)^\beta|\phi(z)||f_n'(z)| + \sup_{|z|>r}(1-|z_n|^2)^\beta|\phi(z)||f_n'(z)| \\ &\le (1+\frac{2}{1-\beta})\|\phi\|_\beta \sup_{|z|\le r}|f_n'(z)| \\ &\quad + \|f_n\|_{B^\alpha}(1+\tfrac{2}{1-\beta})\|\phi\|_\beta \sup\nolimits_{|z|>r}(1-|z_n|^2)^{\beta-\alpha} \\ &< (2+\tfrac{4}{1-\beta})\epsilon. \end{aligned}$$

Next, suppose $\beta > 1$. By [10, Proposition 7], there exists a constant k only depending β such that $\sup(1-|z|^2)^{\beta-1}|\phi(z)| \le k < +\infty$. It follows that, by combing (1) and (4),

$$\begin{aligned} I_1 &\le k\sup_{|z|\le r}(1-|z|^2)|f_n'(z)| + k\sup_{|z|>r}(1-|z_n|^2)|f_n'(z)| \\ &\le k\frac{\epsilon}{\|\phi\|_\beta} + k\|f_n\|_{B^\alpha}\sup_{|z|>r}(1-|z_n|^2)^{1-\alpha} < \frac{2k}{\|\phi\|_\beta}\epsilon. \end{aligned}$$

Finally, suppose $\beta = 1$. By Lemma 2.1, (3) and (4), we have

$$\begin{aligned} I_1 &\le \sup_{|z|\le r}(1-|z|^2)\frac{1}{\ln 2}\ln\frac{2}{1-|z|^2}\|\phi\|_\beta|f_n'(z)| \\ &\quad + \|f_n\|_{B^\alpha}\|\phi\|_\beta\tfrac{1}{\ln 2}\sup\nolimits_{|z|>r}(1-|z|^2)^{1-\alpha}\ln\tfrac{2}{1-|z|^2} \\ &\le \frac{2}{e\ln 2}\|\phi\|_\beta\sup_{|z|\le r}|f_n'(z)| + M\|\phi\|_\beta\frac{1}{\ln 2}\sup_{|z|>r}(1-|z|^2)^{1-\alpha}\ln\frac{2}{1-|z|^2} \\ &\le \frac{2}{e\ln 2}\epsilon + \frac{1}{\ln 2}\epsilon. \end{aligned}$$

Which implies that $\|M_\phi(f_n)\|_\beta \longrightarrow 0$ as $n \to \infty$. Thus M_ϕ is a compact by Lemma 2.3.

Assume that M_ϕ is compact from B^α to B^β, then M_ϕ is bounded from B^α toB^β. It follows that $M_\phi \in B^\beta$ by Theorem A. The proof is complete.

Theorem 3.3. *Let $\beta > 1$, then M_ϕ is a compact operator from B^1 to B^β if and only if $\phi \in H(D)$ and*

$$\lim_{|z|\to 1}(1-|z|^2)^\beta\ln\frac{1}{1-|z|^2}|\phi'(z)| = 0. \tag{6}$$

Proof: Assume $\phi \in H(D)$ and (6) hold. By Theorem A, we know that M_ϕ is bounded from B^1 to B^β, which implies $\phi \in B^\beta$. Let $\{f_n\}$ be a bounded sequence in B^1 which converges to 0 uniformly on compact subsets of D. Let $M = \sup_n \|f_n\|_1 < +\infty$. By [10, Proposition 7] and (6), we have

$$\lim_{|z|\to 1}(1-|z|^2)^{\beta-1}|\phi(z)| = 0.$$

Then, $K = \sup_D (1-|z|^2)^{\beta-1}|\phi(z)| < +\infty$ and given $\epsilon > 0$, there is $0 < r < 1$, we can suppose $r > r_0 = \sqrt{1-\frac{1}{e}}$, such that $|z| > r$ implies

$$(1-|z|^2)^\beta \ln \frac{1}{1-|z|^2}|\phi'(z)| < \frac{\epsilon}{M}, \tag{7}$$

$$(1-|z|^2)^{\beta-1}|\phi(z)| < \frac{\epsilon}{M}, \tag{8}$$

$$\sup_{|z|\le r}|f_n(z)| < \frac{\epsilon}{\|\phi\|_\beta}, \tag{9}$$

$$\sup_{|z|\le r}|f_n'(z)| < \frac{\epsilon}{K}. \tag{10}$$

On the other hand,

$$\begin{aligned}\|M_\phi f_n\|_{B^\beta} &= \sup_z (1-|z_n|^2)^\beta |(\phi f_n)'(z)| \\ &\le \sup_z (1-|z_n|^2)^\beta |\phi'(z)||f_n(z)| + \sup_z (1-|z_n|^2)^\beta ||\phi(z)||f_n'(z)| \\ &\triangleq I_2 + I_3.\end{aligned}$$

Thus for large enough n we have, by (7) and (9),

$$\begin{aligned}I_2 &= \sup_{|z|\le r}(1-|z|^2)^\beta|\phi'(z)||f_n(z)| + \sup_{|z|>r}(1-|z|^2)^\beta|\phi'(z)||f_n(z)| \\ &\le \|\phi\|_\beta \sup_{|z|\le r}|f_n(z)| + \sup_{|z|>r}(1-|z|^2)^\beta \frac{\|f_n\|_1}{\ln 2}\ln\frac{2}{1-|z|^2}|\phi'(z)| \\ &< \epsilon + \frac{2M}{\ln 2}\sup_{|z|>r}(1-|z|^2)^\beta \ln\frac{1}{1-|z|^2}|\phi'(z)| \\ &< \epsilon + \frac{2}{\ln 2}\epsilon,\end{aligned}$$

and, by (8) and (10),

$$I_3 = \sup_{|z|\le r}(1-|z|^2)^\beta|\phi(z)||f_n'(z)| + \sup_{|z|>r}(1-|z_n|^2)^\beta|\phi(z)||f_n'(z)|$$

$$\le K\sup_{|z|\le r}|f_n'(z)| + \|f_n\|_1 \sup_{|z|>r}(1-|z|^2)^{\beta-1}|\phi(z)|$$

$$< K\frac{\epsilon}{K} + M\frac{\epsilon}{M} = 2\epsilon.$$

So $\|M_\phi(f_n)\|_\beta \longrightarrow 0$ as $n\to\infty$. Hence M_ϕ is a compact by Lemma 2.3.

Conversely, if M_ϕ is compact from B^1 to B^β, then it is bounded. By Theorem A, we know that

$$\sup_z(1-|z|^2)^\beta \ln\frac{1}{1-|z|^2}|\phi'(z)| \le C < +\infty,$$

which implies $\phi \in B_0^\beta$. Then, by [10, Proposition 7], we get

$$\lim_{|z|\to 1}(1-|z|^2)^{\beta-1}|\phi(z)| = 0. \tag{11}$$

Now assume that (6) fails. Then there exists a subsequence $\{z_n\}\subset D$ and an $\epsilon_0>0$ such that $|z_n|\to 1 (n\to\infty)$ and

$$(1-|z_n|^2)^\beta \ln\frac{1}{1-|z_n|^2}|\phi'(z_n)| \ge \epsilon_0$$

for all n.

Let $z_n = r_n e^{i\theta_n}$, we take

$$f_n(z) = (\ln\frac{2}{1-|z_n|^2})^{-1}(\ln\frac{2}{1-\overline{z_n}z})^2,$$

We get $\|f_n\|_1 \le 4$ by a directly calculation. Then $\{f_n\}$ be a bounded sequence in B^1 which converges to 0 uniformly on compact subsets of D.

On the other hand,

$$\|M_\phi f_n\|_\beta \ge (1-|z_n|^2)^\beta|f_n'(z_n)\phi(z_n) + f_n(z_n)\phi'(z_n)|$$

$$\ge (1-|z_n|^2)^\beta \ln\frac{2}{1-|z_n|^2}|\phi'(z_n)| - 2|z_n|(1-|z_n|^2)^{\beta-1}|\phi(z_n)|.$$

Noting (11), we have $\liminf_{n\to\infty}\|M_\phi f_n\|_\beta \ge \epsilon_0$. This contradicts the compactness of M_ϕ by Lemma 2.3. This shows that (6) holds. The proof is completed.

Theorem 3.4. *Let $\beta > \alpha > 1$, then M_ϕ is a compact operator from B^α to B^β if and only if $\phi \in B_0^{\beta-\alpha+1}$.*

Proof: Firstly suppose $\phi \in B_0^{\beta-\alpha+1}$. We know M_ϕ is bounded from B^α toB^β by Theorem A. Suppose $\{f_n\}$ be a bounded sequence in B^α which converges to 0 uniformly on compact subsets of D. Let $M = \sup_n \|f_n\|_\alpha < +\infty$ and $K = \sup_z (1-|z|^2)^{\beta-\alpha}|\phi(z)| < +\infty$. Since $\phi \in B_0^{\beta-\alpha+1}$, and by [10, Proposition 7], then given $\epsilon > 0$, there is $0 < r < 1$, we can suppose $r > r_*$, such that $|z| > r$ implies

$$(1-|z|^2)^{\beta-\alpha+1}|\phi'(z)| < \frac{\epsilon(\alpha-1)}{4M}, \tag{12}$$

and

$$(1-|z|^2)^{\beta-\alpha}|\phi(z)| < \frac{\epsilon}{M}. \tag{13}$$

We know that the sequence $\{f_n\}$ converges to uniformly on $|z| \le r$, which follows that the $\{f_n'\}$converges to uniformly on $|z| \le r$. It follows that for large enough n and $|z| \le r$, we get

$$|f_n(z)| < \frac{\epsilon}{\|\phi\|_{B^\beta}}, \tag{14}$$

and

$$|f_n'(z)| < \frac{\epsilon}{K}. \tag{15}$$

On the other hand,

$$\begin{aligned}\|M_\phi f_n\|_{B^\beta} &= \sup_z (1-|z|^2)^\beta |(\phi f_n)'(z)| \\ &\le \sup_z (1-|z|^2)^\beta |\phi'(z)||f_n(z)| + \sup_z (1-|z|^2)^\beta ||\phi(z)||f_n'(z)| \\ &\triangleq I_4 + I_5.\end{aligned}$$

Thus for large enough n we have, by (12) and (14),

$$\begin{aligned}I_4 &= \sup_{|z|\le r} (1-|z|^2)^\beta |\phi'(z)||f_n(z)| + \sup_{|z|>r} (1-|z|^2)^\beta |\phi'(z)||f_n(z)| \\ &\le \|\phi\|_{B^\beta} \sup_{|z|\le r} |f_n(z)| + \sup_{|z|>r} (1-|z|^2)^\beta \frac{2}{(\alpha-1)(1-|z|)^{\alpha-1}} \|f_n\|_\alpha |\phi'(z)| \\ &< \epsilon + M\tfrac{4}{\alpha-1}(1-|z|)^{\beta-\alpha+1}|\phi'(z)| < 2\epsilon,\end{aligned}$$

and by (13), (15),

$$I_5 = \sup_{|z|\le r}(1-|z|^2)^\beta|\phi(z)||f_n'(z)| + \sup_{|z|>r}(1-|z|^2)^\beta|\phi(z)||f_n'(z)|$$

$$\le K(1-|z|^2)^\alpha \sup_{|z|\le r}|f_n'(z)| + \|f_n\|_{B^\alpha}\sup_{|z|>r}(1-|z|^2)^{\beta-\alpha}|\phi(z)|$$

$$< 2\epsilon.$$

So $\|M_\phi(f_n)\|_\beta \longrightarrow 0$ as $n\to\infty$. Hence M_ϕ is compact by Lemma 2.3.

Conversely, if M_ϕ is compact from B^α to B^β, then it is bounded. If $\phi\in B_0^{\beta-\alpha+1}$ fails, by [10, Proposition 7] we have

$$\lim_{|z|\to 1}(1-|z|^2)^{\beta-\alpha}|\phi(z)| \ne 0.$$

Then there exists a subsequence $\{z_n\}\subset D$ and an $\epsilon_0>0$ such that $|z_n|\to 1(n\to\infty)$ and $(1-|z_n|^2)^{\beta-\alpha}|\phi(z_n)|\ge\epsilon_0$ for all n. Set

$$f_n(z) = \frac{(1-|z_n|^2)(z-z_n)}{(1-\overline{z_n}z)^\alpha}.$$

We get $\|f_n\|_\alpha \le 1+2^{\alpha+1}|\alpha|$ by a directly calculation. Then $\{f_n\}$ be a bounded sequence in B^α which converges to 0 uniformly on compact subsets of D.

On the other hand,

$$\|M_\phi f_n\|_\alpha \ge (1-|z_n|^2)^\beta|f_n'(z_n)\phi(z_n)+f_n(z_n)\phi'(z_n)|$$

$$= (1-|z_n|^2)^\beta|\phi(z_n)||f'(z_n)|$$

$$= (1-|z_n|^2)^{\beta-\alpha}|\phi(z_n)|.$$

Let $n\to\infty$, we get $0\ge\epsilon_0$, this is contradict. So $\phi\in B_0^{\beta-\alpha+1}$. The proof is completed.

References

1. J. M. Anderson, J. Clunie, Ch. Pommerenke, On Bloch functions and normal functions, *J. Reine Angew. Math.* **270**(1974): 12-37.
2. L. Brown, A. L. Shields, Multipliers and cyclic vectors in the Bloch space, *Michigan Math. J.* **38**(1991): 141-146.
3. Z. J. Lou, J. D. Guo, D. J. Song, Multipliers and cyclic vectors in Bloch type spaces, *Acta Math. Sin.* **19**(2003):79-88.
4. L. Luo, S. J. Huai, Composition operators between Hardy spaces on the unit ball, *Acta Math. Sin.* **44**(2001): 209-216.

5. K. Madigan, A. Matheson, Compact composition operators on the Bloch space, *Trans. Amer. Math. Soc.,* **347**(1995):2679-2687.
6. S. C. Ohao, R. H. Zhao, Weighted composition operators on the Bloch space, *Bull. Austral. Math.* **63**(2001):177-185.
7. W. Smith, Composition operators between Bergman and Hardy spaces, *Trans. Amer. Math. Soc.,* **348**(1996):2331-2348.
8. X. J. Zhang, Extend cesaro operator on the Dirichlet type spaces and Bloch type spaces of C^n, *Chin. Ann. Math.* **26A**(2005): 138-150.
9. S. L. Ye, Pointwise multiplies in the Bloch-type spaces, *J. Fujian normal univ.* (2005), In press.
10. K. H. Zhu, Bloch type spaces of analytic function, *Rocky Mountain J. Math.* **23**(1993):1143-1177.
11. K. H. Zhu, *Operator Theory in function spaces*, Deker, New York, 1990.

AN IMPROVED INEQUALITY FOR MEROMORPHIC FUNCTIONS WITH FEW SIMPLE ZEROS*

QINGDE ZHANG
Department of Computational Science
Chengdu University of Information Technology
Chengdu 610041, P. R. China
E-mail: qdzhang@cuit.edu.cn

Let $f(z)$ be a non-constant meromorphic function in the complex plane satisfying $N_1(r, \frac{1}{f}) = S(r, f)$ and d be a nonzero constant. Then

$$T(r,f) < 11\overline{N}(r,f) + 11\overline{N}(r, \frac{1}{f' + df - 1}) + S(r,f),$$

except that $f(z)$ has one of the following forms: $(1) Ae^{-dz}$, $(2) \frac{1}{d}(Ae^{-dz} + 1)^2$, $(3) \frac{1}{d}(Ae^{-\frac{1}{2}dz} - 1)^2$, $(4) Ae^{dz}(e^{-dz} + \frac{1}{2dA})^2$, where A is a non-zero constant.

Keywords: Zero, differential polynimial, inequality.

2000 Mathematics Subject Classification: 30D35

1. Main Results

In 2004 we obtained the following result[1].

Theorem A *Let $f(z)$ be a non-linear meromorphic function in the complex plane satisfying $N_{1)}(r, \frac{1}{f}) = S(r, f)$. Then*

$$T(r,f) < 11\overline{N}(r,f) + 11\overline{N}(r, \frac{1}{f' - 1}) + S(r,f),$$

here and henceforth $S(r,f)$ denotes $o\{T(r,f)\}$ $(r \to +\infty)$, except possibly a set of finite linear measure in the interval $[0, +\infty)$. Now we consider the

*Project supported by the National Science Foundation of China (9971097).

linear differential polynomial $f' + df - 1$ instead of $f' - 1$ and obtain the following result.

Theorem 1.1. *Let $f(z)$ be a non-constant meromorphic function in the complex plane satisfying $N_{1)}(r, \frac{1}{f}) = S(r, f)$ where $N_1(r, \frac{1}{f})$ denotes the counting function of the simple zeros of the function $f(z)$ and d be a nonzero constant. Then*

$$T(r,f) < 11\overline{N}(r,f) + 11\overline{N}(r, \frac{1}{f' + df - 1}) + S(r,f), \tag{1}$$

except that $f(z)$ has one of the following forms:

$(1) Ae^{-dz}$, $(2) \frac{1}{d}(Ae^{-dz} + 1)^2$, $(3) \frac{1}{d}(Ae^{-\frac{1}{2}dz} - 1)^2$, $(4) Ae^{dz}(e^{-dz} + \frac{1}{2dA})^2$, *where A is a non-zero constant.*

The above 4 forms of functions satisfy the conditions that $N(r, f) = 0, N(r, \frac{1}{f}) = 0$ and $N(r, \frac{1}{f'+df-1}) = 0$. Therefore (1.1) does not hold for them.

We use the common symbols and notations of Nevanlinna value distribution theory[2]. $N_{1)}(r, \frac{1}{f}), N_{(2)}(r, \frac{1}{f})$ and $N_{(3}(r, \frac{1}{f})$ denote the counting functions of the simple zeros, the zeros of multiplicity 2 and the zeros of multiplicity greater than 2 of $f(z)$ respectively. $\overline{N}_{(2)}(r, \frac{1}{f})$ and $\overline{N}_{(3}(r, \frac{1}{f})$ denote the corresponding functions without counting the multiplicity.

We give some lemmas in Section 2 and the proof of Theorem 1 in Section 3.

2. Some Lemmas

In this section we always assume that the function $f(z)$ is a non-constant meromorphic function in the complex plane satisfying $N_{1)}(r, \frac{1}{f}) = S(r, f)$ and d is a fixed nonzero constant.

Lemma 2.1. *Assume that $f(z)$ is not of the form $Ae^{-dz} (A \neq 0)$. Then $f''(z) + df'(z)$ is not a constant.*

Proof. Assume that $f''(z) + f'(z)$ be a constant. Then $f(z) = Ae^{-dz} + Cz + B$, where A, B and C are constants.

When $AC \neq 0, AB \neq 0$ or $BC \neq 0$, $N_{1)}(r, \frac{1}{f}) = N(r, \frac{1}{f}) \sim T(r, f)$. This is contradictory to the above assumption. Hence $A \neq 0, B = C = 0$. Again it is contradictory to the assumption of Lemma 1. Therefore $f''(z) + f'(z)$ is not a constant. Lemma 1 is proved.

Henceforth we assume that $f(z)$ is not of form Ae^{-dz}.

Set

$$g(z) = \frac{f'(z)}{f(z)}, \tag{1}$$

$$G(z) = f'(z) + df(z) - 1, \tag{2}$$

$$h(z) = \frac{G''(z)}{G'(z)} = \frac{f'''(z) + df''(z)}{f''(z) + df'(z)}, \tag{3}$$

$$F(z) = 18g'(z) + 9g^2(z) - 6g(z)h(z) + 3h'(z) - 2h^2(z) \\ + 6dg(z) - 5dh(z) - 2d^2. \tag{4}$$

Lemma 2.2. *Every zero of $f(z)$ with multiplicity 2 is a zero of $F(z)$.*

Proof. Let $z = z_0$ be a zero of $f(z)$ with multiplicity 2. Then near $z = z_0$ there is an expansion

$$f(z) = K\{(z - z_0)^2 + a(z - z_0)^3 + b(z - z_0)^4 + O(z - z_0)^5\}(K \neq 0).$$

From (2.1) $\sim$ (2.3) and the above expression we obtain

$$g(z) = \frac{2}{z - z_0}\{1 + \frac{1}{2}a(z - z_0) + (b - \frac{1}{2}a^2)(z - z_0)^2 + O(z - z_0)^3\}, \tag{5}$$

$$h(z) = (3a + d) + (12b - 9a^2 - 3da - d^2)(z - z_0) + O(z - z_0)^2. \tag{6}$$

Hence we have

$$g'(z) = -\frac{2}{(z - z_0)^2}\{1 - (b - \frac{1}{2}a^2)(z - z_0)^2 + O(z - z_0)^3\}, \tag{7}$$

$$g^2(z) = \frac{4}{(z - z_0)^2}\{1 + a(z - z_0) + (2b - \frac{3}{4}a^2)(z - z_0)^2 + O(z - z_0)^3\}, \tag{8}$$

$$g(z)h(z) = \frac{2}{z - z_0}\{(3a + d) + (12b - \frac{15}{2}a^2 - \frac{5}{2}da - d^2)(z - z_0) \\ + O(z - z_0)^2\}, \tag{9}$$

$$h'(z) = 12b - 9a^2 - 3da - d^2 + O(z - z_0), \tag{10}$$

$$h^2(z) = (3a + d)^2 + O(z - z_0). \tag{11}$$

Subsituting (2.5) $\sim$ (2.11) into (2.4) we obtain that $F(z) = O(z - z_0)$. Namely $z = z_0$ is a zero of the function $F(z)$. Lemma 2 is proved.

Lemma 2.3. *Assume that $F(z) \not\equiv 0$. Then*

$$\overline{N}_{(2)}(r, \frac{1}{f}) \leq 2\overline{N}_{3)}(r, \frac{1}{f}) + 2\overline{N}(r, f) + 2\overline{N}(r, \frac{1}{f' + df - 1}) + 2N_o(r, \frac{1}{G'}) + S(r, f), \tag{12}$$

where $N_0(r, \frac{1}{G'})$ *does not count the zeros of* $f'(z) + df(z) - 1$ *or zeros of* $f(z)$ *with multiplicity greater than 2.*

Proof. From $m(r, g) = S(r, f)$ and $m(r, h) = S(r, f)$ we have $m(r, F) = S(r, f)$. By Lemma 2 we have

$$N_{(2)}(r, \frac{1}{f}) \leq N(r, \frac{1}{F}) \leq T(r, F) + O(1) \leq N(r, F) + S(r, f). \tag{13}$$

From (2.4) we see that only the poles of $f(z)$,the zeros of $G'(z)$ or the zeros of $f(z)$ of multiplicity not equal to 2 can be the poles of $F(z)$ with multiplicity not greater than 2. Hence

$$\begin{aligned} N(r, F) &\leq 2\overline{N}(r, f) + 2\overline{N}(r, \frac{1}{G}) + 2N_0(r, \frac{1}{G'}) + 2N_{1)}(r, \frac{1}{f}) + 2\overline{N}_{(3}(r, \frac{1}{f}) \\ &\leq 2\overline{N}(r, f) + 2\overline{N}(r, \frac{1}{G}) + 2N_0(r, \frac{1}{G'}) + 2\overline{N}_{(3}(r, \frac{1}{f}) + S(r, f), \end{aligned} \tag{14}$$

since $N_{1)}(r, \frac{1}{f}) = S(r, f)$ by the assumption before Lemma 1. Substituting (2.14) into (2.13) we obtain (2.12). Lemma 3 is proved.

Set

$$k(z) = (3g(z) + h(z) - d)\frac{f'(z)}{f(z)} - 3\frac{f''(z)}{f(z)}. \tag{15}$$

Substitute $\dfrac{f'(z)}{f(z)} = g(z)$ and $\dfrac{f''(z)}{f(z)} = g'(z)+g^2(z)$ into the above expression we obtain

$$k(z) = -3g'(z) + g(z)h(z) - dg(z). \tag{16}$$

Lemma 2.4. *We have the identity*

$$f'(z)F(z) \equiv f(z)\{3k'(z) + k(z)(3g(z) - 2h(z) + 2d)\}. \tag{17}$$

Proof. From (2.15) we have

$$kf = (3g + h - d) - 3f' - 3f''. \tag{18}$$

Differentiating the above identity we have

$$k'f + kf' = (3g' + h')f' + (3g + h - d)f'' - 3f'''.$$

From (2.3) we have $f''' = (h - d)f'' + dhf'$ and substituting it into the above identity we obtain

$$k'f + kf' = (3g' + h' - 3dh)f' + (3g - 2h + 2d)f''.$$

From (2.18) we have $3f'' = (3g+h-d)f' - kf$ and substituting it into the above identity we obtain

$$3k'f + 3kf' = (9g' + 3h' - 9dh)f' + (3g - 2h + 2d)\{(3g + h - d)f' - kf\}.$$

Hence

$$f\{3k'+k(3g-2h+2d)\} = f'\{(9g'+3h'-9dh)+(3g-2h+2d)(3g+h-d)-3k\}.$$

Substituting the expression of $k(z)$ in (2.16) into the right side of the above identity we obtain (2.17). Lemma 4 is proved.

Lemma 2.5. *Assume that* $k(z) \equiv 0$. *Then* $f(z) = A\frac{1}{e^{-dz} - C}$, *where* A *and* C *are constants with* $A \neq 0$.

Proof. From (2.16) we have $3\frac{g'}{g} = h - d$. Hence $g^3 = A_1 G' e^{-dz} (A_1 \neq 0)$. Since $g = \frac{f'}{f}$ and $G' = (f'e^{dz})'e^{-dz}$, the above identity becomes $\frac{f'}{f^3} = A_1 \frac{(f'e^{dz})'}{(f'e^{dz})^2}$. Therefore $\frac{1}{f^2} = 2A_1 \frac{1}{f'e^{dz}} + B_1$ where B_1 is a constant. Hence $\frac{f'}{f^2} = 2A_1 e^{-dz} + B_1 f'$ and $\frac{1}{f} + B_1 f + C_1 = \frac{2A_1}{d} e^{-dz}$ where C_1 is a constant.

Now we assume that $B_1 \neq 0$. From the above identity we see that $f \neq 0, \infty, \alpha$ where α is one of the nonzero root of the equation $B_1 x^2 + C_1 x + 1 = 0$. Hence $f(z)$ is a constant. This is a contradiction. Therefore $B_1 = 0$ and the function $f(z)$ has the form $\frac{1}{\frac{2A_1}{d} e^{-dz} - C_1}$. Lemma 5 is proved.

Lemma 2.6. *Assume that* $F(z) \equiv 0$ *and* $k(z) \not\equiv 0$. *Then* $f(z)$ *is one of the following forms:*

$$(I) A(e^{-dz} + C)^2,\ (II) A\frac{(e^{-dz} - C)^2}{(e^{-dz} - B)},\ (III) A(e^{\frac{d}{2}z} + C)^2,$$

where A, B *and* C *are constants with* $A \neq 0$ *and* $B \neq C$.

Proof. From Lemma 4 we have $3k' + k(3g - 2h + 2d) = 0$. Hence $3\frac{k'}{k} + 3g = 2h - 2d$ and we have

$$f^2 k^3 = A_1 (G' e^{-dz})^2, \tag{19}$$

where A_1 is a non-zero constant. But

$$f^3 k^3 = f^3 g^3 \{-3\frac{g'}{g} + h - d\}^3 = f^3 g^3 \{(\frac{G'e^{-dz}}{g^3})' / (\frac{G'e^{-dz}}{g^3})\}^3. \tag{20}$$

Subsituting (2.20) into (2.19) we obtain $[(\frac{G'e^{-dz}}{g^3})']^3/(\frac{G'e^{-dz}}{g^3})^5 = A_1(\frac{f'}{f})^3$.

Now we assume that there is a point $z = z_1$ in the complex plane such that $f(z_1) \neq 0, \infty$, $f'(z_1) \neq 0$,$G'(z_1) \neq 0$, and $(\frac{G'e^{-dz}}{g^3})|_{z=z_1} \neq 0$. Then near $z = z_1$ we have $(\frac{G'e^{-dz}}{g^3})/(\frac{G'e^{-dz}}{g^3})^{\frac{5}{3}} = A_1^{\frac{1}{3}}\frac{f'}{f}$, where $(\frac{G'e^{-dz}}{g^3})^{\frac{1}{3}}$ is an arbitrary branch. Hence we have $(\frac{G'e^{-dz}}{g^3})^{-\frac{2}{3}} = \frac{2}{3}A_1^{\frac{1}{3}}(\frac{1}{f} + D_1)$, where D_1 is a constant. Therefore $(\frac{g^3}{G'e^{-dz}})^2 = \frac{8}{27}A_1(\frac{1}{f} + D_1)^3$.

This identity holds near $z = z_1$, hence in the whole complex plane. From it we have

$$[(f'e^{dz})']^2/(f'e^{dz})^4 = \frac{27}{8A_1}\frac{f'^2}{f^3(1 + D_1 f)^3}.$$

Now we assume that there is a point $z = z_2$ in the complex plane such that $f(z_2) \neq 0, \infty$,$f'(z_2) \neq 0$,$1 + D_1 f(z_2) \neq 0$, and $(f'e^{dz})'|_{z=z_2} \neq 0$. Then near $z = z_2$ we have

$$(f'e^{dz})'/(f'e^{dz})^2 = (\frac{27}{8A_1})^{\frac{1}{2}}\frac{f'}{\sqrt{(f^3(1 + D_1 f)^3}},$$

where $\sqrt{f(1 + D_1 f)}$ is an arbitrary branch.

Having integrated it we have

$$e^{-dz} = 2(\frac{27}{8A_1})^{\frac{1}{2}}\{f'[\sqrt{\frac{1 + D_1 f)}{f}} + D_1\sqrt{\frac{f}{1 + D_1 f}}] + B_1 f'\},$$

where B_1 is a constant. From it we obtain

$$-\frac{1}{d}e^{-dz} = 4(\frac{27}{8A_1})^{\frac{1}{2}}\{\sqrt{f(1 + D_1 f)} + B_1 f + C_1\},$$

where C_1 is a constant. It is easy to see from this expression that $f(1 + D_1 f) = (B_1 f + C_1 + A_2 e^{-dz})^2$ where A_2 is a nonzero constant. This identity holds near $z = z_2$ and therefore in the whole complex plane. Thus

$$({B_1}^2 - D_1)f^2 + [2B_1(A_2 e^{-dz} + C_1) - 1]f + (A_2 e^{-dz} + C_1)^2 = 0.$$

Assume first that ${B_1}^2 - D_1 = 0$.

(1) $B_1 = 0$. We have $f(z) = (A_2 e^{-dz} + C_1)^2$, namely $f(z)$is of the form (I).

(2)$B_1 \neq 0$. We have $f(z) = -\frac{1}{2B_1}\frac{A_2 e^{-dz} + C_1)^2}{A_2 e^{-dz} + C_1 - \frac{1}{2B_1}}$, namely $f(z)$ is of the form (II).

Now assmue that $B_1{}^2 - D_1 \neq 0$. Then

$$f(z) = \frac{1 - 2B_1(A_2e^{-dz} + C_1) + \sqrt{\Delta}}{2B_1{}^2 - D_1},$$

where

$$\begin{aligned}\Delta &= [2B_1(A_2e^{-dz} + C_1) - 1]^2 - 4(B_1{}^2 - D_1)(A_2e^{-dz} + C_1)^2 \\ &= 4D_1(A_2e^{-dz} + C_1)^2 - 4B_1(A_2e^{-dz} + C_1) + 1.\end{aligned}$$

Since $\sqrt{\Delta}$ is a meromorphic function in the complex plane and $B_1{}^2 - D_1 \neq 0$, it deduces that $D_1 = 0$ (*hence* $B_1 \neq 0$) and $4B_1C_1 = 1$. Hence

$$f(z) = -\frac{A_2}{B_1}e^{-dz} + \sqrt{-\frac{A_2}{B_1^3}e^{-\frac{1}{2}dz}} + \frac{1}{4B_1{}^2},$$

namely $f(z)$ is of the form (III). Lemma 6 is proved.

3. Proof of Theorem 1

Assume that $f(z)$ is not of the form $Ae^{-dz}(A \neq 0)$. By Lemma 1 $f'' + df' - 1$ is not a constant. By Milloux Inequality [2] we have

$$T(r, f) < \overline{N}(r, f) + N(r, \frac{1}{f}) + \overline{N}(r, \frac{1}{f' + df - 1}) - N_{oo}(r, \frac{1}{f'' + df'}) + S(r, f),$$

where $N_{oo}(r, \frac{1}{f''+df'})$ does not count the multiple zeros of the function $f' + df - 1$. Now we assume that the counting function $N_o(r, \frac{1}{f''+df'})$ does not count the multiple zeros of the function $f' + df - 1$ or the zeros of $f(z)$ with multiplicity greater than 1. Then we have

$$T(r, f) < \overline{N}(r, f) + 2\overline{N}_{(2}(r, \frac{1}{f}) + \overline{N}(r, \frac{1}{f' + df - 1}) - N_o(r, \frac{1}{f'' + df'}) + S(r, f), \tag{21}$$

since $N_{1)}(r, \frac{1}{f}) = S(r, f)$ by the assumption of Theorem 1. Obviously we have

$$T(r, f) \geq N(r, \frac{1}{f}) - O(1) \geq 2N_{(2)}(r, \frac{1}{f}) + 3\overline{N}_{(3}(r, \frac{1}{f}) - O(1). \tag{22}$$

Combining (3.1) and (3.2) we have

$$\overline{N}_{(3}(r, \frac{1}{f}) < \overline{N}(r, f) + \overline{N}(r, \frac{1}{f' + df - 1}) - N_o(r, \frac{1}{f'' + df'}) + S(r, f). \tag{23}$$

Now we distinguish three cases.

(1)$F(z) \not\equiv 0$. From the above inequality and (2.12) in Lemma 3 we have

$$\begin{aligned}\overline{N}_{(2}(r,\frac{1}{f}) &= \overline{N}_{(2)}(r,\frac{1}{f}) + \overline{N}_{(3}(r,\frac{1}{f}) \\ &\le 3\overline{N}_{(3}(r,\frac{1}{f}) + 2\overline{N}(r,f) + 2\overline{N}(r,\frac{1}{f'+df-1}) \\ &\quad + 2N_o(r,\frac{1}{f''+df'}) + S(r,f) \\ &\le 5\overline{N}(r,f) + 5\overline{N}(r,\frac{1}{f'+df-1}) + S(r,f).\end{aligned}$$

Substituting the above inequality into (3.1) we obtain (1.1).

(2)$k(z) \equiv 0$. From Lemma 5 $f(z)$ is of the form $A\dfrac{1}{e^{-dz}-C}$ where $A \neq 0$. When $C = 0$ we have $\overline{N}(r,\frac{1}{f'+df-1}) \sim \frac{dr}{\pi} \sim T(r,f)$. When $C \neq 0$ we have$\overline{N}(r,f) \sim \frac{dr}{\pi} \sim T(r,f)$. Therefore (1.1) holds in this case.

(3)$F(z) \equiv 0$ and $k(z) \not\equiv 0$. By Lemma 6 we distinguish three cases.

(3.1)$f(z) = A(e^{-dz}+C)^2$. Obviously (1.1) holds for $AdC^2 \neq 1$. When $AdC^2 = 1$ (1.1)does not hold and $f(z) = \frac{1}{d}(Ae^{-dz}+1)^2$, where $A \neq 0$.

(3.2)$f(z) = A\dfrac{(e^{-dz}-C)^2}{(e^{-dz}-B)}$ where $A \neq 0, B \neq C$. When $B \neq 0$, $N(r,f) \sim \frac{dr}{\pi}$ and $T(r,f) \sim \frac{dr}{\pi}$. Thus (1.1) holds. When $B = 0$ and hence $C \neq 0$ we have (1.1) for $2AdC \neq -1$. When $2AdC = -1$, (1.1) does not hold and $f(z) = Ae^{dz}(e^{-dz} + \frac{1}{2Ad})^2$.

(3.3)$f(z) = A(e^{-\frac{1}{2}dz}+C)^2$. Obviously (1.1) holds for $AdC^2 \neq 1$. It does not hold for $AdC^2 = 1$ and $f(z) = \frac{1}{d}(Ae^{-\frac{1}{2}dz}+1)^2$ where $A \neq 0$.

Theorem 1 is proved.

References

1. Q. D. Zhang, An inequality for meromorphic functions with few simple zeros, *Acta Mathematica,* **47**(2)(2004):285-290(in Chinese). **42**(1967):389-392.
2. W. K. Hayman , *Meromorphic Functions,* Oxford:Clarendon Press, 1964.

THE MÖBIUS INVARIANCE OF BESOV SPACES ON THE UNIT BALL OF $\mathbb{C}^{n*}$

KEHE ZHU
Department of Mathematics
SUNY
Albany, NY 12222, USA
and
Department of Mathematics
Shantou University
Guangdong, 515063 P. R. China
E-mail: kzhu@math.albany.edu

It is well known that, for $1 \leq p < \infty$, the diagonal Besov space B_p of the open unit ball admits a norm or semi-norm $\|\ \|_p$ such that $\|f \circ \varphi\|_p = \|f\|_p$ for all f in B_p and all automorphisms φ of the unit ball. We show here that the same result holds when $0 < p < 1$.

2000 Mathematics Subject Classification: 32A37

1. Introduction

Throughout the paper we fix a positive integer n and let

$$\mathbb{C}^n = \mathbb{C} \times \cdots \times \mathbb{C}$$

denote the n-dimensional complex Euclidean space. For $z = (z_1, \cdots, z_n)$ and $w = (w_1, \cdots, w_n)$ in $\mathbb{C}^n$ we write

$$\langle z, w\rangle = z_1\overline{w}_1 + \cdots z_n\overline{w}_n$$

and

$$|z| = \sqrt{|z_1|^2 + \cdots + |z_n|^2} = \sqrt{\langle z, z\rangle}.$$

*Research partially supported by the US National Science Foundation.

The open unit ball in $\mathbb{C}^n$ is the set

$$\mathbb{B}_n = \{z \in \mathbb{C}^n : |z| < 1\}.$$

We use $\text{Aut}(\mathbb{B}_n)$ to denote the group of all automorphisms, or biholomorphic maps, of $\mathbb{B}_n$.

Suppose $0 < p < \infty$ and N is a positive integer such that $Np > n$. The (diagonal) Besov space B_p consists of holomorphic functions f in $\mathbb{B}_n$ such that the functions

$$(1 - |z|^2)^N \partial^\alpha f(z), \quad |\alpha| = N,$$

all belong to $L^p(\mathbb{B}_n, d\lambda)$, where

$$\alpha = (\alpha_1, \cdots, \alpha_n)$$

is a multi-index of nonnegative integers,

$$|\alpha| = \alpha_1 + \cdots + \alpha_n$$

is the size of α,

$$\partial^\alpha f(z) = \frac{\partial^\alpha f}{\partial z^\alpha}(z) = \frac{\partial^{|\alpha|} f}{\partial z_1^{\alpha_1} \cdots \partial z_n^{\alpha_n}}(z)$$

is a mixed partial derivative, and

$$d\lambda(z) = \frac{dv(z)}{(1 - |z|^2)^{n+1}}$$

is the Möbius invariant measure on $\mathbb{B}_n$. Here we use dv to denote volume measure on $\mathbb{B}_n$ normalized so that that $v(B_n) = 1$.

It is well known that the above definition of B_p is independent of the choice of N; see Section 6.1 of [3]. In what follows we fix a positive integer N such that $Np > n$ and define a "norm" on B_p as follows:

$$\|f\|_p = \sum_{|\alpha|<N} \left| \frac{\partial^\alpha f}{\partial z^\alpha}(0) \right| + \sum_{|\alpha|=N} \left[\int_{\mathbb{B}_n} \left| (1 - |z|^2)^N \partial^\alpha f(z) \right|^p d\lambda(z) \right]^{1/p}.$$

Of course, when $0 < p < 1$, $\|f\|_p$ is not really a norm; B_p is a complete metric space in this case.

The main result of the paper is the following.

Theorem 1.1. *For any $0 < p \leq 1$ there exists a positive constant $C > 0$ such that $\|f \circ \varphi\|_p \leq C\|f\|_p$ for all $f \in B_p$ and $\varphi \in \text{Aut}(\mathbb{B}_n)$.*

The constant C above depends on p, n, and N, but is independent of f and φ. It follows that

$$\|f\|_{p'} = \sup\{\|f \circ \varphi\|_p : \varphi \in \mathrm{Aut}(\mathbb{B}_n)\}$$

defines a Möbius invariant "norm" on B_p, $0 < p \leq 1$, in the sense that $\|f \circ \varphi\|_{p'} = \|f\|_{p'}$ for all $f \in B_p$ and $\varphi \in \mathrm{Aut}(\mathbb{B}_n)$.

When $1 < p < \infty$, there exist unbounded functions in B_p. Therefore, the above theorem cannot possibly hold for B_p when $p > 1$ (because of the presence of the term $|f(0)|$ in the definition of $\|f\|_p$). However, every B_p, $p > 1$, admits a Möbius invariant semi-norm; see [1] and [3].

2. The automorphism group

In this section we gather some information concerning the automorphism group $\mathrm{Aut}(\mathbb{B}_n)$ that is necessary for the proof of the main theorem.

Every unitary transformation U of $\mathbb{C}^n$, or equivalently, every unitary matrix U, is clearly an automorphism of $\mathbb{B}_n$. It is well known that the volume measure dv is invariant under the action of unitary transformations, that is,

$$\int_{\mathbb{B}_n} f(Uz)\, dv(z) = \int_{\mathbb{B}_n} f(z)\, dv(z) \tag{1}$$

whenever $f \in L^1(\mathbb{B}_n, dv)$.

Lemma 2.1. *Suppose $0 < p \leq 1$ and α' is a multi-index of nonnegative integers with size N. Then*

$$|\partial^{\alpha'}(f(Uz))|^p \leq n^N \sum_{|\alpha|=N} |(\partial^\alpha f)(Uz)|^p$$

for all f holomorphic in $\mathbb{B}_n$ and all unitary transformations U in $\mathrm{Aut}(\mathbb{B}_n)$.

Proof. For a unitary matrix $U = (u_{ij})$ and holomorphic f let $g(z) = f(Uz)$. Then by the chain rule, we have

$$\frac{\partial g}{\partial z_{n_1}}(z) = \sum_{j_1=1}^{n} \frac{\partial f}{\partial z_{j_1}}(Uz) u_{j_1,n_1},$$

and

$$\frac{\partial^2 g}{\partial z_{n_1} \partial z_{n_2}}(z) = \sum_{j_1=1}^{n} \sum_{j_2=1}^{n} \frac{\partial^2 f}{\partial z_{j_1} z_{j_2}}(Uz) u_{j_1,n_1} u_{j_2,n_2}.$$

More generally,

$$\frac{\partial^N g}{\partial z_{n_1}\cdots\partial z_{n_N}}(z)=\sum_{j_1}^{n}\cdots\sum_{j_N=1}^{n}\frac{\partial^N f}{\partial z_{j_1}\cdots\partial z_{j_N}}(Uz)u_{j_1,n_1}\cdots u_{j_N,n_N}.$$

Since each entry u_{ij} of a unitary matrix has modulus less than or equal to 1, and since

$$\left|\frac{\partial^N f}{\partial z_{j_1}\cdots\partial z_{j_N}}(Uz)\right|^p\leq\sum_{|\alpha|=N}|(\partial^\alpha f)(Uz)|^p$$

for any choice of $j_1, j_2, \cdots, j_N$, the desired result is then a consequence of Hölder's inequality. □

Combining Lemma 2.1 and (1), we obtain the following.

Lemma 2.2. *If $0 < p \leq 1$, f is holomorphic in $\mathbb{B}_n$, and U is a unitary transformation, then $\|f \circ U\|_p \leq C\|f\|_p$, where C is a positive constant independent of f and U.*

Another class of automorphisms, called involutions, can be defined explicitly. Thus for every point $a \in \mathbb{B}_n$, $a \neq 0$, we define

$$\varphi_a(z)=\frac{a-P_a(z)-\sqrt{1-|a|^2}\,Q_a(z)}{1-\langle z,a\rangle},\qquad z\in\mathbb{B}_n,$$

where

$$P_a(z)=\frac{\langle z,a\rangle}{|a|^2}a,\qquad z\in\mathbb{C}^n,$$

is the orthogonal projection from $\mathbb{C}^n$ onto the one-dimensional subspace $[a]$ generated by a, and $Q_a(z) = z - P_a(z)$ is the orthogonal projection from $\mathbb{C}^n$ onto $[a]^\perp$. When $a = 0$, we simply define $\varphi_a(z) = -z$.

It is well known that each φ_a is an automorphism of $\mathbb{B}_n$. Furthermore, φ_a has the properties that

$$\varphi_a(0)=a,\quad \varphi_a(a)=0,\quad \varphi_a\circ\varphi_a(z)=z, \tag{2}$$

and

$$1-\langle\varphi_a(z),\varphi_a(w)\rangle=\frac{(1-|a|^2)(1-\langle z,w\rangle)}{(1-\langle z,a\rangle)(1-\langle a,w\rangle)}, \tag{3}$$

where z and w are arbitrary points in $\mathbb{B}_n$. See [2] or [3].

If φ is any automorphism of $\mathbb{B}_n$ and $a = \varphi(0)$, then the automorphism $\varphi_a \circ \varphi$ fixes the origin and hence must be a unitary transformation (by a classical theorem of Cartan; see [2] or [3]). This shows that $\varphi = \varphi_a \circ U$ for some $a \in \mathbb{B}_n$ and some unitary U. A similar argument shows that every $\varphi \in \mathrm{Aut}(\mathbb{B}_n)$ can be written as $V \circ \varphi_b$ for some $b \in \mathbb{B}_n$ and some unitary V. We summarize this analysis as the following.

Lemma 2.3. *Every $\varphi \in \mathrm{Aut}(\mathbb{B}_n)$ can be written as*

$$\varphi = \varphi_a \circ U = V \circ \varphi_b,$$

where U and V are unitary transformations, $a = \varphi(0)$, and $\varphi(b) = 0$.

The following result gives an even more detailed decomposition for automorphisms.

Lemma 2.4. *Every automorphism φ of $\mathbb{B}_n$ can be decomposed into the form*

$$\varphi = U \circ \varphi_a \circ V,$$

where U and V are unitary transformations, and $a = (r, 0, \cdots, 0)$ for some $r \in [0, 1)$.

Proof. Fix an automorphism $\varphi \in \mathrm{Aut}(\mathbb{B}_n)$ and choose a unitary transformation U such that $U(a) = \varphi(0)$, where $a = (r, 0, \cdots, 0)$ and $r = |\varphi(0)|$. This is possible, because the unitary group is transitive on any sphere centered at the origin. Since

$$\varphi_a \circ U^{-1} \circ \varphi(0) = 0,$$

an application of Lemma 2.3 shows that there exists a unitary transformation V such that

$$\varphi_a \circ U^{-1} \circ \varphi = V,$$

or

$$\varphi = U \circ \varphi_a \circ V,$$

completing the proof of the lemma. □

3. Möbius invariance of B_p

Let M_n be the dimensional constant defined by $M_n = 1$ for $n = 1$ and $M_n = 2n$ for $n > 1$. The Möbius invariance of B_p exhibits a subtle difference when p crosses this constant.

When $p > M_n$, the Besov space B_p can be given an explicit Möbius invariant semi-norm:

$$\left[\int_{\mathbb{B}_n} |\widetilde{\nabla} f(z)|^p \, d\lambda(z)\right]^{1/p},$$

where $|\widetilde{\nabla} f(z)|$ is the Möbius invariant complex gradient of f at z. This integral diverges for $p \le M_n$ unless f is constant. See Section 6.3 of [3].

When $p = 2$, the Dirichlet type space B_2 admits the following Möbius invariant semi-inner product:

$$\langle f, g\rangle = \sum_{\alpha} |\alpha| \frac{\alpha!}{|\alpha|!} a_\alpha \bar{b}_\alpha,$$

where a_α and b_α are the Taylor coefficients of f and g, respectively, and $\alpha! = \alpha_1! \cdots \alpha_n!$. See Section 6.4 of [3].

When $p = 1$, the space B_1 consists exactly of functions of the form

$$f(z) = c_0 + \sum_{k=1}^{\infty} c_k f_k(z), \tag{4}$$

where $\{c_k\} \in l^1$ and f_k are coordinate functions of automorphisms. Moreover,

$$\|f\|_m = \inf \sum_{k=0}^{\infty} |c_k|,$$

where the infimum is taken over all $\{c_k\}$ satisfying (4), defines a Möbius invariant norm (not just a semi-norm) on B_1. See Section 6.2 of [3].

More generally, for any $p \in (1, \infty)$, the Möbius invariance of B_p follows from complex interpolation between B_1 and B_q, where $q > \max(p, M_n)$. See [1] or [3].

We proceed to show that B_p is also Möbius invariant when $0 < p < 1$. The following integral estimate will be essential for us.

Lemma 3.1. *For $s > -1$ and t real the integral*

$$I(z) = \int_{\mathbb{B}_n} \frac{(1 - |w|^2)^s \, dv(w)}{|1 - \langle z, w\rangle|^{n+1+s+t}}$$

has the following asymptotic properties as $|z| \to 1^-$.

(a) If $t > 0$, then $I(z)$ is bounded.
(b) If $t = 0$, then $I(z)$ is comparable to $-\log(1 - |z|^2)$.
(c) If $t > 0$, then $I(z)$ is comparable to $(1 - |z|^2)^{-t}$.

Proof. See Proposition 1.4.10 of [2]. □

The key ingredient in our proof is the following atomic decomposition for functions in Besov spaces.

Lemma 3.2. *For any $0 < p < \infty$ and $b > \max(0, n(p-1)/p)$ there exists a sequence $\{a_k\}$ in $\mathbb{B}_n$ such that B_p consists exactly of functions of the form*

$$f(z) = \sum_{k=1}^{\infty} c_k \left(\frac{1 - |a_k|^2}{1 - \langle z, a_k \rangle} \right)^b, \tag{5}$$

where $\{c_k\} \in l^p$. Moreover, $\|f\|_p^p$ is equivalent to

$$\inf \sum_{k=1}^{\infty} |c_k|^p,$$

where the infimum is taken over all sequences $\{c_k\}$ satisfying (5).

Proof. See Section 6.1 of [3]. □

Let H^∞ be the space of bounded holomorphic functions in $\mathbb{B}_n$ equipped with the sup-norm $\|f\|_\infty$.

Lemma 3.3. *If $0 < p \le 1$, then B_p is continuously contained in H^∞.*

Proof. If $0 < p \le 1$, then Hölder's inequality shows that l^p is continuously contained in l^1. Since each function

$$f_k(z) = \left(\frac{1 - |a_k|^2}{1 - \langle z, a_k \rangle} \right)^b$$

has sup-norm less than or equal to 2^b, the desired result then follows from Lemma 3.2. □

The above proof actually shows that B_p, when $0 < p \le 1$, is continuously contained in the disk algebra.

We can now prove Theorem 1.1, the main result of the paper.

Fix a function $f \in B_p$, where $0 < p \le 1$. By Lemma 3.3, there exists a constant $C_1 > 0$ such that

$$\sup\{|f \circ \varphi(z)| : z \in \mathbb{B}_n, \varphi \in \mathrm{Aut}(\mathbb{B}_n)\} = \|f\|_\infty \le C_1 \|f\|_p.$$

Combining this with Cauchy's estimates, we can find a constant $C_2 > 0$ such that

$$\sum_{|\alpha|<N} |\partial^m (f \circ \varphi)(0)|^p \leq C_2 \|f\|_p \tag{6}$$

for all $\varphi \in \mathrm{Aut}(\mathbb{B}_n)$.

By Lemma 3.2 (with $b = 1$), there exists a sequence $\{a_k\}$ in $\mathbb{B}_n$ and a sequence $\{c_k\}$ in l^p such that

$$f(z) = \sum_{k=1}^{\infty} c_k \frac{1 - |a_k|^2}{1 - \langle z, a_k \rangle},$$

where

$$\sum_{k=1}^{\infty} |c_k|^p \leq C_3 \|f\|_p^p, \tag{7}$$

and C_3 is a positive constant independent of f and $\{c_k\}$. Since $0 < p \leq 1$, it follows from Hölder's inequality that the integral

$$\int_{\mathbb{B}_n} \left|(1 - |z|^2)^N \partial^\alpha (f \circ \varphi)(z)\right|^p \, d\lambda(z)$$

is less than or equal to

$$\sum_{k=1}^{\infty} |c_k|^p \int_{\mathbb{B}_n} \left|(1 - |z|^2)^N \partial^\alpha (f_k \circ \varphi)(z)\right|^p \, d\lambda(z),$$

where α is a multi-index with size N and

$$f_k(z) = \frac{1 - |a_k|^2}{1 - \langle z, a_k \rangle} = 1 - \langle \varphi_{a_k}(z), a_k \rangle, \qquad z \in \mathbb{B}_n.$$

Note that the last equality above follows from (3).

By Lemma 2.3, we have $\varphi_{a_k} \circ \varphi = U_k \circ \varphi_{a'_k}$, where U_k is a unitary transformation and $a'_k = \varphi^{-1}(a_k)$. It follows that

$$f_k \circ \varphi(z) = 1 - \langle \varphi_{a'_k}(z), b'_k \rangle,$$

where $b'_k = U_k^*(a_k)$.

For u and v in $\mathbb{B}_n$ we consider the function

$$f_{u,v}(z) = 1 - \langle \varphi_u(z), v \rangle, \qquad z \in \mathbb{B}_n.$$

The proof of the theorem will be complete if we can show that there exists a constant $C_4 > 0$, independent of u and v, such that

$$\int_{\mathbb{B}_n} \left|(1 - |z|^2)^N \partial^\alpha f_{u,v}(z)\right|^p \, d\lambda(z) \leq C_4 \tag{8}$$

for all u and v in $\mathbb{B}_n$ and all α with $|\alpha| = N$.

Apply Lemma 2.4 to the automorphism φ_u and then use Lemma 2.2. We conclude that it suffices to prove (8) under the additional assumption that $u = (r, 0, \cdots, 0)$, where $0 \le r < 1$. In this case, we have

$$\varphi_u(z) = \left(\frac{r - z_1}{1 - rz_1}, -\frac{\sqrt{1-r^2}}{1 - rz_1} z_2, \cdots, -\frac{\sqrt{1-r^2}}{1 - rz_1} z_n \right),$$

and so

$$f_{u,v}(z) = 1 - \overline{v}_1 \frac{r - z_1}{1 - rz_1} + \overline{v}_2 \frac{\sqrt{1-r^2}}{1 - rz_1} z_2 + \cdots + \overline{v}_n \frac{\sqrt{1-r^2}}{1 - rz_1} z_n,$$

where $(v_1, \cdots, v_n) = v$.

Consider a mixed partial derivative $\partial^\alpha (f_{u,v})(z)$, where $|\alpha| = N$. It is clear that this partial derivative is zero when

$$|\alpha_2| + \cdots + |\alpha_n| > 1.$$

Therefore, we can assume that at most one of the numbers in $\{\alpha_2, \cdots, \alpha_n\}$ is 1 and the rest are all 0. There are two cases to consider.

The first case is when $\alpha_2 = \cdots = \alpha_n = 0$. Then it is easy to check that

$$\partial^\alpha (f_{u,v})(z) = \frac{\partial^N f_{u,v}}{\partial z_1^N}(z) = \frac{N! r^{N-1}}{(1 - rz_1)^{N+1}} \langle w, v \rangle,$$

where

$$w = (1 - r^2, r\sqrt{1-r^2}\, z_2, \cdots, r\sqrt{1-r^2}\, z_n).$$

Since

$$|w|^2 = (1-r^2)^2 + r^2(1-r^2)(|z|^2 - |z_1|^2) \le (1-r^2)^2 + (1-r^2)(1-|z_1|^2),$$

an application of the Cauchy-Schwarz inequality followed by Hölder's inequality gives

$$|\partial^\alpha (f_{u,v})(z)| \le N! \frac{1 - r^2 + \sqrt{1-r^2}\sqrt{1-|z_1|^2}}{|1 - rz_1|^{N+1}}. \tag{9}$$

Writing rz_1 as the inner product of z with $(r, 0, \cdots, 0)$ and applying part (c) of Lemma 3.1, we see that

$$\int_{\mathbb{B}_n} \left(\frac{(1-r^2)(1-|z|^2)^N}{|1 - rz_1|^{N+1}} \right)^p d\lambda(z)$$

is a bounded function of r. A similar estimate works for the second term on the right hand side of (9), after we make use of the fact that

$$1 - |z_1|^2 \le 2|1 - rz_1|.$$

This proves the inequality (8) when $\alpha_k = 0$ for $2 \le k \le n$.

The second case is when exactly one of the integers in $\{\alpha_2, \cdots, \alpha_n\}$, say α_k, is 1 and the rest are all 0. In this case, we have

$$\partial^\alpha (f_{u,v})(z) = (N-1)! \overline{v}_k r^{N-1} \frac{\sqrt{1-r^2}}{(1-rz_1)^N}.$$

By part (b) of Lemma 3.1, the integral

$$\int_{\mathbb{B}_n} \left(\frac{\sqrt{1-r^2}}{|1-rz_1|^N} (1-|z|^2)^N \right)^p d\lambda(z)$$

is dominated by

$$(1-r^2)^{p/2} \log \frac{2}{1-r^2},$$

which is clearly a bounded function of $r \in [0,1)$.

This completes the proof of Theorem 1.1.

References

1. M. Peloso, Möbius invariant spaces on the unit ball, *Michigan Math. J.* **39** (1992):509-536.
2. W. Rudin, *Function Theory in the Unit Ball of* $\mathbb{C}^n$, Springer-Verlag, New York, 1980.
3. K. Zhu, *Spaces of Holomorphic Functions in the Unit Ball*, Springer-Verlag, New York, 2004.

